THE EXTINC

VANDA FELBAB-BROWN

The Extinction Market

Wildlife Trafficking and How to Counter It

OXFORD
UNIVERSITY PRESS

Oxford University Press is a department of the
University of Oxford. It furthers the University's objective
of excellence in research, scholarship, and education
by publishing worldwide.

Oxford   New York
Auckland   Cape Town   Dar es Salaam   Hong Kong   Karachi
Kuala Lumpur   Madrid   Melbourne   Mexico City   Nairobi
New Delhi   Shanghai   Taipei   Toronto

With offices in
Argentina   Austria   Brazil   Chile   Czech Republic   France   Greece
Guatemala   Hungary   Italy   Japan   Poland   Portugal   Singapore
South Korea   Switzerland   Thailand   Turkey   Ukraine   Vietnam

Oxford is a registered trade mark of Oxford University Press
in the UK and certain other countries.

Published in the United States of America by
Oxford University Press
198 Madison Avenue, New York, NY 10016

Copyright © Vanda Felbab-Brown 2017

All rights reserved. No part of this publication may be reproduced,
stored in a retrieval system, or transmitted, in any form or by any means,
without the prior permission in writing of Oxford University Press,
or as expressly permitted by law, by license, or under terms agreed with
the appropriate reproduction rights organization. Inquiries concerning
reproduction outside the scope of the above should be sent to the
Rights Department, Oxford University Press, at the address above.

You must not circulate this work in any other form
and you must impose this same condition on any acquirer.

Library of Congress Cataloging-in-Publication Data is available
Vanda Felbab-Brown.
The Extinction Market: Wildlife Trafficking and How to Counter It

ISBN: 9780190855116

Printed and bound in Great Britain by Bell and Bain Ltd, Glasgow

To otters, dragons, and other creatures (some human) whom I love.

CONTENTS

Acknowledgements ix
List of Abbreviations xiii

1. Introduction 1
2. The Wildlife Trade and the Drug Trade: Similarities and Differences 31
3. The Character and Scale of the Illegal Wildlife Economy 51
4. Why the Illegal Wildife and Drug Economies Matter 65
5. The Actors 87
6. Policy Response I: Bans and Law Enforcement 103
7. Legal Trade 131
8. Local Community Involvement 161
9. Anti-Money Laundering Efforts 205
10. Demand Reduction 219
11. Conclusions and Policy Recommendations 241

Notes 279
Bibliography 341
Index 391

ACKNOWLEDGEMENTS

This inquiry of necessity spans diverse fields of scholarship and public policy: conservation; biology; wildlife trafficking; criminology illicit economies and their enforcement, including the illegal drug trade and illegal logging and mining; public health issues; and enforcement of and compliance with rules and regulations. Writing this book has therefore required, along with my own extensive fieldwork, that I consult widely with experts and activists across these various domains of research, ecology-relevant industry and community behaviors, policymaking, and law enforcement.

Special thanks go to the intrepid team of biologists and nature-lovers who over the years have been associated with Tropical Birding, including Keith and Yvonne Barnes, Christian Boix, Ken Behrens and Robert White, as well as other environmentalists who helped me get deep into forests and high into mountains around the world to study illegal economies, understand their impact on nature, and see and appreciate the planet's magical fauna and flora.

My fieldwork has often involved interviews with individuals whose names I cannot reveal—poachers, wildlife and drug smugglers, members of criminal groups, illegal loggers—for they spoke to me under the condition of complete anonymity. Indeed, some of them met and talked with me at considerable personal risk, including those trying to enforce laws and regulations and those trying to avoid or subvert them. I want to thank them here, since their insights and opinions, and sometimes self-justification, immeasurably enrich the analysis offered in this book.

ACKNOWLEDGEMENTS

Given this somewhat unorthodox empirical research, I am especially and deeply appreciative of the support of The Brookings Institution and the encouragement of its President, Strobe Talbott; Executive Vice President, Martin Indyk; and Vice President for Foreign Policy, Bruce Jones. My thanks also go to Charlotte Baldwin and other colleagues at Brookings who were an important part of that institutional support. My resourceful senior research assistant, Bradley Porter, as well as a team of highly skilled interns over the years deserve special credit.

I am grateful to the anonymous reviewers who looked at the manuscript for the Brookings Institution, and for Hurst Publishers and Oxford University Press. Their comments and suggestions not only enhanced the scholarly quality of the book, but pointed me to new lines of inquiries and new fields of study. I particularly want to highlight the penetrating and detailed comments of Professor Jonathan Caulkins of Carnegie Mellon University, which allowed me to improve the manuscript significantly. I am also thankful to my colleague Dr Michael O'Hanlon, a Brookings Director of Research on Foreign Policy Studies (himself a prolific writer on public policy issues), for his most useful suggestions regarding the book's expository style and organization.

Additionally, the suggestions of Ken Williams of the Air Force history office have not only improved the quality and readability of the book but also provided supportive coaching during the entire research, writing, and publishing process and beyond. I am very grateful for his friendship as well as advice.

Such invaluable friendship and support has also come from many dear friends, including Dr Olivier Guillard, a prominent French expert on Asia as well as a man of plants and rocks; Ambassador Ronald Neumann, the President of the American Academy of Diplomacy; and Ambassador Cameron Munter, the President and CEO of The East-West Institute. A close interlocutor on the book and many other issues, Cameron generously read multiple iterations of the manuscript, stimulated my thinking, and gave me much appreciated input.

Dr Seyom Brown, key counselor on all dimensions and phases of the project—from the vantage (or disadvantage) point of being my husband—has been a frequent fieldwork companion as well as the principal source of the emotional encouragement, critical commentary, and multidimensional backing indispensable for the often challenging, and sometimes risky aspects of the work.

ACKNOWLEDGEMENTS

As with all of my ventures, my mother Jelena Felbábová deserves my thanks for putting up with my research in dangerous places and on dangerous topics. In this book, I want to especially acknowledge my father, Josef Felbáb. Though he died when I was fifteen, he profoundly and in many dimensions shaped who I am today. Exemplifying courage, solidarity, perseverance, and a commitment to freedom and civil liberties in the face of life-threatening risks and hardships, he also instilled in me the love of nature. From when I was very little, he would take me to bird watch in the forests and hills of southern Bohemia and visit the zoos of 1980s Czechoslovakia, and was willing to read for hours my beloved dry-as-can-be encyclopedias of animals. In addition to imparting profound respect for wild creatures, he also made me aware of man-made destruction of the world's ecosystems, manifested vividly near the village in southern Bohemia where we lived, and the need to do all that is possible to preserve nature and species.

My research has been generously supported by the Open Society Foundations and the Government of Norway. Brookings recognizes that the value it provides is in its absolute commitment to quality, independence and, and impact. Activities supported by its donors reflect this commitment. The views contained in this book are my own.

LIST OF ABBREVIATIONS

AML	anti-money laundering
ASEAN-WEN	Association of Southeast Asian Nations Wildlife Enforcement Network
CBNRM	community-based natural resource management
CIFOR	Center for International Forestry Research
CITES	Convention on the International Trade in Endangered Species, 1974
EIA	Environmental Investigation Agency
FAO	United Nations Food and Agriculture Organization
FATF	Financial Action Task Force
FinCEN	Financial Crimes Enforcement Network
FSC	Forest Stewardship Council
IFAW	International Fund for Animal Welfare
IIED	International Institute for Environment and Development
ICRAF	World Agroforestry Center
IUCN	International Union for the Conservation of Nature
NTFPs	non-timber forest products
PES	payments for ecosystem services
REDD+	reducing emissions from deforestation and forest degradation
SDGs	Sustainable Development Goals
SIUs	special interdiction units
TCM	Traditional Chinese Medicine
TTP	Trans-Pacific Trade Partnership

LIST OF ABBREVIATIONS

UNDP	United Nations Development Programme
UNEP	United Nations Environment Programme
UNESCO	United Nations Educational, Scientific, and Cultural Organization
UNODC	United Nations Office on Drugs and Crime
USAID	United States Agency for International Development
VDC	village development committee
WWF	World Wide Fund for Nature
WRI	World Resources Institute

1

INTRODUCTION

The planet is currently experiencing alarming levels of species loss caused in large part by intensified poaching stimulated by a greatly expanding demand for animals, plants, and wildlife products. The rate of species extinction, now as much as 1,000 times the historical average and the worst since the dinosaurs died out 65 million years ago, deserves to be seen, like climate change, as a global ecological catastrophe meriting high-level policy initiatives to address its human causes. In addition to irretrievable biodiversity loss, poaching and wildlife trafficking pose serious threats to public health. Diseases linked to wildlife trafficking and consumption of wild animals have included SARS and Ebola. Wildlife trafficking can trigger global pandemics. Wildlife trafficking can also undermine human security of forest-dependent communities, cause local, national, and global economic losses, and even pose threats to national security. But conservation policies to preserve species can equally undermine human security when they constrain the access of poor populations to the natural resources on which they depend for basic livelihoods. The causes and modalities of poaching and wildlife trafficking and the complexities of policies for dealing with them are the subject of this book.

Much of the killing of wildlife is illegal, as is the illegal selling and buying—trafficking—of the captured animals and their body parts. As such, countering poaching and wildlife trafficking is viewed by many experts and policymakers as analogous to the effort to combat the trafficking of another commodity of global concern: addictive drugs. While the primary focus of the book is on wildlife issues, my analysis distills and applies lessons from the experience of dealing with illegal drugs. The goal is to inform the conservation community of the analogous challenges and the usefulness, limitations, and outcomes of the same policy tools applied in the drug field. Central to my analysis is the fact that in both wildlife and drug domains, legal production and legal trade are intertwined in complex ways with the illegal production and trade. In both fields, the design and governance of legal production and trade structure in important ways their illegal aspects, and vice-versa, and this dynamic influences the effectiveness of the various policy measures.

Drawing lessons from decades of drug policies and applying them to poaching and wildlife trafficking, I analyze the outcomes and effectiveness of five policy tools: bans and law enforcement, including interdiction; permitting legal trade; involving the local community through alternative livelihoods programs, compensation mechanisms, and, in the case of wildlife, also community-based natural resource management approaches; anti-money-laundering efforts; and demand-reduction strategies. For each of these tools I consider the regulatory theory behind its use—that is, what its proponents promise the particular approach will deliver—and I analyze the actual outcomes, assessing the theory's advantages and limitations. Throughout the book I also bring in examples from other illegal economies or relevant public policy experiences, including illegal logging and mining, to elucidate what might work under which circumstances for addressing poaching and wildlife trafficking. In exploring how to reduce demand for illegally-sourced wildlife products, the final policy tool chapter also taps into lessons from public health campaigns. These provide additional perspectives on the usefulness and limitations of analogies with the drug-control problem, especially efforts to induce compliance and change behavior.

Although there are many analogies between the illegal drug and wildlife economies, including the often very decentralized and very diverse structures of the illegal trade in both commodities, there are some important differences. Beyond the obvious fact that drugs, unlike wildlife, are addictive, illegal drugs are a nondepletable commodity:

INTRODUCTION

whether synthetic or plant-based, they can be produced indefinitely. Wildlife, however, is highly depletable. While in theory some wildlife species can be replenished through breeding—whether in the wild or in captivity—in practice, depletion is often far faster than the rate at which breeding can replenish a species. And once a species is extirpated, it cannot be brought back (unless through some future black magic of genetic engineering). Drugs, however, will be produced again and again.

Because of the nondepletable characteristic of the drugs at issue, many drug-policy reformers have come to emphasize as the most appropriate focus of policy the reduction of the harm associated with drug use, trafficking, and counter-narcotics policies themselves, and have given up on the hope of eliminating the use of harmful drugs and the illegal drug trade. Quite different is the crucial objective of environmental conservation efforts to preserve biodiversity and hence suppress poaching and wildlife trafficking. Although not all species can be saved, saving even a proportion of species from extinction is the goal. Consequently, to deal with poaching and wildlife trafficking, policies must operate on timelines that are highly constrained—a fundamental difference with drugs. The conservation community often talks about transforming wildlife trade from the currently high-profit, low-risk venture to a low-profit, high-risk enterprise. But drugs, by contrast, are not only high-risk; they also remain high-profit because the more law enforcement there is, the more profitable the illegal trade becomes. Yet, eliminating prohibition and enforcement will boost demand. Moreover, the typical efficacy levels of the various tools used to address illegal drug production and trade will often be insufficient in efforts to counter the illegal wildlife trade, while the emerging new objectives of drug reform policies will be inadequate in the case of the illegal wildlife trade.

Drugs and wildlife also differ in terms of policy saliency. Combating poaching and wildlife trafficking currently receives unprecedented attention by governments, but drugs for the most part still greatly surpass wildlife in the intensity and level of governments' interest. Consequently there is a vast disparity in the amount of law enforcement, economic, and other resources devoted to suppressing and managing the two illegal economies. Around the world, governments annually devote tens of billions of dollars to combat the illegal drug trade, while even now, at the height of unprecedented policy attention, they devote only tens of millions of dollars to combat poaching and wildlife trafficking.[1] Thus, it is all the more crucial that policymakers assign resources strategically, and that in policy design and implementation, programs avoid measures that will not work or that are based on overly optimistic estimates of how much can actually be accomplished.

From assessment of the effectiveness of the five tools used in the illegal wildlife and drug economies, I derive policy implications and recommendations for dealing with poaching and wildlife trafficking. I show that although there is great urgency for correctives—given the explosion in poaching and wildlife trafficking and the precipitous decline of numerous wildlife species, far beyond the currently iconic ones: elephants, tigers, and rhinos—each of the policy tools comes with many complexities, contradictions, and limitations. One of the fundamental problems is that conservation policies that promote biodiversity and seek to save species can hurt poor marginalized populations dependent on natural resource extraction for their basic livelihood. If such conservation policies are employed, such as particular forms of law enforcement relying on forcible displacement or lethal force, they will often be rejected by local populations as contradictory to their interests and rights. This is all the more so if conservation is associated with painful colonial legacies. Local populations can mobilize against such conservation policies or embrace poaching and wildlife trafficking. What is illegal—that is, poaching, illegal grazing or other forms of encroachment in protected areas and their intersections with global wildlife trafficking—may be seen as perfectly legitimate by local populations. I describe how the drug trade too is permeated by the contradiction between legality and legitimacy: dependent on illegal drug cultivation for basic livelihoods and any form of economic advancement, many poor marginalized farmers of illegal drug crops regard growing what they want as legitimate and resent and mobilize against countermeasures. Since basic human security can be severely undermined by efforts to suppress the illegal production and trade in drugs or wildlife, those who sponsor such illegal economies can develop substantial political capital. Law enforcement can prevail in circumstances when many reject the regulations and norms that

INTRODUCTION

enforcement defends, but under such circumstances law enforcement becomes very costly—normatively politically, as well as in terms of resources. Law enforcement is easiest where the vast majority of people have internalized the laws and norms, and believe that obeying laws and norms best advances their interests and aspirations. Thus giving local communities and other actors material as well as non-material stakes in conservation enhances it. However, alignments of competing interests cannot always be achieved.

The debates about how best to design drug and wildlife policies to combat illegal production and illegal trade are to an important extent debates about means. But these debates are also about values, including local versus global values and the rights of individuals and communities versus the imperatives of states and the logics of global regimes. Moreover, an element of the controversy over how best to deal with poaching and wildlife trafficking is also part of the debate about human welfare versus animal welfare and biodiversity preservation.

The goal of policy should be to minimize these tradeoffs and sacrifices among and within those values. Yet as this book shows, there is no easy solution for either accomplishing this or eliminating poaching and wildlife trafficking. All the policy tools discussed come with limitations, downsides, and a vast variation in outcomes. Thus rather than elevating one tool, policy needs to remain flexible as to what tool is deployed in a particular place and at a particular time. What works today in one place may not work tomorrow in the same place, and it may already not work today in another place. Policies to counter poaching and wildlife trafficking need to be specifically tailored to local contexts, even while elements of such policies will inevitably have global impacts. There is no policy silver bullet to stop the bullets of the poachers.

A Wildlife Market

In the sweltering heat of Indonesia, hundreds of cages containing birds, lizards, bats, and mammals were stacked upon one another, with scores of specimens often crammed into one cage. Several dozen white-eyes (a bird genus) were squeezed into a cage appropriate for one canary. At least a hundred bats were stuffed into another container. In a cage atop this stack were more than fifty green agama dragon lizards, some dead, with their bodies rotting amidst those still alive and desperately competing on the ceiling of their container for a sliver of space. Two baby civets, on sale for 400,000 Indonesian rupiah each (about $40), were shoved into an adjacent box.[2] Like the rest of the unfortunate animals—squirrels, chipmunks, black-naped orioles, drongos, leafbirds, shamas, mynas, partridges, and the highly prized and highly threatened lorises—the civets had no water and no protection from the full blast of the sun, which was literally baking them alive. Many of the animals would die in this (in)famous Yogyakarta bird market before they were sold.[3]

The Yogyakarta market, like other wildlife markets in Indonesia and East Asia, serves as a perfect incubator for diseases that can mutate and jump between species, like avian influenza and SARS. Such zoogenic diseases could potentially set off a catastrophic pandemic, killing millions of people. The spread of the viruses to domestic animals and humans is exacerbated by the trade in roosters for cockfights, also on sale in the market amidst the birds and animals caught in the wild. Even the animals still active enough to be bought by customers often do not survive as household pets.

The inhumane treatment of animals in the many wildlife markets I visited during my research across the Indonesian archipelago was to me as heart wrenching as the devastation this unmitigated trade in wild birds and other animals has wreaked upon Indonesia's ecosystems. Orange-headed thrushes and white-crested laughing thrushes, available in cages to eager buyers, are now exceedingly rare in the remnants of Indonesia's forests. In Jakarta's Jatinegara, wildlife, tiger parts, rhino horns, and even live orangutans and Komodo dragons can be bought.[4] In Labuan Bajo on Flores, in the eastern part of the country, all kinds of illegally and legally caught fish are on sale, including many highly threatened species, such as sharks, manta rays, and Napoleon wrasse.[5]

While I find the state of these wildlife markets deeply disturbing, for many Indonesians it is perfectly acceptable. Keeping pet birds, often caught in the wild and regardless of whether they are endangered or not, is deeply ingrained in the culture and has a long tradition. For many Indonesians, efforts to change the market in birds involves an imposition of Western values, carrying with it overtones of the injustices of the colonial era and the imposition of environmental policies at the expense of local people.[6] Even the Indonesian animal rights move-

INTRODUCTION

ment that came out of the *reformasi* movement that emerged around the time of the end of the Suharto dictatorship backed off from taking on bird markets.

Such wildlife markets as these also exist in Thailand, Myanmar (Burma), China, and Vietnam. In Myanmar's key wildlife smuggling hubs, such as Mong La and Tachilek, tiger penises, pangolin scales, and potions from a huge diversity of animals are hawked.[7] In different forms, wildlife markets and the poaching and problematic consumption of legally and illegally hunted wildlife exist as well in South Asia, Africa, Latin America, the Middle East, the United States, and Europe.

The current demand for highly prized items such as ivory, rhino horn, and tiger parts threatens to extirpate these iconic species, as well as many much less known and prized ones, thus undermining the accomplishments of the early 1990s that halted some of the most dangerous and egregious poaching. Increased buyer power, population growth, and globalization have led to the global proliferation of wildlife markets, now flourishing in developed, emerging, and developing countries.[8]

Specialized wildlife markets supplied by illegal hunting exist throughout the world. However, East Asia currently contains the areas of the most intense legal and illegal trade in wildlife and wildlife products, with China the biggest consumer of both, including ivory.[9] Also a very large consumer of legal and illegal wildlife products, Vietnam surpasses China in the consumption of rhino-horn products from rhinos poached in Africa and Asia. The increasing affluence of the expanding middle classes in places such as Vietnam, China, and Thailand, and their greater connectedness to global trade, lie behind this exploding demand for wildlife. China in particular has become like a great vacuum cleaner, sucking natural environments empty of wildlife—not only in China and its neighbors, but also in Africa and elsewhere where elephants, rhinoceros, and many other species are illegally and legally hunted at devastating and often unsustainable rates.[10] Tens of millions of wild animals are shipped each year to southern China for food or to East and Southeast Asia for use in Traditional Chinese Medicine (TCM).[11] Many species—such as tigers, Asian and African species of rhinoceros, various species of pangolin, Tibetan and saiga antelopes, and freshwater and marine turtles—are now on the verge of extinction as a result of commercial exploitation and the increase in consumer demand for these luxury goods.[12] Although wildlife consumption has deep and long historic roots in Asia, the level of cross-border trade between China and neighboring countries, and increasingly also distant regions, has reached a level unmatched in history and is decimating the planet's wildlife populations.[13]

East and Southeast Asian diaspora communities often spread the taste for wildlife to new areas, expanding local habits of exploiting wildlife, whether as pets or for food or other uses. Traditional markets and demand for wildlife exist everywhere, but globalization and the increasing purchasing power of large segments of the world's population have expanded and intensified traditional demand. Significant markets for wildlife exist in Africa and Latin America, as well as in the United States and Europe.[14] Disturbingly, the United States—despite having one of the world's strongest regulatory regimes regarding the importation of wildlife, and some of the most extensive and intensively enforced regulations criminalizing the illegal trade in it—ranks as the second largest market, after China, for illegal wildlife products. Demand for TCM within the United States, often linked to Asian communities in the country, accounts for an important portion of that demand.[15] Demand and supply markets also exist in Latin America. Some are new and some have existed for a long time. They include, for example, the illegal trade in parrots in Brazil and the illegal trade in reptile skins that supplies the affluent in Mexico, who love boots made of exotic materials. Demand for wildlife, both sustainable and environmentally problematic, also exists in other places often ignored in the story of global poaching, including various East and West African countries, which are often characterized as merely source countries.[16]

Other environmental threats such as climate change, deforestation and other habitat destruction, industrial pollution, and competition between indigenous and invasive introduced species often impact on ecosystems on a large scale.[17] Particularly given those threats and vulnerabilities, poaching and wildlife trafficking (as well as badly managed legal trade) can drive species into local and worldwide extinction.[18] For endangered species with slow reproductive rates and already numbering in the low thousands, the existence or absence of poaching and trade can mean either their extinction or survival. From iconic species (such as tigers and elephants) to those which less frequently capture the attention of publics

INTRODUCTION

around the world (such as reptiles, frogs, and insects), effectively managing wildlife trade and curbing its illegal dimension are often required for species preservation and biodiversity conservation.

Fighting against this irretrievable destruction of species and ecosystems are many determined individuals. They often work against all odds and are sometimes found in the most surprising places. In Bali, Indonesia, Jatmika, a very impressive young female Muslim veterinarian, has been supervising and establishing rescue shelters so that law-enforcement officials have an incentive to retrieve illegally owned and captured species, even though raising resources is difficult. In the past, the Indonesian police often used the small number of available animal shelters as an excuse for not undertaking raids, claiming that they could not care for the rescued animals. Indeed, some 95 percent of animals confiscated from wildlife markets or private collections are too sick and damaged to be returned to the wild. With few releases possible, because they might introduce new diseases that could devastate wild populations, most of the recovered animals will have to be treated at shelters for the rest of their lives or be put down.

In Somalia, on the dangerous border of Somaliland and Puntland where al-Shabaab militants and criminal gangs regularly operate, a determined wildlife conservationist, Abdi Jama, is trying to establish ecotourism and protect local species. He even bought an open jeep from an ecotourism company in South Africa and drove it all the way up the continent to Somalia to take the few tourists and biologists who come to Somalia to see the precious local wildlife on safaris. Yet his effort is an immense struggle against overwhelming odds.[19] One day, Abdi bought a live pregnant cheetah in a market and released her into the wild. A few weeks later, the cheetah was again on sale in the market, this time with her newborn cub. Although less than 7,000 cheetah remain in the wild globally, many are trafficked from Somaliland and the Horn of Africa to the Middle East, where they are sought for their fur and as pets. Most will survive less than two years in captivity. Rich Saudis and Emiratis also come to this part of Somalia to hunt its rare antelopes and bustards; local people who desperately need income eagerly escort them, and forest rangers gladly take the customary bribe (equal to about $100). In response, Abdi pays a water station guard, a seventy-year-old man who lives alone in a hut, to keep a lookout for poachers. When he sees them and passes on the message, Abdi tries to mobilize his subclan to prevent poaching.[20] Abdi is a controversial figure, and his efforts are not always supported by all the nearby local communities. Often very poor local people understandably, and by necessity, frequently see local animals through the prism of economic benefits or costs, rare as the animals might be: as elsewhere in Africa, meat from antelopes and leopard skins are sold in Somaliland's capital, Hargeisa. Constraining such hunting and trade in wildlife hurts the people's livelihoods and opportunities for their children, as much as letting the trade be hurts the environment and can extirpate species.

There are many admirable individuals committed to conservation: honest park rangers, wardens, and managers, such as those of the Mara Conservancy in Kenya, who, despite the odds, try to implement better policies, save animals from poachers, and give local communities a stake in conservation; a police officer in Brazil who took it upon himself to create a manual of endangered species to motivate his lackadaisical colleagues to conduct seizures of trafficked animals and arrest wildlife smugglers; and a Nepalese colonel who devotes himself to protecting the country's reserves and the tigers, rhinos, and elephants that live in them. The dedicated include conservation biologists and social scientists, without whose fundamental work conservation would be impossible and whose courage sometimes consists of challenging established political, scientific and advocacy stances that fail to produce effective environmental results, or of standing up for the people who are severely hurt by conservation policies. There are committed environmental NGOs who advocate for endangered species and ecosystems, and do so in thoughtful ways based on actual evidence and outcomes rather than just basing their work on ideology. Especially valuable is the sometimes dangerous work of investigative journalists who expose the corruption and criminal networks involved in wildlife poaching and global trafficking. In doing research for this book, I was fortunate to speak with many of these concerned champions of wildlife, and in the chapters to come I recount their policy experiences and cite their work. There are, of course, many more whom I have not met.

In the eyes of some of the affected communities and individuals, not all of these actors are heroes. Many poor communities, often historically marginalized, are dependent on hunting animals and natural resource extraction for basic livelihoods. Sometimes their own poaching can be separated from the poaching that supplies the global trafficking networks, while other times it cannot. For them, exclusion from protected areas, especially due to coerced displacement and the use of force, hurts their most immediate needs and is associated with the conservation policies of the colonial era. They deeply resent the rangers who seek to keep them from hunting in protected areas and prevent their cattle from grazing there, as well as the NGOs that advocate conservation policies that clash with their interests, needs, and rights. They complain that such "Western" conservation puts the wellbeing of animals ahead of their human needs. To align their needs and interests with efforts to oppose unsustainable poaching and wildlife trafficking requires that they are given economic, political, and normative stakes in conservation. But doing so, as my narrative shows, is not easy.

Many consumers of wildlife products similarly see it as their right to consume and utilize wildlife regardless of environmental consequences and see environmental advocacy against it as an intolerable encroachment on their lifestyle. If their material well-being can be assured from a legal supply, such as of captive-bred animals, then their interests and the interests of international conservation can be synchronized. But when a legal supply is not available, because some animals are too difficult to breed in captivity or because the legal supply serves principally as a means of laundering illegally sourced wildlife (that is, making it possible to hide the fact that the animals were illegally caught in the wild by falsely labeling them as having been bred in captivity or caught with a legal permit), reducing people's consumption patterns involves both changing norms and law enforcement.

The Global Poaching Crisis

Beyond elephants, rhinoceros, and tigers, many other animals are captured, slaughtered, and trafficked for trinkets, for TCM substances, or for the global pet trade. In combination with habitat destruction and global warming, hunting and poaching might eliminate entire genera of species, further weakening remaining ecosystems. The poaching numbers are staggering and devastating: Between 2010 and 2012, almost 100,000 elephants were killed for their tusks in Africa.[21]

According to a 2016 report by the International Union for the Conservation of Nature, African elephants have thus experienced a drastic net population decline of some 111,000 since 2006, leaving a likely current population of 415,000. (In parts of Africa that could not be surveyed, an additional 135,000 elephants could possibly exist.)[22] Estimates of recent poaching range from 20,000 to 50,000 a year,[23] though in 2014 a smaller number of elephant carcasses—13,511—was discovered in Africa.[24] Between 2009 and 2015, Tanzania lost between 50 and 60 percent of its elephant population (from 109,000 in 2009 to 43,000 in June 2015) and Mozambique some 48 percent.[25] By contrast, in India 217 Asian elephants were illegally killed between 2009 and 2014.[26] The smaller number reflects the much smaller population of elephants in India and the difficulty of locating them in forests. For many global trafficking networks, poaching elephants in Africa is easier. Indeed, many of the elephants illegally killed in India are killed by local people who are trying to defend their crops.

The five species of rhinos in the world are faring even worse as a result of poaching over the past decade. Some 98 percent of Africa's two species of wild rhinos exist in just four countries: Namibia, Kenya, Zimbabwe, and South Africa. In South Africa, over two-thirds—some 20,000 animals—are concentrated. Yet in South Africa alone, 1,215 rhinos were killed for their horns in 2014, up from 13 in 2007. In 2014, a critical tipping point was reached in South Africa, where the number of deaths was believed to have surpassed the number of births (a poaching rate that could wipe out all of Africa's rhinos by 2022 if not reduced).[27] In 2015, the total number of poached rhinos in South Africa and those that died as a result of poaching (such as orphaned infants) was estimated as being between 1,160 and 1,500.[28] The total number of rhinos poached since 2010 in that country alone is over 4,828, with the surviving population of rhinos believed to be about 20,000.[29]

The northern subspecies of the white rhino is already almost extinct in the wild. The last four specimens that could potentially breed were transported from a Czech zoo to the Ol Pejeta Conservancy in Kenya. Their horns were removed so poachers are not tempted to kill them, and they are guarded by armed wildlife enforcement officials day and night.[30] All the Asian rhino species—the Indian, Sumatran, and Javan—are also critically endangered and poached. In India, 141 rhinoceros

INTRODUCTION

were killed by poachers between 2009 and 2014.[31] The Vietnamese subspecies of the Javan rhino was extirpated in 2010 when the presumed last specimen was shot in Vietnam by a poacher for its horn.[32]

Both rhino horn and ivory are highly valuable products. A kilo of rhino horn currently fetches some $65,000 in China and Vietnam.[33] A kilo of ivory meanwhile fetches between $3,000 and $6,000,[34] ivory prices having doubled in recent years.[35]

Tiger products are also extremely profitable, with poaching putting huge pressure on the world's remaining tigers. Some 171 tigers were poached in India between 2010 and 2015, while 19 very rare Siberian tigers are known to have been killed in Russia in 2012 and 2013.[36] This has taken place despite the commitment of more than $100 million over five years by those countries who are home to tiger populations, with the aim of reducing poaching and doubling tiger populations by 2020.[37] Although a total of 171 animals may seem low in comparison with the tens of thousands of elephants poached, given that the worldwide tiger population is under 4,000, with particular subspecies numbering in the hundreds, such poaching rates are devastating and can easily drive species to extinction.

Other species are traded in greater numbers. For example, the eight species of pangolin, scaly anteaters whose bodies are covered in keratin scales, are declining precipitously throughout their range in Africa and Asia (with 94 percent of the Asian population wiped out in China). They are perhaps the world's most poached and trafficked mammals.[38] More than one million pangolins are believed to have been taken from the wild over the past decade,[39] while in 2011, between 40,000 and 60,000 were poached in Vietnam alone.[40] Meanwhile, the illegal trade in reptiles, which is also driving species to extinction, numbers millions of specimens per year for many species.

This global poaching crisis and the illegal trade in wildlife coexist with a large and equally expanding legal trade. Although sometimes the legal wildlife trade enables habitat conservation and contributes to species recovery, at other times it facilitates the illegal trade by enabling the laundering of illegally sourced wildlife and boosting demand for wildlife products, including illegal ones.

THE EXTINCTION MARKET

Policy Urgency and Opportunity

At this moment of acute crisis, wildlife trafficking is attaining unprecedented policy attention. Halting wildlife trafficking was the key topic of the September 2016 Conference of the Parties to the Convention on the International Trade in Endangered Species (CITES), whose goal is to facilitate the legal trade in wildlife and protect endangered species. The United Nations Security Council has also discussed the issue, and the goal of eliminating wildlife trafficking has been incorporated into the United Nations Sustainable Development Goals (SDGs). Wildlife trafficking has also been an important subject in the negotiation of the Trans-Pacific Trade Partnership (TTP) treaty. The governments of the United States, the United Kingdom, and Germany have increased spending on combating wildlife trafficking around the world, expanding the menu of diplomatic, law enforcement, alternative livelihoods, and demand-reduction efforts.

Perhaps most importantly, the United States and China have committed themselves to fighting global poaching and wildlife trafficking. For years indifferent and even outright complicit in global wildlife trafficking, Chinese law-enforcement agencies have confiscated and destroyed illegally sourced ivory and seized other illegal wildlife products, such as rhino horn and Asian black bear paws, though as yet systematic enforcement of wildlife regulations is hardly frequent or prioritized by the Chinese authorities.[41] Crucially, after years of international lobbying, in 2015 China promised to declare its ivory market—a crucial driver of elephant poaching and a facilitator of ivory laundering—illegal. At the 2016 Conference of Parties, China also strongly supported an approved non-binding resolution that all domestic ivory markets be closed down, a decision opposed by Japan—like China, a country with a large domestic ivory market.

The decision to ban the global ivory trade was made at the 1989 Conference of Parties after the massive hunting of elephants decimated their numbers. In the 1970s and 1980s, demand exploded in Europe, North America, and Japan, and started taking off in China. Demand vastly surpassed the legal supply, and poaching halved Africa's elephant population from 1.3 million in 1979 to 600,000 in 1989.[42] Notably, the 1989 Conference of Parties adopted the ban despite objections from nine African countries, along with Japan and China.[43]

INTRODUCTION

In a policy experiment to provide resources for conservation in Africa, Japan was allowed in 1999 to legally buy 55 tons of ivory from Botswana, Namibia, and Zimbabwe from stocks that had accumulated from deceased elephants. But Japan was not able to maintain strict control of the distribution of the ivory within its borders, and smuggling increased.[44] Nonetheless, in 2008, China too was given a license to buy 62 tons of ivory, while Japan bought an additional 39 tons.[45] As long as Chinese stores can produce a license (often fake), they are permitted to sell the ivory officially acquired in 2008, as well as mammoth ivory and ivory from before the 1989 ban. Like Japan, China has been unable to prevent the laundering of poached ivory in its legal market. Demand for ivory has skyrocketed and poaching greatly intensified.

In December 2016, China promised to end its legal ivory market by the end of 2017. In the original announcement in June 2015, China's top official responsible for CITES matters stated that a total ban on ivory processing and sales would come "very quickly," adding "one year, two years, three years, four years, 10 years. Is that quick or not quick compared to the history of the world?"[46] In fact, there was widespread expectation that the market would be closed down by the end of 2016. In 1993, China similarly announced a ban on the commercial breeding of tigers at its tiger farms and the selling of tiger products—without, as of the end of 2016, ever properly implementing it. Instead, Chinese authorities have been quietly issuing licenses to farms to sell tiger products,[47] even though such trade facilitates the laundering of poached tiger products. Demand for legal and illegal tiger products has not abated in China and seems to be increasing, and the smuggling of poached tiger products into China continues. Nor has the wild tiger population in China (which is in single digits) recovered since farms were established. And at the 2016 Conference of Parties, China objected to a resolution calling on all countries to close down their tiger farms, which are also present in Thailand, Vietnam, and Laos, and from which poached as well as farmed tiger products are also often smuggled. In saying domestic markets in tiger products were not the business of CITES since CITES only covers international trade, China contradicted its position on domestic ivory markets.

Meanwhile, China imposed a temporary one-year halt on the importation of ivory trophies from legal hunts in Africa, as trophy importation can facilitate the smuggling and laundering of ivory, as well as the products from rhinos and other poached animals.[48] In February 2015, China also imposed a one-year ban on the importing of ivory carvings, but it once again excepted ivory from before the 1989 ban.[49]

In December 2016, Hong Kong announced that it would phase out illegal trade in ivory by 2021.[50] Since Hong Kong too has long served as a major hub of ivory carving and laundering, the declaration was welcomed by wildlife conservation NGOs, though many call for faster implementation. At least 400 licensed sellers in Hong Kong are permitted to trade in ivory from mammoths or stocks predating the 1989 ban on the ivory trade. As in mainland China, ivory from poached elephants is widely mixed into the legal ivory supply, and is likely to significantly outstrip in volume actual legal stocks.

How the prohibition and, crucially, its enforcement are designed in mainland China and in Hong Kong, and what demand-reduction policies accompany it, will crucially influence its effectiveness. In the concluding chapter, I sketch what these designs could look like.

Often considered the second largest consumer of wildlife, including trafficked wildlife, the United States has similarly tightened its policies. There has been a near-total ban in the United States on commercial ivory since 2014, and new restrictions were added in 2015, prohibiting previously allowed imports of pre-1989 ivory and ivory from elephants that have died of natural causes.[51] Commercial importation of antique ivory has also been banned. Restrictions have been imposed to limit the types and number of wildlife trophies that can be brought into the United States, with trophy imports from Tanzania suspended temporarily until poaching there has been reduced, while from Zimbabwe they have been suspended indefinitely.[52] After intense lobbying by environmental NGOs, many airlines now refuse to transport ivory and other trophies. Only antique ivory can be traded across state lines within the United States, with New York and New Jersey having completely banned ivory trading. California has also passed tougher anti-ivory legislation than the federal government in order to limit smuggling and laundering. Perhaps as much as 77 to 90 percent of ivory sold in San Francisco and Los Angeles in 2014 was illegal, faked to look antique.[53] Not everyone has weighed in on the same side: the National Rifle Association opposes a total federal ban on trade in ivory, supporting trophy hunting and the trade in weapons with inlaid ivory.[54]

INTRODUCTION

Beyond ivory, another major global battle underway concerns whether a country should be allowed to sell its stocks of rhino horn in a one-off sale, and whether this move would facilitate conservation by increasing resources for conservation or fuel poaching and wildlife trafficking. South Africa signaled that it would request permission for this at the 2016 Conference of Parties. However, it did not; Swaziland did instead. Much to the frustration of several southern African countries, that permission was denied, as were Namibia's and Zimbabwe's requests to be allowed again to sell their ivory stocks.

In 1977, all five species of rhino were listed in CITES Appendix I, meaning that the international commercial trade in rhinos or their body parts was fully prohibited. The white rhino population subsequently recovered, while the numbers of black rhino continued to plummet. Over the next twenty years, the black rhino went extinct in eighteen African countries.[55] Efforts to tighten regulations and restrictions included repeated calls for governments to destroy all stockpiles of rhino horn.

However, in southern Africa, white rhino populations rebounded, and licensed, legally permitted, trophy hunting became a big business in South Africa, with exports of the trophies permitted by parties to CITES. The 2008 Conference of Parties also granted permission to both South Africa and Namibia to export annually trophies from up to five black rhinos, a critically endangered species,[56] and in 2014, a license to shoot a rhino in Namibia was auctioned to a US hunter for $350,000.[57]

The recovery of rhinos in southern Africa has been enabled by giving local ranch owners and communal conservancies economic incentives for conservation, and by effective local *in situ* enforcement. Thus, since 1997, South Africa has repeatedly requested a license to sell off its rhino-horn stocks. But because of the lack of established controls on legal trade, the burgeoning demand for rhino horn in Asia, particularly Vietnam and China, and the poaching and global trafficking crisis it set off, as well as corruption in South Africa's trophy-hunting sector and government regulatory and enforcement agencies, it has been denied the license each time. In 2015, however, a South African judge lifted a domestic ban on rhino-horn trade after two large rhino breeders in South Africa sued the government for infringing their right to trade in sustainably utilized wildlife codified in South Africa's constitution.[58]

Actions are being taken elsewhere. In February 2014, forty-six countries agreed to the London Declaration on Illegal Wildlife Trade, calling for poaching and wildlife trafficking to be considered "serious crimes" in the context of the United Nations Convention against Transnational Organized Crime.[59] That same month, the US executive branch released the *National Strategy for Combating Wildlife Trafficking*, calling for strengthening of domestic and global enforcement, reducing of demand for illegally traded wildlife at home and abroad, and building of international cooperation and public–private partnerships to combat wildlife poaching and trafficking.[60] In November 2015, the US House of Representatives passed the Global Anti-Poaching Act, placing wildlife crime in the same category as drug and gun smuggling, increasing penalties and making wildlife trafficking a liable offense for money laundering and racketeering. An equivalent bill was introduced in the US Senate. The new legislation also authorized the US Department of State to deny assistance to countries that fail to uphold international commitments to protect threatened species, replicating the highly controversial and rather ineffective drug-certification legislation. Countries that fail to comply with counter-narcotics commitments are subject to aid cutoffs unless they obtain a presidential national security waiver. In the 2016 Intelligence Authorization Act, the House of Representatives also mandated the director of national intelligence to report on wildlife trafficking networks and targets for disruption. The European Commission has established a €350 million (about $380 million) trust fund to promote conservation in Africa. The African Union is also drafting a regional strategy for halting wildlife poaching.

The Search for Effective Policy: Lessons from Trafficking in Drugs and Wildlife

The struggles over ivory and rhino-horn poaching are emblematic of a larger policy search for how to mitigate unsustainable hunting, design anti-poaching policies, and suppress wildlife trafficking. Although maintaining a ban on the sale of rhino products and declaring domestic ivory markets illegal makes good sense in the current context of the massive laundering of poached animals through the legal trade in wildlife, the overall policy lesson is not and should not be to end all legal sales of

INTRODUCTION

wildlife. Under some circumstances, legal sales from hunting or farming crucially underpin and enable wildlife conservation in a way that bans, prohibition, and law enforcement will not able to accomplish because they fail to give key actors an economic stake in conservation. Indeed, even the recovery of the white rhino throughout the 1990s in southern Africa was crucially underpinned by such market mechanisms. Allowing the legal trade of farmed crocodilians also resulted in the recovery in the wild of several crocodilian species. However, allowing legal trade has not always produced such desirable conservation outcomes. A legal trade can, and often does, allow for the laundering of poached animals through the legal trade. It may also boost overall demand, including undesirable demand for poached animals and their products.

Given the precipitous and irretrievable collapse of species, there is desperation in the conservation community to find policy silver bullets. Calls for bans on the legal wildlife trade and hunting, and tougher legislations, enforcement, and penalties, are often accompanied by unrealistic expectations of how effective interdiction and law-enforcement policies can be and how easy it might be to coerce those involved in wildlife poaching, including poor marginalized local communities dependent on natural resource extraction, to abstain from hunting. Characterizations of contemporary poaching and wildlife trafficking as something that does not involve poor communities and is solely conducted by organized criminal gangs and terrorist and militant groups, though widespread today, are often deeply flawed and easily lead to counterproductive policies. Radical proposals calling for shooting poachers on sight (often the *de facto* policy in various African countries, and at various times in India) are highly undesirable, both normatively and in terms of their efficacy.

Alternative livelihoods and other efforts to obtain conservation support from local communities, such as through community-based natural resource management (CBNRM), also have highly varied results. Sometimes they have worked very well; at other times they have not produced satisfactory conservation outcomes.[61] Thus the conservation community is increasingly looking at the successes and failures from decades of global drug policies to learn from them and enrich its own discussions and debates.

These debates are highly polarized in both the drug and wildlife fields. Many environmental NGOs advocate strict bans and call for tougher law enforcement.⁶² Indeed, this is the policy flavor *du jour*. For some, no legal trade in wildlife, particularly if it involves killing wild animals, should be allowed. Others only oppose legal trade under specific circumstances and in specific wildlife commodities, such as rhino horn and ivory. Some conservation biologists support the calls for far tougher law enforcement. Law enforcement authorities often also support bans, not merely because that serves their budgetary interests (as is sometimes alleged by opponents of bans), but because the existence of a legal trade alongside an illegal one significantly complicates the law-enforcement task of having to sort through what is legal and what is not, and because the legal trade provides loopholes for the illegal one to exploit.

Many conservation biologists and conservation economists oppose bans and support market-based mechanisms to promote conservation. They point out that bans have often failed, and that governments, businesses, and local communities need to be given material stakes in conservation or conservation will fail and species will be lost. Their adage is: Wildlife stays if wildlife pays. For some of them, the global destruction of ivory stocks that has taken place in China, Kenya, the Philippines, Sri Lanka, the United States, Gabon, and elsewhere symbolizes how emotions and passions get the better of effective conservation.

Many so-called critical conservationists, including some anthropologists, geographers, sociologists, political scientists, as well as some biologists, may also oppose global bans. In particular, however, critical conservationists are opposed to conservation policies that hurt poor, marginalized populations. Thus they reject law-enforcement-heavy policies and emphasize the historic injustice of colonial conservation policies that forcefully evicted and brutalized native populations around the world. They lament what they see as the disproportionate and unfair power of international environmental NGOs advocating bans and the establishment of protected areas from which local communities are prohibited from accessing resources. Critical conservationists thus fear that environmentalism is swinging back to the discriminatory approaches of exclusion, marginalization, and force from which it sought to distance itself in the 1990s. Not all critical conservationists endorse market-based mechanisms that allow the legal trade in wildlife. Particularly those of neo-Marxist or postmodernist persuasion see the market as another means of dispossession of local communities. Most critical conservationists, however—as well as some conservation biologists—call for CBNRM approaches that give local communities rights to local land and its wildlife, and empower them to make their own decisions over local resources.

These debates are hardly simple ones, taking place between different philosophies and ideologies or along North/South divisions. Different constituencies within a country—whether it is a supply, transshipment, or demand country—support and advocate different policies. Moreover, within countries at the same point along the production and trade chain there are differences. Kenya and South Africa are both supply countries for wildlife trafficking, and both are experiencing massive poaching rates. They share colonial legacies of environmental conservation often rejected by local populations. Yet Kenya opposes a legal trade in ivory, and in 1977 banned all hunting in the country, making next to no exceptions, even for subsistence hunting and CBNRM. South Africa, on the other hand, has repeatedly lobbied for allowing the sales of its ivory and rhino stocks, and has made economic incentives for conservation, including trophy hunting and the trade in wildlife, a key hallmark of its conservation policies.

In the drug-policy field, the lineup of who supports what approach is rather different. With the exception of marijuana, which is a mild drug in terms of intoxication and addiction, legalizing the drug trade is mostly promoted by NGOs and free-market and libertarian economists. Many drug-policy scholars, including criminologists, social scientists, psychologists, and medical experts on addiction, emphasize that advocates of drug legalization oversell its benefits and underestimate its costs and downsides. Even while such scholars point out the significant failures and costs of the war on drugs, opposing highly punitive approaches such as the incarceration of users or drug-crop eradication policies without effective and adequate alternative livelihoods being put in place, they do not necessarily support legalization. They are concerned that legalization, particularly if accompanied by intense commercialization, will increase addiction. They also argue that making the drug trade legal does not inevitably reduce the violence of criminal groups or bankrupt them. Often these scholars instead support decriminalization and the depenalization of use, harm reduction measures (also promoted by NGOs), differently sequenced eradication policies, and differently designed interdiction and policing of retail markets. Yet there remain many committed supporters of tough approaches toward drugs—not only among counter-narcotics officials in the West, but also among many NGOs and religious groups in Latin America and the United States. Even more so among most of the governments and publics of East Asia, the Middle East, and Russia, the drug-policy reform efforts underway in the United States, Latin America, and Western Europe find very few supporters.

In the wildlife trade and conservation domain, the debates are over means: whether bans and law enforcement or the market and legal trade are the best mechanisms to assure species survival and biodiversity, and whether the state or local communities are the most effective locus of decision-making over environmental policy. But the debates are also about values: For some environmental NGOs, killing animals is unacceptable. Conversely, advocates of local communities hurt by conservation policies point out that this sentiment puts animals ahead of people. Proponents of CBNRM efforts often argue that environmental conservation should not contradict efforts to empower local communities and promote their economic development, and in fact CBRNM is the best mechanism of advancing environmental conservation. Sometimes that is the case. But at other times, a local community, just like the state and business, may not have an interest in conservation. It may want to make money quickly and as much as possible by participating in logging, the conversion of forests into agricultural land, or poaching and wildlife trafficking. Even so, maximalist versions of CBNRM hold that an affected local community should be not merely one of the environmental policy stakeholders, but rather the principal authority for deciding how local natural resources are managed so that it can ensure that its security, wellbeing, and rights are not ignored by the preferences of globalist conservationists and local vested lobbies.

As with drugs, all of the policy options have worked sometimes, just as they have often not worked. Both fields are characterized by an enormous variation in policy effectiveness and outcomes, affected by local cultural and institutional contexts as well as exogenous factors. Proponents of particular policies often believe that if only more of their approach were adopted, then better and more consistent out-

INTRODUCTION

comes would follow. Supporters of CBNRM approaches thus point out that despite the growth of the size of national parks and protected areas and the increased militarization of wildlife enforcement policies advocated by many environmental NGOs, ecosystem degradation continues unmitigated, as do poaching and wildlife trafficking. Advocates of bans respond that more intensive and tougher enforcement is needed. Meanwhile, when CBNRM efforts do not bring about desired conservation outcomes, their proponents argue that the failures are not the result of a faulty concept but merely of inadequate and incomplete implementation of that concept, and that greater rights should be transferred to local communities.

Yet a major part of what can be learned from drug policy, and what can usefully inform conversations of how to deal with poaching and wildlife trafficking, is the absence of silver bullets, the limitations of the various drug policies, their failures and downsides, and the harm and costs of the policies themselves. Reducing demand is crucial, but difficult, and appealing only to altruism hardly ever works sufficiently. Interdiction, tough law enforcement, and tough penalties have not eliminated drugs and not dramatically reduced global flows. Anti-laundering efforts have not yet defunded illegal trading. Rarely have alternative livelihoods succeeded in weaning drug farmers off growing illicit crops on a large scale. Some of the failures are the outcome of poor design and insufficient resources, but in the case of many policies there are inherent limitations concerning the levels of effectiveness that can be achieved. Interdiction is simply hard. Anti-money-laundering efforts are like looking for a needle in a haystack.

The illegal drug and wildlife economies share many structural characteristics that enable and constrain policy effectiveness. But drugs and wildlife are not fully analogous, and in Chapter 2, I begin the analysis by comparing and contrasting the key similarities and differences between the two realms and unpack the meanings of "legality," "illegality," and "trafficking." I then highlight the structural characteristics of illicit economies that privilege interventions in supply, transshipment, or demand components of the illegal trade in each. These include the technological and institutional requirements for production of the illicit commodity or service; the dispersion of supply; the labor intensiveness of production and transshipment; the technological and institutional requirements for smuggling; the detectability of flows; and the nature and dispersion of consumption and demand. However, I also emphasize how the "social embeddedness" of illicit practices—that is, their local institutional and cultural context—shapes each illegal economy, and thus also needs to shape the particular thrust of policy interventions. I further analyze the differences between the illegal drug and wildlife economies, the variations in drivers and impacts within each, the uneven distribution of the threats, costs, and impact across and within countries, and of policies to mitigate them. I describe also the historic legacies and baggage of conservation policies, their intersection with colonialism, and the evolution of conservation approaches since the 1990s, including CBNRM.

In Chapters 3 through 5, I analyze the size and structure of the illegal wildlife economy and its actors, and contrast them with the size and organizational structures of the illegal drug economy. I also analyze the threats that poaching and wildlife trafficking and the illegal drug economy pose, and the costs that policies to mitigate them can generate, including by undermining the human security of poor marginalized populations. Chapters 6 through 10 explore in detail the various policy tools and their varied outcomes and effectiveness for both drugs and wildlife, and sometimes also for other illegal economies or in other public policy domains.

The analysis and policy implications presented are the product of years of studying various illegal markets in drugs, timber, and wildlife and their connections, or lack thereof with organized crime and militant groups. In addition to comprehensive reviews of the relevant literature, I have conducted extensive field research on the drug trade, militancy, and illegal logging, mining, and wildlife trafficking in Indonesia, China, India, Afghanistan, Nepal, Kenya, Tanzania, Ethiopia, Somalia, Djibouti, Namibia, Morocco, Brazil, Peru, Colombia, and Mexico. During fieldwork, I interviewed militants, illegal drug producers and traffickers, poachers, miners and loggers who worked illegally, consumers of wildlife products, law-enforcement officials, park rangers and wardens, operators of ecolodges and tourism industry representatives, environmental NGO activists, and conservation biologists. I draw on this research and bring in vignettes and illustrations throughout the analysis.

INTRODUCTION

The concluding chapter pulls together the policy implications and offers guidelines and recommendations for applying—or avoiding—the various elements and design of the policy tools analyzed in the preceding chapters. As a specific illustration, the conclusion uses those guidelines to sketch what the design, implementation, and enforcement of the upcoming ivory bans in China and Hong Kong could look like.

Here, anticipating the discussion of the last chapter but broadly outlined, are the key policy implications of the analysis and the recommendations that flow from them. I want to put my cards on the table—that is, to fully disclose to readers at the outset which basic policies I have found to be most effective, as well the values and philosophical predispositions underlying these findings. That said, I hope readers will find that I have been scrupulous in presenting the best arguments for and against the contending approaches and policies. I also admit that arriving at my own judgments as to what policies are most efficacious and just at this particular time for dealing with particular elements of the illegal wildlife trade has not been easy, as the evidence and outcomes are so mixed and contingent.

A Summary of the Key Findings

Given the increasing threat wildlife trafficking poses to the survival of vulnerable species and the current scale of elephant poaching, Beijing should implement its promise to close down its ivory market, and should do so as soon as possible, with enforcement measures in place to interdict and discourage the buying, selling, and trafficking of ivory. Other countries, such as Japan, should also do so. South Africa and other countries should not be given licenses to sell either ivory or rhino horn until both demand and poaching significantly decline and better monitoring and enforcement structures are in place in supply and demand countries.

However, if demand levels off and law enforcement is tightened so that a lot fewer illegally sourced products leak into the legal supply chain, then serious consideration should again be given to permitting some tightly controlled legal supply in both ivory and rhino horn. Providing economic incentives for conservation to as many stakeholders as possible greatly facilitates wildlife regulation enforcement.

Indeed, allowing sustainable hunting even in endangered species, such as for trophy hunting, as long as strict monitoring and controls can be put in place, should be the policy predisposition so that local stakeholders have material interests in promoting conservation. My preference is often toward permitting legal trade. Yet, when levels of corruption, law evasion, and the laundering of wildlife are pervasive, policymakers need to be willing to move to temporary and locale-specific bans, and sometimes even outright global blanket bans, for specific species. Any policy will need to be reevaluated and possibly changed if its outcomes are not positive.

Hunting can be sustainable and can benefit the environment, such as by keeping overpopulated species in check. This is particularly so if the numbers of keystone predator species are down, whether as a result of poaching, habitat destruction, or other causes. Poaching and unsustainable hunting of apex predators is particularly problematic as it has large-scale repercussions throughout an ecosystem. Indeed, not all poaching, just as not all legal trade, threatens a species' survival and health. Thus in buffer zones and even in core parts of protected areas, it makes good sense to allow limited hunting of non-endangered species and the limited sustainable extraction of natural resources to mitigate food insecurity and income losses of poor local communities. Thus it is crucial to make a case-by-case assessment of what type of hunting is sustainable at what levels. Careful monitoring and reassessments need to be conducted regularly and repeatedly to reassess whether wildlife populations and ecosystems are bearing up well, since the impact of limited exploitation may change over time.

Whenever possible, local communities should be given rights to land in conservation areas and to proceeds from sustainable wildlife utilization. Marginalized communities should receive assistance to secure their rights and should have a strong voice in the determination of land use and protection to achieve environmental equity and sustainability. The rights conferred should be limited, with restrictions applied on use to ensure the preservation of biodiversity: communities should not be allowed to destroy valuable biodiversity areas. This does not preclude sustainable logging or the hunting of non-endangered species for either subsistence or trophy hunting, or of limited grazing in protected areas.

INTRODUCTION

Such conditional rights should also include the community's entitlement to 100 percent of the revenues derived from sustainable wildlife management. However, the revenues should be taxed. That way, external government support for local community efforts can be established, and the state may have fewer incentives to collude in external poaching and deforestation or leave the community high and dry when other actors, such as wildlife traffickers or the logging industry, threaten its resources. All such arrangements need to be monitored and reassessed on a regular basis, with a strong input from local communities.

If allowing the exploitation of environmentally significant areas by local communities results in the significant degradation of natural ecosystems, then consultative and compensated resettlement may well be necessary. At a minimum, compensation must ensure that the local community is economically no worse off than when it lived in the protected area and, ideally, the level of compensation will also reduce poverty. That means, however, that proponents of exclusive protected areas must help in raising money to offset the consequence of resettlement and/or placing restrictions on land use. Indeed, conditional cash transfers should become a standard element of the policy conservation toolbox, particularly in circumstances where legal trade in a species is not viable and where ecotourism or alternative livelihoods simply cannot generate enough resources to offset the benefits foregone by local communities who were going to convert natural ecosystems, such as forests, into agricultural land or hunt threatened species.

Effective monitoring and law enforcement are crucial for the success of any of the tools analyzed in the book. Strengthening law enforcement regarding wildlife and reducing corruption among its agencies is critical. But human rights abuses by law-enforcement authorities should never be tolerated, and ineffective and reprehensible policies, such as shoot-to-kill and shoot-on-sight, should not be adopted. *In situ* law enforcement is by far the most effective element of trying to prevent poaching. Dismantling wildlife trafficking networks should focus on apprehending as much of their middle operational layers in one sweep as possible to minimize their regeneration capacities, a policy far more effective than arresting wildlife "kingpins" (as morally desirable as that may be) or flooding prisons with low-level poachers. But interdiction needs to be carefully designed so as not to produce the counterproductive effect of greater seizures leading to greater poaching. That is why *in situ* law enforcement and targeting of middlemen, as opposed to *en route* seizures, is so important.

Finally, retail and demand markets need to be reshaped—both through persuasion and messaging campaigns and through enforcement, though penalties should differentiate between small-level buyers of an ivory statue, for example, and speculators and heavy users of prohibited wildlife products. This recommendation rejects the premise that all values are relative. My visceral reaction to the Indonesian bird market was not one of opposing the hunting of all wild animals for the pet trade, though I certainly oppose the unlicensed hunting of endangered species. It was very much about the cruel conditions in which the animals were kept. That may well be a Western view, but if so, that should not discredit it. Some values ought to prevail universally: avoiding unnecessary cruelty is chief among them, just as stopping slavery and human trafficking should be promoted as important global values. Thus while emotionalism and ideological commitments should not cloud judgments about policy efficacy, minimizing avoidable harm, such as unnecessary cruelty toward animals or humans, should be an essential criterion in deliberations over and choice of conservation policies.

This study is not only about minimizing harm to the environment. It is equally about minimizing the cost and harm that remedial policies themselves cause, and this needs to be a crucial part of the conversation. Inevitably there will be difficult trade-offs and there will be disagreements. The goal should be to minimize losses as much as possible across the range of values and objectives, and tough choices will have to be made along the way.

Perhaps the most important finding and guideline of this book is that we must expect huge variation in policy outcomes. This is so for both drug policy and wildlife policy. What works well for a species in one locale may not work well for the very same species in a neighboring locale. What works well in suppressing demand at a given time may not work ten years later. How much a legal or illegal market can be shaped or reduced through supply-and-demand measures depends on their elasticities, which can change over time, as well as a host of other factors, such as local institutional and cultural settings.

INTRODUCTION

Policy thus should allow for experimentation. But the flexibility to experiment with policies, such as in specific pilot projects, needs to be combined with diligence and regular reevaluations and not depend solely on one-off assessments. Allowing for policy failure and thus policy adaptation is key. Otherwise, assessments will be obfuscated and species and ecosystems will irretrievably disappear. Sometimes avoiding policy failure may involve merely tinkering with design and implementation; at other times it may require a complete overhaul of policy, from introducing bans to allowing legal trade or revoking community management schemes. This, the central finding of the book, may be the one that very few readers will want to hear. Who wants to accept that their favorite approach may be inadequate at times? Maintaining flexibility and honesty so as to revise policies that are inadequate in particular cases can be very hard when emotions, philosophies, and interests have become vested.

In many ways, the book is a depressing story. Not only is the scale of species destruction catastrophic, there are no easy solutions and few positive outcomes. All of the five policy responses analyzed in this book—bans on wildlife trade, allowing legal trade, mobilizing local community buy-ins for conservation, going after the smugglers' money, and reducing demand—have failed far more often than they have succeeded. Successful applications of these approaches have been highly contingent on specific settings. This book does not seek to provide a universal blueprint for ending poaching and wildlife trafficking or identify the best solution. In fact, it is a call against such thinking. It analyzes both the failures and successes of policies that have been proposed and tried out, and it endeavors to derive best practices for each approach as well as realistic assessments of what kind of challenges and effects for each must be anticipated. For regulatory designs to have maximum effect in stopping poaching and wildlife trafficking, and thus saving species, they will have to be highly specific to a particular species and cultural and institutional setting as well as cognizant of the best practices and lessons derived from the efforts analyzed in this book and presented in the concluding chapter.

However, the fact that policy tools are highly imperfect, that there is no one-size-fits-all design, and that policy choices are highly contingent on specific settings does not mean that we should give up on efforts to stop poaching and wildlife trafficking. It is our moral imperative to do all we can to preserve the planet's species and biodiversity, and to do so as much as possible in ways that also improve the welfare of humans. It is in our own species' self-interest to do so, even to prevent disease pandemics, to preserve our natural heritage and revenue options for future generations, and to minimize the negative impacts of wildlife trafficking in supporting organized crime and militancy and in undermining governance. We must struggle with the policy complexities, conduct honest assessments of policy effectiveness, deficiencies, and alternatives in specific settings, and be ready to adjust policies in the service of environmental conservation and human welfare so as to avert future tragedies.

2

THE WILDLIFE TRADE AND THE DRUG TRADE

SIMILARITIES AND DIFFERENCES

The illegal trades in wildlife and drugs are transactional crimes. By violating procedures and laws, they supply a black market in prohibited or semi-prohibited products for which there is a demand. They are sometimes erroneously called "victimless crimes" in the sense that the buyers and sellers are working together, which makes them harder to thwart than predatory crimes, such as rape or robbery, that have identified victims who can serve as eye witnesses.

Poaching and wildlife trafficking and illegal drug production and trafficking are similar in terms of the threats they pose to states, societies, peoples, and nature. The two illegal economies also have similar structures. In both, there are some vertically integrated production and smuggling organizations, as well as (in fact, most of the time) some very loose ones, with many small traders and operators.

The two illicit economies share many other structural characteristics that influence the selection and effectiveness of the policies employed for countering them. Indeed, many of the policy tools available for dealing with drug trafficking and wildlife trafficking are the same.

However, the outcomes of the policies exhibit considerable variation due to other characteristics they share: their social embeddedness. The internationalization of norms against drug use and the consumption of wildlife products is vastly different in different communities across the world. The norms and regimes that govern the wildlife trade and the drug trade, and that establish illegality and prohibit production and trade, are seen by some populations, particularly poor, marginalized people mostly located in the global South, as illegitimate Western/ Northern constructs imposed on those who prefer to hunt animals or cultivate drugs. Thus policies to suppress both illicit economies also generate threats to the human security of marginalized populations. Moreover, in both illegal economies, the costs, harm, and threats posed by permitting production and trade or prohibiting them are very unequally distributed across the world. But the North/South dichotomy is simplistic and can obscure the different constituencies within the North and within the South that are pushing for radically different policies. The costs and benefits of these illegal economies and the various policies that have been developed to manage them are also unequally distributed within particular societies.

A very important difference between the two economies is that the commodity traded in the wildlife black market is highly and rapidly depletable whereas drugs are a non-depletable resource. Policy baselines and trends in the drug and wildlife economies also differ significantly. Yet much can be learned about how to combat wildlife trafficking by looking at what has and what has not worked in combating drug trafficking.[1]

Definitions of the Illegal Trade in Wildlife ... and in Drugs

Wildlife trade, which can be both legal and illegal, includes all sales or exchanges of wild animal and wild plant resources. Under the 1974 Convention on the International Trade in Endangered Species of Wild Fauna and Flora (CITES), to which governments voluntarily adhere, only trade that does not threaten the survival of a species is permitted. The basic presumption is that extracting natural resources and trading in them is allowed, except where and when the data show that those actions threaten species survival.

Today, CITES accords varying degrees of protection to more than 35,000 species of wild animals and plants, whether these are traded as live specimens for the pet trade or for body parts and products derived from them. Implementing enforcement for such a vast number is complex and resource-demanding. Indeed, when CITES was set up, neither its administrators nor signatories anticipated that protection of more than a few species would be involved.[2]

The trade in some species is altogether prohibited; in others, it is permitted with restrictions. All international commercial trade in species categorized as "endangered" and listed in Appendix I, such as tigers and orangutans, is illegal under CITES. Exceptions for scientific or conservation purposes, such as the export of licensed trophies, can be granted and require both export and import permits. Many other species, potentially at risk and vulnerable but not endangered, are listed in Appendix II. International trade in species listed in Appendix II may be authorized by the granting of an export permit or reexport certificate. No import permit is necessary for these species under CITES (though it can be required under national laws). Such permits and certificates should only be issued if the relevant authorities are satisfied that the trade will not be detrimental to the survival of the species in the wild.

Species listed in Appendix III are protected by national legislation of at least one country, and that country has requested the assistance of others in controlling trade in that species.

However, the fact that a species is not listed under CITES, and hence its trade is not illegal under international law, does not imply that the levels of trade in that species are sustainable and do not cause environmental damage. Indeed, often a species is added to a CITES list precisely because previously ill-regulated culling and trade have decimated it.[3]

As of 2016, there were a total of 183 country signatories to CITES who are referred to as CITES "parties." The parties meet every two or three years at the Conference of the Parties to make decisions and pass resolutions to interpret and implement the convention, such as assessing in which appendix a species should be listed or whether countries are complying with their treaty obligations. Between Conferences of the Parties, the CITES Standing Committee acts on behalf of the conference. It meets annually and assesses the implementation of the conference's resolutions and decisions and makes recommendations for parties to improve their implementation. In addition, the CITES regime also has a Secretariat in Geneva. The Secretariat facilitates meetings, provides evaluations, assistance, and technical training, sometimes undertakes scientific studies, and makes recommendations regarding the implementation of the convention. Neither the Secretariat nor the Standing Committee has an independent enforcement capacity. Implementation and enforcement are solely carried out by parties.

Illegal drugs are consumable chemicals banned in the jurisdiction in question or by international treaties. The principal international treaties that regulate the international drug trade—the 1961 UN Single Convention on Narcotic Drugs and the 1971 UN Convention on Psychotropic Substances—define which substances are fully or partially prohibited on the basis of their medical usefulness, addictive properties, and recreational highs. How prohibition and restriction on production, sale, and distribution are enforced is further specified in the 1988 UN Convention Against Illicit Traffic in Narcotic Drugs and Psychotropic Substances.[4]

Currently there is largely uniform agreement that so-called hard drugs, such as heroin and cocaine, are illegal. But such agreement is lacking for other drugs, notably alcohol (banned in Muslim countries) and regional drugs such as qat. Whereas the trend in drug policy for decades was toward further criminalization and punitive policies, the loosening of drug policies is under way in parts of Western Europe, Latin America, and the United States. In 2013, Uruguay became the first country to fully legalize cannabis (marijuana). Long before Uruguay's initiative, the Netherlands had effectively legalized the possession and retail sale of cannabis, but not its production or wholesale trade in it. However, this trend toward reform is not shared by China, East Asia, Russia, and the Middle East, which remain staunch supporters of prohibitionist and punitive regimes.

Like wildlife, drugs are also categorized and listed on international and domestic schedules. These schedules are supposed to detail the various threats and risks of abuse versus any medical benefits of a drug. There are international disagreements about how specific drugs are classified, such as about whether chewing coca leaf has, in fact, a high chance of creating addiction or whether cannabis has medical properties.[5] Whereas CITES lists species on the basis of threats to the survival of a species, illegal drug schedules are based on the level of threat that the consumption of the substance poses to the health of the individual, including the likelihood of addiction.

As in the case of drugs, there are disagreements about what constitutes illegal trade in wildlife. In this book, I apply the term "illegal" to any trade in wildlife (animals, plants, or their products) in violation of national or international laws, such as CITES. I use the same principle—that is, some violation of either national or international laws—to define illegal drugs. As with wildlife, illegality can apply to the production, trafficking, or use of such drugs. In other words, illegal commodities—whether drugs or wildlife—encompass commodities, the production and marketing of which are either completely prohibited by governments and/or international organizations, or partially proscribed unless the production and marketing comply with special licenses, certification, taxation, and other economic and political regulations. (Of course, some countries have laws against genetically modified crops. I do not discuss those in the book.)

Both the illegal trade in wildlife and illegal trade in drugs have three elements, any or all of which can be illegal: production, trade, and consumption. In the case of wildlife, one can thus distinguish between poaching (illegal hunting), which may be for local or international markets or for consumption. Wildlife trafficking is the illegal trade element. When I want to refer to both poaching and trafficking, I sometimes use the term "illegal wildlife economy." Similarly, I apply the term "illegal drug economy" to the combination of illegal drug production and illegal trade.

The Structural Characteristics of the Illegal Wildlife, Drug, and Other Economies ... And the Crucial Importance of Local Contexts

Certain structural characteristics of illicit economies, including drugs and wildlife, determine their varied susceptibility to countermeasures aimed at the supply, transshipment, or demand aspects of those economies. Because of these different structural characteristics, suppression of particular illicit economies places different requirements on government enforcement capacity and points of intervention. Optimal strategies for dealing with these illicit economies thus vary substantially, putting a different premium on demand, transshipment, and supply strategies and their subcomponents. The determinative structural characteristics include the following.

The Technological and Institutional Requirements for the Production of the Illicit Commodity or Service

This factor determines the ease of entry by new producers, the ability of the illicit economy to shift to other areas in response to supply-side countermeasures, and the visibility of production and the ability to detect it. In the case of nuclear smuggling, for example, the ability to produce highly enriched uranium or steal radioactive substances is largely restricted to a small and select group of actors. The ability to monitor and restrict supply is thus greatly facilitated. By contrast, the technological know-how and physical requirements needed to cultivate illicit crops and process them into illegal drugs or to poach animals are ubiquitous, so that supply-side suppression policies frequently result only in the relocation of production to new locales.

The Dispersion of Supply

This factor determines the resource-intensiveness of supply-side countermeasures and the ability of producers to escape suppression through generating numerous sources of supply. In the case of nuclear smuggling, supply is concentrated in relatively few well-known locations, even though the United States and the international community do not have easy access to all of them. Still, the concentration of supply facilitates monitoring and control mechanisms. In contrast, illicit crops and wildlife poaching are dispersed around the world, with few limits on areas into which actors can move (often only the fact that no wild animals are left there prevents them), making detection, interdiction, and the restriction of supply inherently difficult.

The Labor-Intensiveness of Production and Transshipment

Like the dispersion of supply, this structural characteristic affects the ease of detection. It also affects the state's ability to generate legal sources of livelihood for participants in the illicit economy as well as smugglers' and other sponsors' ability to generate political support by protecting the illicit economy. The production of illicit crops is by far the most labor-intensive illicit economy, employing hundreds of thousands to millions of people in particular locales. Nuclear smuggling networks, by contrast, are highly restrictive, with few participants. The smuggling of conventional small arms falls in between. Most of the time, poaching and illegal logging, mining, and grazing employ fewer people in any particular locale than illicit crop cultivation. However, these other illegal economies can still employ thousands or tens of thousands of people in a particular locale. Globally, over a billion people are dependent on animals and animal products for basic livelihood, and quite a large portion of environmental regulations restrict access to animal habitats. The greater the number of people poaching—for subsistence or for the international wildlife trade—the greater are the resource requirements and the challenges for law enforcement, and for efforts to provide the affected communities with alternative livelihoods.

The Technological and Institutional Requirements for Smuggling the Illicit Commodity

This factor determines the resource-intensiveness of transshipment countermeasures and the ability of the illegal producers to escape suppression by diversifying smuggling routes. The ability of technologically unsophisticated pirates to shift their area of operation on the open seas is extensive, while the resource-intensity demands on maritime interdiction by coastguards and navies are large. Similarly, the technological requirements for human trafficking are small; this is one of the reasons why there is not only a great prevalence of the phenomenon, but also why there is great diversity in the organizational structure of human-trafficking enterprises. Smuggling the body parts of dead animals is much easier than smuggling radioactive materials.

The Detectability of Flows

This structural characteristic influences the efficacy of interdiction policies. In the case of maritime piracy and drug smuggling, such criminal actions and shipments are comparatively visible, also due to the sheer volume of the illicit commodity or service, thus permitting at least the modest efficacy of interdiction policies. In the case of smuggling wildlife or nuclear materials, on the other hand, the quantities smuggled are frequently so small and the geographic areas so large that often only inside information from informers can result in successful action against those involved.

There are of course variations not just across illicit economies, but also within them: smuggling live animals, for example, is more complicated than smuggling animal body parts. Such differences can also facilitate detectability: a baby orangutan looks like an orangutan, and customs officials should be able to identify transporting them as illegal if there is no valid license; but a law-enforcement officer may not easily know whether a piece of meat comes from an endangered reptile or from a sheep.

Commodities also vary by their value-to-weight ratios. The higher the value-to-weight ratio, the easier it is to smuggle a commodity. Thus smuggling rhino horn or cocaine is much easier than smuggling exotic meats.

The Nature and Dispersion of Consumption and Demand

This is the least purely structural variable, and the one most intensely interactive with social embeddedness, as analyzed below. Structurally, demand can be either widely dispersed or highly concentrated among a few users. The concentration as well as different social drivers of demand determine the effectiveness of particular policy measures to suppress it. The nature of demand also varies widely in terms of the commitment of buyers. Demand groups seeking to buy nuclear materials look very different from demand groups seeking to buy ivory products. For example, terrorist groups trying to acquire nuclear or radiological capabilities are a highly limited group of very determined, dangerous actors. Most criminal and terrorist groups will be self-deterred from participating in nuclear smuggling or contemplating such weapons.[6]

In the case of the wildlife trade, on the other hand, as in the case of illegal drugs, consumers are a very large and diverse group. In Europe, for example, where wildlife consumption is a luxury and ecological sentiments strong, consumers can be sufficiently influenced through demand-reduction messages raising environmental awareness. The demand for wildlife products in China is also part of the market in luxuries, but here many consumers do not exhibit great attachment to conservation and see wildlife and nature only from a consumption standpoint. Demand reduction for them will have to be supplemented with other messages and measures. In large parts of Africa, Asia, and even Latin America, forest communities still hunt to obtain protein or the income they need to access basic services such as health care. Such needy consumers will be far less susceptible to normative messaging than "luxury" consumers. Efforts to reduce their demand for wildlife will instead need to be addressed through policies that provide them with legal protein, alternative sources of income, and regular health care. Unlike nuclear smuggling, the smuggling of conventional small arms more closely resembles wildlife trafficking in that there are many highly diverse and dispersed sources of supply and transshipment, and a widespread demand for them among militants and criminal groups.

However, while the structural characteristics of an illicit economy determine broadly how resource-demanding and easily executable the policy interventions are along the supply, transshipment, or demand vectors, the local institutional and cultural context—the social "embeddedness" of the illicit economy—determines the optimal shape of a particular policy intervention. For example, strategies to reduce illegal logging will need to be shaped differently if logging is driven by a demand for hardwood timber in external markets, or a demand for cleared land, or a demand to meet energy needs through firewood and charcoal consumption among a local (sometimes poor) population. The social embeddedness of an illegal economy and undesirable behavior also influence what policies are accepted by local communities and what rules they internalize.

The Differences between the Illegal Drug Economy and the Illegal Wildlife Economy

There are, of course, important differences between the illegal drug and wildlife economies, with significant implications for policy.

Addiction: A Further Difference in Demand

An obvious difference is that the consumption of illegal wildlife products is not addictive whereas the consumption of most illegal drugs can be.

The vast majority of drug users, perhaps 80 percent, are not addicts and have the capacity to stop using drugs either as a result of their own volition, or because of court orders or other forms of public monitoring. But 20 percent or so of users do become physiologically—or psychologically—dependent, with their habit being a chronic disease. Unable to stop, even with repeated treatment or despite (inappropriate) imprisonment, these "hooked" individuals will use drugs for decades, cumulatively consuming an estimated 80 percent of the volume of illicitly trafficked drugs, with a non-trivial impact on production and smuggling. These addiction effects are not present in the consumption of wildlife products or the illegal possession of wild animals.

Depletable versus Nondepletable Resources

The most significant difference between the drug and wildlife economies is that illicit drugs are a non-depletable resource. They can be grown and regrown or produced synthetically more or less indefinitely. Many wildlife species, by contrast, are rapidly depletable. Even though in theory wildlife species can reproduce in captivity, in practice captive breeding is frequently challenging, and ultimately dependent on wild stocks, with those that are released vulnerable to poaching.

In addition to the nondepletable character of illicit drugs, the great adaptability of drug producers and traffickers in response to law-enforcement policies, and the persisting demand for drugs, are key defining features of the drug trade. There is an increasing recognition that not only drug use and the drug trade generate intense harm and threats, but so do many antidrug policies. Slowly though not uniformaly around the world, drug policies in Western Europe and Latin and North America are moving away from focusing simply on the suppression of production and trade and toward minimizing harm and threats, including those caused by antidrug policies themselves.[7] In particular, reformers emphasize the need to decrease the violence, mortality, morbidity, and social disruption associated with the drug trade and drug use, and the need to reduce the vast numbers of people around the world imprisoned on drug charges. Decriminalization of drug use and public health approaches such as needle exchange and methadone maintenance, the legalization of marijuana production, and the reduction of penalties for nonviolent drug distribution are increasingly in vogue. For decades, the chief architect and enforcer of the global prohibitionist and punitive drug-policy regime, the United States, now wobbles between some reforms, such as state-led marijuana legalization, and some old tendencies—with variations state by state.

The speed of species depletion not only sets different objectives for the conservation of wildlife but also drives a policy urgency not paralleled in the effort to combat drug trafficking and use.

Policy Baselines and Trends

Unlike the case with drugs, the policy trend regarding wildlife has been increasingly toward greater criminalization of the illegal trade and much more severe penalties for poachers and traffickers, toughening CITES and domestic legislation, and augmenting law-enforcement resources. Such trends in environmental conservation policies, particularly increasing penalties and imprisonment, thus resemble drug policies in the United States in the 1980s and 1990s.

Increasing toughness is a very common response to much antisocial behavior, including cruelty toward animals. Even for drugs and street crime, it is only some countries that overemphasize law enforcement and incarceration, where the scale of incarceration is now triggering a backlash, chiefly the United States.[8]

The unsatisfactory results of drug policies also reveal why increased seizure is an incomplete and potentially misleading measure of policy effectiveness. Yet environmental NGOs and wildlife law-enforcement officials repeatedly report seizures of smuggled wildlife goods as indicators of the implementation of CITES policies. Reports by Chinese officials, for example, of the amount of ivory they have seized and destroyed are invoked to indicate the new motivation in China to crackdown on its illegal wildlife markets. Yet in the absence of good baselines, more seizures can indicate an increased level of smuggling. Wild assumptions are often made about whether interdiction captures 10 percent of illicit flows in any particular place or 5 percent or 40 percent.

In the case of wildlife, seizures are a problematic indicator for other reasons as well. Increased levels of seizure still mean that animals are being illegally killed or removed from the wild, thus signifying a fundamental policy failure.

Indicators such as seizures and the number of effective prosecutions of poachers and wildlife traffickers are at best indirect proxies of a government's determination to combat poaching. The real measures of policy success ultimately are an increasing stable population of wild animals and decreasing numbers of animals illegally killed or removed from the wild.

There are other differences between the markets in drugs and wildlife. Among them are: the different levels of violence in the two markets (typically much higher in the illegal drug trade than in the illegal wildlife trade); the different effects of drug consumption compared to wildlife consumption on children's productivity and social success, and the related, politically potent dynamic of parental lobby groups present in drug policy but absent in wildlife policy; and the level of public outcry against ill-looking addicts in public places (typically high across all countries) versus the general public acceptance of even gruesome wildlife markets in Asia.

Variations in Drivers and Impacts Within the Two Illegal Economies

There are also variations within particular illegal economies, not just between them. Wildlife consumption driven by a desire to ostentatiously display products from rare species to demonstrate status, power, and wealth is often very pernicious. Scarcity due to overharvesting and unsustainable exploitation of species will only increase the value of the illegal commodity; the market will not stabilize and will require intervention to correct it. Under such circumstances, a ban that reduces such visible displays of consumption may well reduce overall demand (not just its visibility) and is likely to be a highly valuable policy tool. On the other hand, merely banning the consumption of bushmeat driven by food security needs will likely only push demand underground without reducing demand overall. Thus there is a profound difference whether illegal bushmeat is consumed as an exotic luxury or for subsistence.

Both types of demand for bushmeat also differ from the demand for illegal drugs, which are consumed mostly for the purpose of getting high or for medicinal purposes, such as medical marijuana. Illegal opiates such as opium or heroin may be consumed as painkillers if legal pain medication is not available, as is often the case in developing countries. Suppressing people's desire to get high is different from changing people's perceptions of prestige and fun, or reducing subsistence-based consumption of an illegal commodity. And consumer responses to bans, enforcement, and demand reduction efforts will differ accordingly.

Another way to think of the differences within the two illicit economies is to consider the varied impact and harm that different elements of the illegal trade pose. For example, limited subsistence-based hunting of non-endangered species and limited grazing in protected areas may be fully consistent with sustaining biodiversity and healthy populations of species. Devoting resources to such low-impact violations of laws could thus be inefficient (enforcement should be dedicated to high-impact unsustainable industrial poaching), counterproductive (alienating local communities who could provide intelligence on industrial-scale poachers), and unethical (threatening food security and income sources among marginalized communities). However, even subsistence hunting that has a low impact at a particular time could be a gateway to industrial-scale high-impact hunting at a later date. For example, wildlife trafficking networks can take advantage of the knowledge of local individuals and employ them as trackers and hunters.

Once again, there are analogies with the drug trade. Most pharmacological and drug-policy experts believe that cannabis is much less addictive than cocaine or methamphetamines, and risks of lethal overdose are much smaller. These experts, and increasingly publics, in the United States and Latin America support permitting the legal production and trade in medical and recreational marijuana. But even marijuana is a performance-degrading drug.[9] Moreover, opponents of marijuana legalization see its use as a gateway to hard drugs.

A Brief History of the Evolution of Environmental Conservation Policies

The trend toward toughening law enforcement in the environmental field is not new: In some ways, the pendulum has merely swung back.[10]

Early environmental policies—in the United States and those promoted by the British and French powers in Africa and Asia—were colonial projects, carried out at the expense of local populations. The establishment of national parks in the United States, including

Yellowstone and Death Valley, often involved the forcible and harsh displacement of Native Americans so that white tourists and hunters could enjoy the parks without having to encounter and fear the presence of indigenous populations.[11] In the name of environmental conservation, such forcible displacement of local communities, often without compensation, and restriction of their access to protected areas became common throughout Africa.[12] National parks were promoted and advertised as devoid of people, making tourists and hunters averse to seeing local pastoralists and farmers inside protected areas.

For example, the iconic Kruger National Park in South Africa, the locus of today's rhino wars, was created by white men without regard for the economic and social consequences for African populations. Named after Paul Kruger, the tough Boer president of the Transvaal Republic, the park came to symbolize racial discrimination and white economic and social domination as local communities were pushed out from the protected land.[13] The park's first warden between 1902 and 1946, Lieutenant-Colonel James Stevenson-Hamilton, often credited with expanding and developing the park, also became notorious for clearing the park of black settlements and presenting it as an exclusive preserve for white South Africans. Stevenson-Hamilton's nickname, still the name of the largest tourist camp in Kruger—Skukuza—is derived from the Zulu verb *khukhuza*, meaning to scrape clean.

While poor black Africans were prevented from subsistence hunting in protected areas, rich white men could indulge in hunting in them for fun. African farmers and pastoralists were often portrayed as the sole agents of soil erosion and deforestation. Some scholars even argue that colonial overlords sought to humanize animals and dehumanize Africans in order to justify violence in the name of conservation.[14] At times, black populations were allowed back in to national parks when colonial authorities realized that they could not run the parks without them. These so-called black tenants had to pay rent in labor or in cash.[15]

Wildlife control policies also became inextricably linked to the general British colonial policy in the 1930s of concentrating African populations into dense and controlled settlements without economic opportunities and social amenities.[16] Environmental policies thus came to be strongly associated and directly overlapped with colonial oppression in the minds of many African and Asian populations. Not surprisingly, they felt morally justified and economically empowered by illegally hunting and exploiting protected areas. Poaching became not only a means of subsistence but also a form of rebellion against colonial rule.

When African communities mobilized to reduce or eliminate protected areas, they were mostly overruled. In Tanzania in the mid-1950s, for example, the Maasai mobilized to cut the famous Serengeti in half so that they could move through and graze their cattle in the area. The international conservation community, led by the famous conservationist and president of the Frankfurt Zoological Society, Bernhard Grzimek, mobilized against this move and the size of the park remained unchanged.

Postcolonial regimes in Africa and Asia centralized control over resources and land and largely maintained existing policies prohibiting the utilization of protected areas. This was particularly so in cases where ecotourism could bring valuable resources to the central government and elites. Regimes also centralized control over resources and land. Unlike Indonesia and Brazil, for example, which have derived large revenues from logging and mining and relatively small revenues from ecotourism, and which have thus at times been sluggish and unmotivated when it comes to enforcing environmental regulations in protected areas, many African countries have sought to enforce anti-poaching and deforestation laws in national parks.

Nonetheless, even in Latin America and Asia after the end of colonialism, ethnic minorities and indigenous populations often bore the economic and social costs of environmental conservation, with losses of food security, dwellings, incomes and livelihoods, and ways of life.[17] When insurgencies spread in Asia between the 1950s and 1980s, control over forests and the displacement of minorities meant environmental conservation and counterinsurgency efforts became intertwined in places such as India and Thailand.[18]

Even today, the purchasing of land for private reserves in Africa often entails, if not the outright forcible displacement of local people, the loss of jobs for farm laborers, many of whom have lived in the area for generations. New parks such as the Tarangire and Manyara National Parks in Tanzania, the Gir Forest National Park in India, and others in South Africa, Ethiopia, and Mozambique have led to evictions, inadequately compensated relocations, and the undermining of human security and economic and social wellbeing of local people.[19]

There are wide disagreements among scholars as to how many people around the world have been negatively affected by environmentally led evictions, displacement, and restrictions. Estimates for Africa alone range from 900,000 to 14.4 million people.[20] Some critical scholars speak of an "environmental colonialism" that seeks to save Africa from Africans, keeping it for animals and white tourists only.[21] They often blame national governments as well as large powerful environmental NGOs for perpetuating the displacement of people in the creation of new protected areas.[22]

In opposition to these exclusionary environmental policies dubbed "fortress conservation,"[23] a new environmentalism emerged in the 1990s aiming to put "community into conservation."[24] An important strand of this new environmentalism is so-called community-based natural resource management (CBNRM), which puts the social, political, and economic needs and rights of affected communities at the front and center, and which embraces three pillars—"benefits, empowerment, and conservation."[25] Coinciding with broader developments in the social sciences in the 1990s calling for the decentralization of power and emphasizing local communities' ability to govern the commons,[26] advocates of CBNRM argue that African pastoralists or forest-dwelling communities should not be viewed as criminals and degraders of the environment, but instead as citizens with rights, including rights to hunt or extract resources, and should be given authority for managing a country's conservation estate.[27]

Even before the new environmentalism was theoretically conceptualized, community-based conservation schemes began emerging from the ground up in Africa. In the 1960s and 1970s, for example, local communities around Kenya's Amboseli National Park were granted permission to charge fees for tourists hunting outside the park, as well as being given a portion of the park's revenues and granted access to social benefits such as water distribution systems. These compensatory measures were adopted as a way to mitigate the local community's anger at the park's establishment, which had led the community to slaughter many wild animals in and around the park in protest.[28] Similar initiatives emerged in Namibia, South Africa, and Zimbabwe. In South Africa, the new initiatives emerged in response to community demands to have their rights restored after the end of apartheid. In

Namibia, these approaches were pioneered in the mid-1980s in response to an acute poaching crisis, particularly of elephants and black rhinos. A general trend toward the decentralization of power and devolution or rights to local communities, including those over natural resources, was in vogue in the 1990s.

Indeed, environmental policies, like other public policies, often reflect broad political trends and institutional designs favored at particular times. Environmental policy is not merely about the physical condition of natural ecologies, but also about dominant notions of the role of the state, markets, and communities.

Not all conservationists have embraced the new environmentalism. Some conservation biologists are skeptical about the extent of the harm conservation is charged with having inflicted on local communities. They also question whether it is an appropriate or too ambitious goal to make environmental conservation policy responsible for poverty alleviation and the social and political empowerment of communities, and whether such a move will inappropriately raise expectations and involve conservation in difficult political struggles.[29]

Just how much of the new conservationism should be embraced is also an issue among environmental NGOs. Some, such as Birdlife International, Conservation International, Nature Conservancy, the Wildlife Conservation Society, and the Worldwide Fund for Nature, formed the Conservation Initiative on Human Rights to promote collaborative learning and support the practical implementation of human rights frameworks within conservation. Nonetheless, issues such as the presence of communities in protected areas continue to be difficult and divisive. In particular, as CBNRM outcomes have hardly been uniformly positive and as industrial-scale poaching has escalated, many environmental NGOs have demanded a return to tougher law enforcement. Some environmental NGOs also oppose the legal trade in wildlife, an attitude that advocates of CBNRM lament as they often promote the utilization and marketing of wildlife.

The Range of Policy Options and Variations in Outcome

The structural characteristics of the illegal economies in drugs and wildlife and their social embeddedness give rise to the same range of policy options: banning trade in particular commodities and enforcing prohibitions on trade and production; permitting legal trade, though perhaps with some restrictions; providing alternative livelihoods to those unjustly hurt by conservation or drug policies; shrinking or shaping the illegal economy through anti-money-laundering efforts; and discouraging demand for illegal products while strengthening demand for no-harm products.

One formulation of CBNRM which transfers all rights over natural resources to local communities goes beyond alternative livelihoods. It often involves permitting legal hunting and the legal trade in wildlife—in this, too, there are some analogies with the drug trade, such as permitting coca farmers in Bolivia to grow coca leaves for traditional use rather than for the cocaine trade. However, what is different about this CBNRM formulation is that it *de facto* puts human wellbeing clearly ahead of environmental conservation. The local community is not just one of many policy stakeholders but the actual primary policy decider.[30] The objectives of policy can thus become reordered: a community's economic wellbeing can take precedence over environmental conservation.

But neither drug nor environmental policies are made on the basis of purely rational analysis. First of all, the evidence is often ambiguous and subject to different interpretations. Moreover, in the case of both wildlife- and drug-policy design, the same policy (whether it involves bans, legal trade, or CBNRM) often produces vast variations in outcomes even in cases of the same animal or plant species or genus. For example, drug decriminalization (which some countries refer to as depenalization) and discretionary laws reduced the number of people imprisoned on drug-use charges in Italy and Portugal, but increased it in Brazil.[31] Bans on poppy cultivation and eradication drives suppressed the level of cultivation in Myanmar in the late 1990s and early 2000s, but the level increased again toward the end of the latter decade.[32] Arrests of drug traffickers and the dismantling of drug-trafficking groups in the United States has produced almost no changes in levels of violence in the drug market since 2000. On the other hand, in Mexico, between 100,000 and 150,000 people have died since 2006, when interdiction provoked a major turf warfare among drug-trafficking groups.[33] In the Mexican case, other factors, such as the hollowing out of the effectiveness of Mexico's

THE WILDLIFE TRADE AND THE DRUG TRADE

law-enforcement agencies due to corruption and criminal infiltration, also contributed to the scale of carnage.[34]

Second, the policies for countering these illegal economies are only one factor, and perhaps not even the most important factor, influencing their location and strength. The illegal trade in drugs is one or two orders of magnitude more violent in Latin America than in East Asia, despite the fact that the volume of illegal drugs produced, trafficked, and consumed in East Asia is no smaller than in Latin America. Drivers internal to the illegal market itself, such as fads and fashions among users, marketing strategies among drug dealers, balances of power among criminal groups, and behavioral socialization among drug traffickers, often operate independently from, and are able to elude, policy interventions.

Third, as in the case of CBNRM, policy evaluations and choices entail not only judgments about policy instruments, but sometimes also highly contentious judgments about policy objectives, their rankings, and the difficult (and often profoundly varying) ethical judgments that different people will make. Moreover, emotions are part of such judgments. A prominent drug-policy expert with whom I discussed this book suggested, for example, that the big advantage that strategies toward wildlife have over those toward the drug trade is that:

> we have no objection in principal to legal production. If it were cost effective to farm herds of a million elephants and flood the market with cheap ivory, there would be no particular objection to that sort of farming, as long as it was done humanely. By contrast, legal production of dependence-inducing intoxicants still creates enormous social harm. There is no important harm caused by consumption of the poached products; all the harm is in the supply chain. By contrast, the very reason we ban drugs is because they sometimes harm the users.[35]

In addition, they also harm the families of those addicted, and possibly the communities around them, such as through petty crime and disorderly behavior.

Of course, many proponents of animal rights will find the above attitude absolutely unacceptable. Some object to any killing of animals with intelligence levels close to humans, such as primates or elephants. Even if the elephants were not killed, and the ivory collected merely from animals that die of natural causes, some would object to keeping wild animals on farms—and in fact, some animal rights and environmental NGOs oppose legal farming of even crocodiles and reptiles. Many in the environmental community have strongly argued against the farming of tigers and any large cats, a policy preference endorsed at the 2016 Conference of Parties. Some take this position because legal farming, such as of tigers in China and Laos, can facilitate the trafficking of illegally caught wild animals. Other environmental advocates do not believe as a matter of principle that wild animals should be kept in captivity. For example, a decade ago, the prominent Canada-based International Fund for Animal Welfare (IFAW), which has some 1.2 million supporters worldwide, mounted an intense campaign to prohibit the use of elephants for tourist rides into the African bush since IFAW considered it of no conservation value and cruel, even though local communities wanted to develop this kind of tourism.[36]

People have similarly vastly different philosophical precepts toward how to deal with vices and addiction. Many proponents of drug legalization, for example, argue that not only will addicts get better access to treatment if drugs are legal, but also that they have the right to become addicts (and destroy their lives) if they choose to do so, and that the state has no right to regulate such personal decisions.

Nations as well as individuals differ in these judgments about liberties and what the state or international community has a right to regulate. Thus the Scandinavian countries, in which the state paternalistically regulates a wide range of social behavior (mandating, for example, that people take a minimum amount of vacation, and discouraging people from working overtime) tend to have very restrictive and prohibitionist policies toward drug use. By contrast, in drug-violence-plagued Colombia, drug consumption is not illegal and the state takes the position that if drug users want to ruin their lives, that is their right and business, even as health care and support for drug addicts are vastly underprovided in the country.

In sum, policy outcomes are highly contingent, and policy objectives, implementation strategies, and tools need to be tailored to local cultural settings as well as to the structure of the particular economy and ecological realities.

3

THE CHARACTER AND SCALE OF THE ILLEGAL WILDLIFE ECONOMY

As with the illegal drug economy, at the core of the illegal wildlife economy is a strong demand. Unfortunately, the demand for illegal wildlife products and indeed also for many legal wildlife products has expanded dramatically over the past two decades and at rates that generate unsustainable hunting. These rates have had a devastating environmental impact, and the numbers of harvested animals can be staggering in their volumes.

The Varied Demand Behind the Illegal Wildlife Economy

For many people, wildlife is an important source of protein, and for particular marginalized and forest-dwelling communities it is the most important source of protein. Many forest-dependent communities are among the world's poorest and they are economically reliant on and culturally accustomed to hunting.[1] In Central Africa, where the hunting of wildlife, including endangered primates, has long been a hot and difficult environmental policy topic, bushmeat can provide up to 80 percent of a family's protein needs and up to 70 percent of household income, and be the dominant cash source.[2] The fact that high rates of cattle disease in these areas prevent cattle rearing complicates finding legal and sustainable protein replacements. Cutting off the access of such communities to bushmeat can undermine their food security and further impoverish them, even as their bushmeat consumption can have disastrous effects for highly endangered species. In Uganda, a country with high population growth and large, poor, forest-dwelling communities, for example, even endangered primates are hunted for subsistence. (Uganda is also an important smuggling hub for ivory and wildlife illegally sourced in South Sudan and the Democratic Republic of the Congo.)

Even as local communities may be dependent on wildlife and forests for basic livelihoods and have practiced natural-resource extraction for centuries, their consumption rates can surpass sustainable levels, such as when birth rates and life expectancy increase. Sometimes, even as a result of environmental conservation efforts, the increased affluence of such communities can paradoxically result in greater bushmeat hunting rather than the substitution of other forms of protein with the improvements in income worsening, rather than reducing, poaching.

Indeed, a very large part of the global market for wildlife and illegal wildlife products is about affluence. Such markets are extensive in Asia, where much of wildlife food consumption is of expensive exotic meats. Turtles, civets, pangolins, salamanders, wild populations of wrasses, groupers, and sharks are literally eaten away by Asian consumers. Anything can be served (and purchased) in restaurants specializing in exotic items: the rarer, the more appetizing and pricier.[3]

Sellers and markets specializing in the exotic are particularly problematic from a sustainability perspective because in rarity markets, demand goes up with price instead of declining with it. If demand for a depletable good actually increases with price, that wipes out the tendency of markets to approach a stable equilibrium. Economic pressures on species extinction grow substantially.

Wildlife products also include curios, trophies, collections, and accessories, be they Japanese ivory *hankos* (personal name seals) or other ivory carvings, turtle carapaces, coral, beetles, horns, and antlers. Some rhino horn consumption falls into this category, with rhino horn libation cups considered a symbol of great prestige. Such vessels have been produced and prized around the world for millennia: in ancient Greece, rhino horn was believed to purify water; centuries later in Europe, the Middle East, and Asia, rhino horn cups were assumed to detect and sometimes neutralize poisons.

THE CHARACTER AND SCALE

Skins from reptiles (particularly crocodiles and snakes) and from pangolins and muntjacs (an Asian deer species), furs (such as from snow leopards, clouded leopards, and tigers) and wool and hair from many animal species are used in the production of boots, clothing, footwear, and shawls (including the shahtoosh shawl from Tibetan antelopes).[4] Some are used for traditional costumes, such as in Tibet, where this practice is centuries old and the costumes are highly valued. Some go to the modern fashion industry, catering, for example, to the newly emerged market for furs, such as in Russia. Boots and leather products with snake and reptile skin accessories are loved in Mexico and Latin America and among Hispanic communities in the United States.

Data about demand are not only incomplete, but also inadequate for effective policy design. Such data paucity is acute for non-Asian markets. Fewer systematic studies of the consumption of ivory and other wildlife products have been conducted in the United States than in China, for example; there is little understanding of which groups or types of individuals in the United States are key consumers of what products, and how much wildlife consumption has expanded beyond particular subgroups, such as ivory and rhino horn among East Asian Americans, reptile skins among Latino Americans, or illegal trophies among rich white Americans. Even fewer studies have been conducted on existing or expanding demand for illegal wildlife in Latin America.

The pet trade is dominated by reptiles and birds, such as parrots and songbirds, but includes also tropical fish and mammals, including tigers, lions, and jaguars, orangutans, and other primates. A large part of that trade is international, such as the trade in songbirds between Indonesia and China or parrots between Central America and the United States, but much remains, often with little scrutiny, within countries, such as the illegal pet trade involving parrots and monkeys in Brazil.

Underemphasized demand also exists in Africa. In West Africa, for example, the cane rat, along with elephants and crested mangabay monkeys a major source of crop damage, is also a prized delicacy. Demand for its meat is driven by relative wealth in urban centers as well as the poverty and protein-deficiency of forest-dwelling populations. In Zambia, local communities often illegally set savannahs on fire to stimulate short green growth for livestock. As the scared animals flee the fire, entire villages line up to hunt the fleeing wildlife. Meanwhile, in the Mediterranean and North Africa, captured songbirds are widely sought out as a delicacy as well as pets.[5] In Kenya and Tanzania, poaching takes place not just for the international trade in ivory and rhino horn but also as retaliation against lions and leopards who killed livestock, and against elephants and wildebeest for trampling crops. In much of Kenya's Tsavo National Park as well as the Maasai Mara National Reserve, game-meat poaching is common, with the meat from large and small antelopes, hippopotami, and other animals sold to regional markets in Kenya, Tanzania, Uganda, and Rwanda.[6]

A very large part of the global demand for wildlife arises out of the practice of Traditional Chinese Medicine (TCM), which uses natural plant, animal, and mineral-based materials to treat a variety of illnesses, maintain vitality and longevity, and enhance sexual potency. Dating back at least 3,000 years, TCM practice is deeply ingrained in the culture of East and Southeast Asian countries, especially those with large Chinese populations and influence, and practiced by hundreds of millions of people. For centuries, tiger penis and snake blood have been prescribed as elixirs and aphrodisiacs. Tiger bone has been used to suppress arthritis, rhino horn to treat fever, convulsions,[7] delirium, and "oppressive ghost dreams,"[8] and bear bile to cure infections and inflammations.[9] Although effective medicinal alternatives are now available to many (though not all) people in East Asia, and although many TCM potions fail to cure anything,[10] so-called *ye wei*, "wild taste" (that is, the prized consumption of wild species) continues to expand greatly. With globalization and the expansion of Chinese diasporas around the world, demand for TCM has spread worldwide.

Much of TCM is quite untraditional. Demand is not static with respect to either consumer groups or promoted goods. Traders invent the desirability of products, sometimes attributing TCM qualities to products that historically have not been part of TCM, other times aggressively promoting new attributes of desirability. For example, rhino wine and rhino alcohol (ground-up rhino horn mixed with wine or alcohol) were invented during the past two decades as symbols of wealth, status, power, and prestige.[11] Gifting such products is also a mechanism of establishing friendships and business partnerships, cementing weddings, and buying loyalty or favor.

The top ten importers of TCM from China are the United States, Germany, Hong Kong, Japan, Korea, Vietnam, Malaysia, Taiwan,

Singapore, and Indonesia.[12] Many are key markets for other illegally sourced wildlife, including fish. Japan, for example, is a critical player not only in illegal fishing and the global overexploitation of the oceans, but also in preventing the tightening of fishing regulations, such as stricter quotas on the collapsing population of bluefin tuna; it also insists on maintaining whaling, and imports whale meat from the two other countries that still practice whaling, Norway and Iceland.[13] European countries too are heavily implicated in illegal fishing. The European Union, for example, is believed to be a key poaching and smuggling center for European eels, with between 20 and 40 percent of this trade derived from poachers and unlicensed fishermen.[14] The EU is also one of the main markets for illegal caviar, with more than 16 tons impounded between 2001 and 2010.[15] Between 20 and 32 percent of the weight of wild-caught seafood imported to the United States in 2011 was likely either illegal or unreported.[16]

In many East Asian countries, including China, little non-consumptive value, such as the pleasure of watching or photographing wild animals without killing or removing them from the wild, is attributed to wild animals or natural ecosystems. Rather, nature and wild animals and plants are seen through the prism of their utilization as sources of income, food, and prestige, and they are thus prized for their consumptive value. Unlike in India, where many animals are considered sacred or at least deserving of protection, and where many people abstain from meat consumption for religious reasons, in China and among overseas Chinese communities normative and cultural habits push in the opposite direction. As Li Zhang, Ning Hua, and Shan Sun note in their survey of wildlife consumption prevalence and consumer preferences in southwestern China, "[f]rom a traditional Chinese perspective ... wild animals are a resource to be exploited, not something to be protected for their intrinsic value."[17] Although this is changing to some extent as a result of international lobbying and environmental-awareness campaigns, similar attitudes tend to be prevalent throughout East Asia. Wildlife is seen primarily in terms of consumption, with wildlife conservation often considered a Western imposition and luxury. That is not to say that nowhere in Asia or Africa do people intrinsically appreciate nature and wild animals. Some indigenous communities have traditions and taboos which enhance conservation efforts, and many rural Asians and Africans want their grandchildren to be able to enjoy seeing (and hunting) wild animals.[18] However, environmental consciousness tends to be much lower than in the West.

Consumption of wildlife to indicate status is very important in East Asia. A significant consumption group in China is that of young males with a good income and, disturbingly, high education levels. (At the same time, the percentage of people who believe it is not right to consume wild animals is far lower among people with primary education and below than among other groups.)[19] Overall, according to surveys, some 30 percent of respondents in China consume wild animals while some 40 percent think no wild animals should be consumed.[20] In addition, 57.5 percent of those who consume wildlife voluntarily started doing so as a result of word of mouth, the media, introductions by friends, relatives, and acquaintances, or the influence of authorities, indicating peer-pressure dynamics. Heavy consumers are least likely to forgo consumption—despite awareness campaigns, and even while they contribute money to conservation efforts.

In addition to growing demand, other factors have contributed to the expansion of the illegal wildlife trade by facilitating supply. Over the past two decades, various countries in East Asia, including crucially Vietnam and China, have opened up their economies and strengthened their international legal and illegal trade connections. Infrastructure development has connected previously inaccessible wild and rural areas to regional and global markets. Commercial logging has also increased access to wilderness areas for wildlife exploitation. Similarly, the growing presence of Chinese businesses in Africa and Latin America has enabled the establishment of new supply chains for legal and illegal wildlife products from these regions to East Asia and East Asian communities worldwide, setting off a new global poaching wave. Although wildlife-law-enforcement agencies in Africa have become intensely focused on wildlife trafficking involving East Asia, Latin America's focus on this issue remains low.

The Scale of the Illegal Wildlife Economy and Comparisons with the Illegal Drug Trade

Measuring illegal economies is extremely hard. Scholars, practitioners, and NGO policy advocates strongly disagree about which measures are appropriate and which numbers are correct. Sometimes data are lacking to credibly resolve policy debates, and pointing out problems and discrediting the validity of particular estimates is easier than providing undisputable authoritative estimates.

For example, some reports estimate that the illegal drug economy is worth between $300 billion and $500 billion annually, a very wide range to start with, but even that range and order of magnitude are questioned by others.[21] The most recent estimate by the United Nations Office on Drugs and Crime (UNODC) is now more than a decade old. In 2005, UNODC put it at $322 billion annually, representing 0.9 percent of global GDP.[22] It is possible that the illegal drug economy has grown over the past decade as new users in China, East Asia, the Middle East, Brazil, and Argentina have entered the market. The overall quantity of drugs consumed has very likely increased, but it is not clear that the overall amount of money spent on drugs has also increased. Many of the dollar figures come from countries where prices are high, and prices have fallen a lot in places where drugs used to be expensive. However acute or grossly exaggerated or underestimated this number is, the illegal trade in drugs is widely believed to be the world's largest illegal economy. Some 246 million people, or 1 in 20 people between the ages of 15 and 64 years, were believed to use an illicit drug in 2013.[23]

In terms of dollar values, the illegal wildlife trade is likely an order of magnitude smaller. Combining illegal trade in animals and their parts with non-timber plants, illegal fishing, and illegal logging, the United Nations Environment Programme (UNEP) estimated in 2014 that the total illegal wildlife trade was between $50 billion and $150 billion per year. Of that, UNEP valued illegal fishing at $10 billion to $23.5 billion a year; illegal logging at between $30 billion and $100 billion; and the illegal trade in animals and non-timber plants at $10 billion and $26.5 billion.[24] Working in collaboration with the INTERPOL, UNEP revised its estimation in 2016. It put the value of all environmental crimes at between $91 billion and and 258 billion annually, a 26 per cent increase from its 2014 estimate. Arguing that environmental crime is rising by some 5 to 7 per cent yearly, it also assessed that annually the illegal wildlife trade amounts to between $7 billion and $23 billion, forestry crimes to between $51 billion and $152 billion, and illegal fishing to between $11 billion and $24 billion.[25] A World Bank article assessed natural resource crimes such as poaching, illegal logging, and wildlife trafficking to amount to a "US$213 billion industry."[26] A WWF 2012 report asserted that the global wildlife trade (not counting illegal fishing and timber) was worth $19 billion, a number widely cited,[27] while a US Congressional Research Service report assessed the global size of the illegal wildlife trade at $5 billion to $20 billion.[28] Defenders of Wildlife, another prominent environmental NGO, estimated in 2015 that the value of illegal wildlife trade in the United States was $2 billion annually, compared with a US legal wildlife trade worth three times as much or $6 billion annually.[29]

Surprisingly, the size of the illegal wildlife trade in Vietnam, one of the largest source, transshipment, and consumer countries for wild animals and plants, was estimated only at $66.5 million annually in the late 1990s and early 2000s,[30] which could be due to the fact that prices for most legal and illegal wildlife products are much lower there than in the United States. That differential does not hold for every desirable wildlife product, such as rhino horn, which tends to be most expensive in places such as Vietnam and China. Indeed, Vietnam has even surpassed China as the largest consumer of rhino-horn products.[31] Thus it is likely that the $66.5 million figure is a significant underestimate of the current size of the Vietnamese market.[32] Similarly, the global illegal ivory trade was estimated at $264 million from 2000 to 2010,[33] but its value today could be higher. As the next chapter shows, the social costs, the threats and harms that the illegal wildlife economy poses to states and communities are much higher and more complex than is expressed merely by the overall estimated monetary value.

As with all illicit economies, estimating the value of trade in illegal wildlife is inherently difficult as much of it is clandestine, hidden, or minimally monitored. Many numbers put forth—whether about the illegal drug trade or illegal wildlife trade—may be no more than guesses and myths.[34] For example, a widely cited statistic is that an estimated 350 million plants and animals are sold on the black market every year. Although this figure is often presented as "current" and attributed to a large variety of sources or cited without specific attribution, its provenance is in fact unclear and goes at least as far back as 2001, when a WWF report used it.[35]

THE CHARACTER AND SCALE

Estimating the size of the illegal wildlife economy is difficult not only for global aggregate dollar values, but also for specific species. The difficulties also pervade estimating the poaching, trafficking, and sustainability rates of particular species. There are no regular and consistent mechanisms for monitoring many wildlife populations throughout the world. Unlike the illegal drug trade, the illegal wildlife trade globally or in specific regions was not necessarily regularly monitored until 2016. In May 2016, UNODC released the first issue of *The World Wildlife Crime Report: Trafficking in Protected Species*,[36] the equivalent of its yearly *World Drug Report*. Like the *World Drug Report*, it is based mostly on data that other entities—governments, NGOs, or UN-sponsored monitoring groups such as WWF-TRAFFIC—report to UNODC. Such data come with severe limitations, including underreporting or overreporting for political purposes, or simply due to a lack of capacity. Some countries, for example, choose not to monitor their drug production or trafficking at all. Nonetheless, the yearly UN "World Drug Report" and other regular UNODC drug-crop monitoring estimates provide at least one consistent, if highly imperfect, monitoring mechanism, and generate time series.

For many wildlife species, only NGO monitoring, such as by TRAFFIC, or sporadic scientific studies provide data on wildlife population levels or poaching levels. Such data tend to be limited, with time series lacking. For some species in particular regions, data might be a decade or two old. While species population trends do not express how many animals are illegally poached, they indicate the most fundamental issue: whether the species' survival and sustainability are being maintained, regardless of how much poaching goes on. (The flourishing or decline of a particular species are also critically dependent on its habitat preservation as well as other factors.) A crucial reason for the paucity of consistent monitoring is a lack of funding, often dependent on the vagaries of donors. Only after almost two decades of a major poaching crisis has a systematic attempt been made to monitor elephant populations in Africa. This is now being done under the aegis of the Great Elephant Census, with funding from Microsoft's cofounder Paul Allen.

Even with highly imperfect data likely underestimating the size of the illegal wildlife trade, the amount of wildlife removed from the forests and waters of the world is shocking. For example, the annual hunting rates of bushmeat are estimated at 1 to 5 million tons in Central Africa and 67,000 to 164,000 tons in the Brazilian Amazon.[37] In Malaysia's Sarawak province alone, 23,000 tons or 2.6 million animals were shot yearly for bushmeat and exotic meat during the 1990s, while in neighboring Sabah, a staggering 108 million animals were killed annually.[38]

Southeast Asia overall is one of the global hotspots of poaching and wildlife trafficking. Between 1998 and 2007, over 35 million CITES-listed animals were exported from there, including 16 million seahorses (a prized delicacy in East Asia) and a further 100,000 other fish, 17.4 million reptiles, 1 million birds, 400,000 mammals, and 300,000 butterflies.[39] Out of these, over 30 million (approximately 300 species) were wild-caught, with the rest derived from captive-breeding programs.[40] In addition, 18 million pieces, or 2 million kg, of live coral was exported.[41] Out of the total exports, the proportion of illegal to legal trade in wildlife was relatively low, involving less than a quarter of a million specimens over the decade.[42]

Nonetheless, the legal trade often serves to launder illegally captured animals, with illegally caught ones labeled as captive-bred or legally traded.[43] In the early 2000s, an estimated 20 million seahorses were taken annually from the South China Sea and Gulf of Thailand, of which 95 percent were destined for China via Hong Kong.[44] In 2011, 20 tons of dried seahorses were seized globally, with half of the global seizures made in Peru alone, suggesting a much greater illegal trade in other parts of the world with weaker enforcement.[45] Apart from the enormous scale of the legal and illegal wildlife trade, these numbers also show that the "size" of the market depends substantially on whether one measures it by weight, monetary value, or other measures, such as damage to the environment, economic potential, or human subsistence.

However, since many of the above numbers were obtained from official documents governing trade in CITES-listed species, and thus represent mainly legal trade or seizures of illegally traded items, the actual volumes of illegally sourced and traded wildlife are likely much higher. For example, in the 1990s and 2000s, some 2 million box turtles were exported from Indonesia annually, greatly exceeding the official Indonesian quota of 18,000 and making 99 percent of the trade illegal.[46] Similarly, the trade in Tockay gecko from Java amounted to some 1.2 million individuals a year, enormously exceeding the Indonesian quota of 25,000,[47] and implying that 98 percent of the trade was illegal. In 1999 and 2000 alone, 25 tons of wild freshwater turtles and tortoises were caught and exported each week from northern Sumatra to China, amounting to about 1,300 tons a year just from one small region.[48] Over 50 percent of Asia's freshwater turtles (forty-five species) are now considered in danger of extinction as a result of overexploitation.[49] Almost 30,000 items made from the critically endangered marine hawksbill turtle were found on sale in Vietnam in 2002, signifying the death of thousands of turtles.[50] Other illegally traded reptiles include cobras, pythons, monitor lizards, and crocodiles. The carapaces and skins are prized as decoration and luxury objects, such as for clothing accessories; their meat, allegedly containing aphrodisiacal or curative properties, is considered a delicacy. Sometimes they are kept as illegal pets.

Sea turtles too are poached for their eggs, meat, and shells, or illegally caught as by-catch. Between 1980 and 2014, over 2 million sea turtles were captured globally. Around the world, at least 13,900 sea turtles per year are illegally taken, with poaching hotspots including Indonesia, Fiji, and Mexico, where between 2000 and 2014 some 65,000 were illegally taken.[51] Some estimates of the poaching of marine turtles are much higher: according to one report, 100,000 green turtles are killed in Indonesian and Australian waters alone each year.[52] Although no worldwide data on the total population size of green turtles exist, based on beach monitoring the population size of breeding females is believed to be between 85,000 and 90,000.[53]

In critically endangered species, the numbers of killed animals can be far smaller, but the detrimental effects on species survival are immense. Feeding the strong Chinese market, fifty-one tigers, for example, were illegally killed in Sumatra between 1998 and 2002, out of a population of around 800 individuals before 1998.[54] The total tiger population in Indonesia is now estimated at only 371.[55] In northeastern Laos, seven tigers were killed during 2003 and 2004, and their bones sold for about $50,000. There, as well as in Vietnam and China, wild tigers number in single digits. A shocking 2,200 tigers were estimated to have been killed by poachers in India over the last decade, sharply reducing its population to today's 2,226 and critically threatening three decades of tiger conservation efforts.[56] Positive news of tiger conservation came out in April 2016, when the total population of tigers left in the wild was estimated at 3,890,[57] up from a 2013 estimate of 3,200.[58] The latest survey is believed to be more accurate, but it does not imply that the tiger population actually increased by almost 700 in four years, only that more animals were recorded in the population surveys.

Similarly, some 200 leopards have also been poached in India yearly between 2002 and 2012.[59] Due to the international trade in its gallbladder bile and paws, the population of the Asian black bear has collapsed by some 50 percent.[60] Over 22,000 great apes were estimated to be trafficked globally between 2005 and 2011, of which 60 percent were chimpanzees. Other extensively smuggled apes include orangutans, bonobos, and gorillas.[61]

Among mammals, the trade in endangered and iconic species, such as bears, tigers, rhinos, and elephants, often receives the greatest attention, but the most commonly traded ones are macaques, leopards, and pangolins. For example, 270,000 individual macaques and 91,000 leopards were traded legally between 1998 and 2007, with less than 1 percent of the total number reported as caught in the wild after 2004. China and Malaysia were the principal exporters, and the European Union and Singapore the principal importers. However, these statistics do not necessarily indicate correctly the size of the legal and illegal trade as traffickers often list wild-caught animals as captive-bred.

Between 1993 and 2003, over 80,000 pangolin skins were illegally exported from Laos to international markets, primarily in the United States and Mexico. Another 15,000 were confiscated in Thailand in 2002 en route to China from Laos and Indonesia. Since 2006, over one million specimens of the eight species of pangolin were likely taken from the wild.[62] Regarded as highly nutritious, pangolin meat sells for $45 per pound in China and $60 a pound in Vietnam. Pangolin scales are prescribed to cure everything from skin ailments to lack of milk in breastfeeding mothers.[63] Meanwhile, over 8,500 water snakes of five species were estimated to be harvested daily during the 1990s from Cambodia's Tonle Sap, an ecological hotspot and a UNESCO-designated biosphere—possibly the most intense harvesting of snakes anywhere in the world.[64]

THE CHARACTER AND SCALE

In Latin America, birds (such as parrots and macaws), monkeys (such as lion tamarins and marmosets), and many reptiles (including caiman and various lizards) are among the most often illegally trafficked wildlife species. They are illegally hunted as pets or for their skin. Other poached species include totoaba fish, sea cucumbers, sharks, and coral. Considered a delicacy in China, totoaba bladders sell for as much as $20,000 there, and for between $7,000 and $14,000 in Mexico, where the species occurs.[65] In 2013, Mexican authorities seized an estimated $2.25 million worth of trafficked totoaba bladders.[66] Illegal totoaba fishing also has consequences for other species. The critically endangered vaquita porpoise, for example, is being driven to extinction because of becoming ensnarled in the nets of people fishing for totoaba. Only 30 vaquitas remain in the wild off Mexico, down from a population of some 570 in 1997, prompting desperate conservation efforts, including possibly a captive breeding program.[67] China's and Hong Kong's sea cucumber market is estimated at $60 million a year, with a kilo selling for as much as $600 there, and is increasingly supplied not just from Asia, but also from the protected and sensitive Galapagos Islands, Mexico, and the Caribbean.[68] Extensive illegal fishing takes place in the waters of Peru, Chile, and Argentina. In May 2015, Ecuador's authorities seized some 100,000 illegal shark fins from an estimated 30,000 sharks.[69] Globally, an astounding and shocking 100 million sharks are killed every year.[70]

Many other species of animals and vast numbers of plants not listed in this chapter are increasingly funneled into the global legal and illegal wildlife economy. Particularly if they are not charismatic megafauna, such as elephants, rhinos, lions, or tigers, and are instead insects or invertebrates, the legal and illegal trade in them may be unmonitored and fail to capture public attention. Demand for legal and illegal wildlife products continues to evolve and mostly expand—not only in overall volume and consumption rates, but also in the number and types of species consumed. New species, historically not significantly affected by human consumption, are being introduced by poachers and traders into the global wildlife market. Often they are promoted after the supply of another species has collapsed due to overexploitation. At other times, they are advertised as curiosities.

4

WHY THE ILLEGAL WILDLIFE AND DRUG ECONOMIES MATTER

The attention paid by policymakers to illegal economies and resource expenditures to combat them are shaped by the perceived threat those economies pose. That is why funding to suppress drug trafficking is far greater than funding to address wildlife trafficking. Equally, people's attitudes toward wildlife preservation or consumption are influenced by the harm and benefit poaching gives them. For many conservationists, the severe threats to species and ecosystems caused by poaching and wildlife trafficking are a major threat that needs to be mitigated. For some policymakers, it is a low priority issue. But linkages to other threats, such as terrorism, whether real or purported, can gain great policy attention.

Policy attention and funding are also critically shaped by the intensity of lobbying that surrounds a particular illicit economy. As the prominent social scientist Mancur Olson explained long ago, if costs are concentrated and visible, powerful lobbies form to mitigate those costs, particularly if the benefits of an alternative policy are diffuse.[1] Not all actors have equal capacity to lobby effectively: those hurt by conservation approaches, such as local indigenous communities or forest-dwelling people, often do not have as powerful a lobbying voice as global conservation NGOs representing white, affluent, educated young and middle-class people or large economic and politically connected businesses.

THE EXTINCTION MARKET

This chapter examines the threats posed by wildlife trafficking and poaching, and contrasts them with threats posed by the illegal drug and other illicit economies. In discussing threats to national security, I also dissect the often exaggerated and simplistic linkages between terrorism and poaching and wildlife trafficking.

I also analyze why suppression of the illegal wildlife and drug economies, and the law-enforcement policies involved in that suppression, sometimes pose severe threats to local populations. For some, participating in poaching or drug production might be the only way to assure their food and human security. Under such circumstances, it is those who protect the community from law-enforcement action and the illegal economy from suppression who accumulate hefty political capital.

Environmental, Human Security, Health, and Economic Costs

The threats posed by illegal and unsustainable legal trade in wildlife are serious and multiple. Foremost among them is the irrevocable loss of species and biodiversity. Although the number of our planet's species—estimated to be between 3 and 100 million, with 10 million often accepted as the most likely number[2]—may seem enormous, rates of species extinction are also dramatic. For example, if current extinction trends in Southeast Asia alone continue, between 13 and 42 percent of Southeast Asian animal and plant species will be wiped out this century. At least half are endemic species, and thus species losses in the region will in many cases represent global extinctions.[3]

Poaching can also disturb delicate habitats and generate cascade effects, devastating entire ecosystems, such as when it eliminates keystone species. A keystone species has a disproportionately large effect on its environment relative to its abundance.[4] Such species play a vital role in maintaining the structure of an ecological community, affecting the presence and population size of other organisms. Keystone species include apex predators, such as lions, tigers, and sharks. Prey species can also be keystone species. For example, extirpation campaigns (by the state and farmers, not poachers) of prairie dogs in the United States, plateau pika in Tibet and China, and the European rabbit in Europe all had critical consequences both for their predators and the

WHY ILLEGAL WILDLIFE & DRUG ECONOMIES MATTER

broader grasslands ecosystems of which they are a part. Their declines have led to precipitous collapses of brown bears, Tibetan foxes, steppe polecats, Pallas's cats, saker falcons, and upland buzzards on the Tibetan plateau, black-footed ferrets, prairie falcons, eagles, badgers, and bobcats in the United States, and Iberian lynxes in Europe. Other species may not depend on them for food, but for creating nesting sites. Thus the extermination campaigns of the plateau pikas have led to the disappearance of birds such as Tibetan ground-tits and snow finches.[5] Blamed for degrading ranchers' and herders' pastures, all three species actually help regulate rainwater absorption and prevent flooding.

Linkages between species effects can be complex, multistage, and hidden, and often not known *a priori*. Moreover, traffickers tend to move from one species they hunt out in an area to another species, leaving behind empty forests and ecosystems.

For forest-dependent peoples, the knock-on effects can precipitate the unraveling of fresh water supplies and food production.[6] Such threats to basic subsistence easily amount to severe threats to the human security of local populations. Numerous studies have noted the importance of wild food products for marginalized communities, especially women and children.[7] By some estimates (which nonetheless vary widely), 1.6 billion people rely on forest resources for all or part of their livelihood.[8] According to another estimate, the collection—including hunting—of non-timber forest products (NTFPs) represents as much as 50 percent of the income of forest-dependent communities.[9] Many are among the world's poorest people. In Central Africa, where poaching for subsistence threatens endangered primates, some 30 to 60 million people in forested environments are poor.[10] In Indonesia and Malaysia, that number is also in the tens of millions. Beyond forest-dependent communities, in sixty-two developing countries several billion people use wild meat, fish, and insects for 20 percent of their protein needs, while forest fruits and vegetables supply their vitamin needs.[11] Moreover, access to forests and forest-based products can provide critical safety nets when other food sources are unavailable, such as during periods of drought, famine, or war. In Zimbabwe, for example, wild resources were estimated to constitute some 35 percent of household income during the 1990s, with the number rising for the poorest and in times of distress.[12] In the Sahel, those most vulnerable to hazards are also those most dependent on access to the natural commons for food, firewood, and medicines.[13]

Indigenous people as well as other communities can also derive cultural and spiritual values from access to natural ecosystems.[14] These can range from exploiting natural resources, to fearing wild animals, to considering them sacred and worthy of protection. The Bathangyi people in the Rwenzori Mountains National Parks in the Democratic Republic of the Congo, for example, consider chimpanzees kinsfolk deserving of respect and protection, whereas other local communities hunt them for food or avoid them as they are thought to possess magical powers. Yet many Western environmental activists, whom community-based natural resource management advocates blame for unfairly excluding forest-dependent communities from protected areas, can also claim that they suffer cultural and spiritual losses by seeing natural environments and biodiversity destroyed.

Such threats to the human security of local vulnerable populations come not only from poaching by outsiders, but also from law-enforcement agencies intent on halting poaching and natural resources extraction by local communities in protected areas. Even when such approaches deliver the desired environmental outcomes, the subsistence, way-of-life, and human security costs to local communities can be severe. Sometimes, local communities may not know that hunting bushmeat is prohibited. A United Nations Environment Program study in the Democratic Republic of the Congo, for example, found that only 0.5 percent of individuals selling bushmeat acknowledged the illegality of the trade, with meat sold openly in local markets.[15] Other times, communities may well be aware of relevant laws and have encountered forceful and punitive enforcement of them, but because of economic dependence on hunting or a normative rejection of the laws, they refuse to comply. Merely designating a species as endangered affects people living in close proximity to it, by restricting its extraction or limiting access to its range.[16]

Moreover, the environmental benefits of excluding local communities from protected areas may not materialize if a local community continues to intrude into—or, in its view, justly access—a protected area. Even today much poaching is for basic subsistence.[17]

Severe food loss can threaten more than local populations. The depletion of global fish stocks by illegal fishing and unsustainable legal fishing can have disastrous effects on food security globally. In 2013, fish accounted for about 17 percent of the global intake of animal protein and 6.7 percent of all protein.[18] In developing countries, the latter number rises to 20 percent for 2.6 billion people and in small island nations to over 50 percent.[19] Policies to sustain fish production can also prevent some terrestrial habitat degradation, such as deforestation, which is often caused by land conversion for cattle ranching or agriculture.

Rich biodiversity increases the chances of nature's adaptation to radical environmental shocks, such as climate change. Better adaptability in turn enhances food security and agricultural resilience.[20] Although few studies have systematically examined biodiversity's role in ecosystem service delivery, such as erosion mitigation and hydrological functions,[21] greater biodiversity likely also improves these functions. Short of major shocks, biodiversity may contain substantial as yet unknown economic and health benefits for mankind, including poor populations: bio-prospecting in Guatemala, for example, has generated between $4 million and $6 million annually for 6,000 families living in or adjacent to protected areas.[22]

Threats to wildlife and biodiversity are of course also substantially caused by habitat degradation resulting from illegal or environmentally unsound logging and mining. Tropical forests host 50 to 90 percent of all of our planet's biodiversity, and yet approximately 17 million hectares are cleared annually.[23] In the Congo and Amazon basins, illegal logging and mining decimate some of the world's last rain forests, contribute to carbon release and global warming and species loss. In addition to destroying acacia forest ecosystems, illegal logging in East Africa leads to further soil erosion and desertification.

The drug trade too generates environmental harm. Coca cultivation in Latin America results in deforestation, and the processing of coca leaves into cocaine leaks highly toxic materials, such as kerosene, into the waterways of some of the richest ecosystems in the world. In the United States, illegal marijuana cultivation has negatively affected US national and state parks.[24] In California, legal cultivation of medical marijuana as well as illegal cultivation for recreational purposes has worsened water scarcity and depletion in the drought-stricken state.[25] Yemen's water scarcity has been aggravated by widespread cultivation of the thirsty qat.[26] However, drug-eradication policies, such as in Colombia or Peru, often only push coca cultivation deeper into previously untouched forests and national parks, thus exacerbating deforestation. Herbicide spraying of coca fields also negatively affects natural habitats, particularly if the herbicides enter waterways, such as in rainforests.

If the illegal wildlife trade depletes a species to such an extent that sustainable legal trade in that species is no longer possible or ecotourism collapses, severe economic losses in a particular area or for an entire country can follow.[27] For example, in 2015, ecotourism revenues represented 14 percent of Tanzania's GDP, of the order of several hundred million to over $1 billion.[28] In Botswana, similarly, tourism is the second-highest revenue earner, after diamonds, bringing in 15 percent of GDP and creating 30,000 formal and informal jobs.[29]

Yet both the illegal wildlife and drug economies have more complex economic effects, not all of which are negative. Significant portions of a country's economy can depend on illegal drug production. For example, for three decades now, opium poppy has been Afghanistan's leading cash crop, with the country supplying most of the world's market in illegal opiates. The United Nations Office on Drugs and Crime estimates that opium production represents 4 percent of Afghanistan's GDP in terms of farm-gate prices,[30] while valued at the border, it represents about 10 to 15 percent.[31] But these numbers are misleadingly low. A focus solely on farm-gate values does not take into account value-added in Afghanistan or economic spillover effects, such as the fact that much of the consumption of durables and non-durables, as well as construction, is underpinned by the opium poppy economy. Drug revenues thus likely constitute between a third and a half of Afghanistan's economy.[32]

Crucially, drug cultivation and processing generate employment for poor populations. Farmers of illegal crops can number in the tens or hundreds of thousands in a particular country, such as in Afghanistan, Colombia, Peru, or Mexico. Globally, millions of people, often the poorest, most marginalized, and indigenous, are dependent on the cultivation of illegal drug crops. Not only does the drug economy allow the poor to make ends meet, it can also facilitate upward mobility. Ninety-nine percent of illicit crop farmers will remain caught in the trap of poverty, marginalization, and illegality. They will not replicate the rags-to-riches stories of some of the world's most famous drug traffickers such as Pablo Escobar in Colombia or Joaquín Archivaldo

WHY ILLEGAL WILDLIFE & DRUG ECONOMIES MATTER

Guzmán Loera (also known as El Chapo) in Mexico, or translate their riches into social respectability and official political power as Du Yuesheng in China and Lo Hsing Han in Myanmar did. But cultivating illegal drugs will enable many poor farmers to send their children to school, access health care, and buy durable goods such as mopeds or electricity generators which they could otherwise not afford. When their access to the illegal drug economy is cut off, such as through cultivation bans or eradication schemes, the poor face even greater economic immiseration and deprivation. Without compensation, their basic livelihood and human security will be compromised.

The same is true about poaching. Although global poaching and trafficking have become more organized,[33] many poor individuals and communities willingly participate in them and do not embrace conservation. For them, hunting, sale and consumption of animals, and the conversion of natural habitats to agriculture or resource exploitation are means of economic survival and social advancement. Ignoring this uncomfortable truth, as has become a fad in some parts of the conservation community, including many environmental NGOs, will produce unsustainable and ineffective policies.

Very large-scale illicit economies, more often involving drugs than wildlife, also contribute to inflation, and hence can harm legitimate, export-oriented, import-substituting industries. They can encourage real-estate speculation and rapid rises in real-estate prices, and undermine currency stability. They can also displace legitimate production. Since the drug economy is economically superior to legitimate production—not only in profitability, but also in requiring less developed infrastructure and not imposing large sunk and transaction costs—the local population is frequently uninterested or unable to participate in a different form of economic activity. The existence of a large illicit economy thus complicates efforts at local development and crowds out legitimate economic activity. The illicit economy can lead to the so-called Dutch disease, where a boom in an isolated sector of the economy causes or is accompanied by stagnation in other core sectors since it causes rises in land and labor costs.[34]

Illegal and unregulated trade and in consumption of wildlife can spread viruses and diseases, endangering local species and food supplies, introducing harmful invasive species that generate ecological and further economic losses of the order of tens of billions of dollars annually,[35] and facilitating the cross-species transfer of diseases from animals to humans. The outbreak of SARS, for example, may have been augmented by the consumption of civets in China, even though horseshoe bats were likely the original carrier of the virus progenitor.[36] The transfer of avian influenza from wild birds to humans likely originated in China's wildlife markets.[37] Outbreaks of Ebola have been linked to bushmeat consumption in Central Africa, such as via the hunting and consumption of chimpanzees, wild pigs, as well as, once again, bats. In the case of highly infectious diseases, such as Ebola and SARS, public health crises represent potential threats to national and even global security, killing not just tens of thousands of people in a particular region but potentially tens of millions of people around the world and devastating the economy and public order in the most severely affected countries. The 2003 SARS outbreak in China infected 8,000, killed 800, and cost more than $40 billion.[38] West Africa's 2014 Ebola outbreak infected 30,000 people, killed 11,000, and cost more than $2 billion in Sierra Leone, Liberia, and Guinea.[39] Moreover, it caused GDP contraction of 8.2 percent in Liberia and 13 percent in Sierra Leone; in Guinea, the 2 percent contraction wiped out all growth and caused stagnation.[40]

Drug use also poses public health dangers—not only from overdoses and increased mortality and morbidity, but also from the spread of communicable diseases such as HIV/AIDS, hepatitis, and drug-resistant tuberculosis. Unsafe needle sharing and other undesirable practices by drug users fuel epidemics of such diseases. Yet there is overwhelming evidence that criminalizing drug use only compounds those public health dangers, and that harm-reduction practices such as the provision of clean needles and anti-overdose medication, embraced by some countries in Western Europe, greatly improve public health. Maligned as encouraging drug use, harm reduction policies are opposed elsewhere, including Russia, East Asia, and the Middle East. The United States is only now starting to experiment with them as it faces a new heroin epidemic. Addiction also has negative social effects on families and communities, sometimes tearing neighborhoods apart. But excessive law enforcement against local drug retail markets, and worse, against drug use itself such as through the imprisonment of drug users, also generates highly negative social effects.

WHY ILLEGAL WILDLIFE & DRUG ECONOMIES MATTER

Political and Rule-of-Law Costs

The illegal drug trade and other large-scale illicit economies, including wildlife trafficking, can also threaten the state and societies politically by providing an avenue for criminal organizations and corrupt politicians to enter the political space, undermining democratic processes. With financial resources from illicit economies, these actors can secure official positions of power as well as wield influence from behind, compromising the legitimacy of a political system. The more corrupt the system, the easier it is for a wide range of illicit economies, including poaching and wildlife trafficking, to thrive and their sponsors to penetrate the official political domain.

Linkages between the illegal drug trade and wildlife trafficking, and between wildlife trafficking and organized crime more broadly, are increasingly being investigated.[41] Sometimes these linkages have in fact emerged in reality, not just hypothetically. For example, the leader of a Mexican drug-trafficking organization, Samuel Gallardo Castro, was killed for failing to pay another smuggler $1 million for a shipment of totoaba swim bladders destined for Asia, leading Mexican officials and various analysts to speculate that Mexican drug-trafficking groups were diversifying their activities into wildlife smuggling into Asia.[42] Certainly Mexican drug gangs have enlarged their crime portfolio well beyond drugs. Many Mexican drug groups are today involved in a myriad of other smuggling rackets and generalized extortion. Bach Mai "Boonchai," identified as a major wildlife trafficker along with his brother Bach Van Limh, is also linked in Thai police documents to importing drugs to Thailand from Laos.[43] The two Vietnamese brothers are alleged to operate a global wildlife trafficking network from sites in Thailand and Vietnam, and are also said to organize poaching in Africa.[44]

Preexisting organized crime groups that have specialized in the smuggling of other goods can sometimes be motivated to diversify into the illegal wildlife trade and be strengthened by it by finding new ways to generate resources, reinforce their assets, and perhaps even expand their logistical networks. One example of this kind of diversification includes Asian criminal gangs operating in Cape Town, South Africa. Historically, they supplied methaqualone and precursor agents for the production of methamphetamines to local gangs in a racially mixed area in Cape Town known as Cape Flats. Initially, they would export abalone (a mollusk) back to China and East Asia, where it is coveted for its presumed aphrodisiacal qualities. Later on, they also branched out to exporting lion bone and ivory, and ultimately also rhino horn.[45]

Particularly in areas where wildlife trafficking is new but other kinds of organized crime have long overwhelmed law enforcement and corrupted the political and judicial systems, preexisting organized crime networks can have important competitive advantages, such as established corruption networks extensive know-how in evading law-enforcement agencies, as well as large amounts of financial capital and power to intimadate. They can sometimes also become motivated to take over wildlife trafficking directly. Most of the time, however, large organized crime groups merely tax other criminal activities in their areas of influence. Even the highly diversified Mexican drug-trafficking groups do not directly run all prostitution rings, human trafficking, migrant smuggling, illegal logging, and even drug production and retail in Mexico. Oftentimes they only provide cover and tax local criminal outfits and franchises. They might also kill off or force out local criminal gangs that refuse to pay them.[46]

Such linkages and diversification are not inevitable. A European Parliament report, for example, diligently looked for the linkages between organized crime and wildlife trafficking in Europe and could find little by way of evidence beyond reptile and caviar smuggling.[47] Indeed, much of the evidence for linkages between general organized crime groups and wildlife trafficking has been weak, tangential, and subsequently questioned.[48] It is also certainly possible that over time, drug-trafficking and other criminal groups will learn that dabbling in the illegal wildlife trade can bring high profits with little extra risk. But so far, most wildlife trafficking networks, whether highly organized or highly disorganized, have been their own special beast.

Casting the poaching and wildlife smuggling problem as one of large-scale groups involved in organized crime or global drug cartels, just like overemphasizing the linkages between militancy and wildlife trafficking, is politically advantageous even if often inaccurate. Such claims generate policy attention and resources. They also reduce focus on vulnerable communities dependent on natural resource extraction, diverting attention from the fact that stopping them poaching or encroaching into protected areas causes serious hardships for them.

WHY ILLEGAL WILDLIFE & DRUG ECONOMIES MATTER

Overemphasizing the organized crime element can also serve to skew policy toward one's smuggling rivals when government officials, including wildlife enforcement officials, are implicated in poaching and wildlife trafficking. They can direct law-enforcement action against alleged global organized crime groups even as their own smuggling networks remain intact and gain a competitive advantage.

While the detrimental effects of large-scale illicit economies on democratic processes are apparent, it is often an inappropriate analytical leap to assume that the emergence of extensive illicit economies will always necessarily challenge political stability and threaten ruling elites. To the extent that external traffickers make alliances with local out-of-power actors, traffickers develop a conflictual relationship with the state, and political instability may well follow. But if the governing elite captures new crime rents, a symbiosis or merger between external (and internal) traffickers and ruling elites may develop.

Large illicit economies, including wildlife trafficking, can also eviscerate judicial systems and the rule of law. First, as the extent of an illicit economy rises, the investigative and prosecutorial capacity of law-enforcement and judicial systems decreases. Impunity grows, undermining the credibility and deterrence of judicial systems and the authority of the government. Second, powerful traffickers frequently turn to violent means, killing or bribing prosecutors, judges, and witnesses, to deter and avoid prosecution.[49]

Even when large-scale traffickers are absent and the illicit economy involves small-scale poachers and traders in illegal products, this can have profound effects on the rule of law. The more people violate the law, the greater the costs of law enforcement and judicial prosecution. Both can easily become overwhelmed, and poaching or crime can become an internalized way of life. On the other hand, extensive incarceration and pervasive violations can be a profound indicator that the law itself is fundamentally incongruous with the needs of large segments of society, and it may become unenforceable and unsustainable.

Of course, efforts to suppress illegal economies often lead to substantial increases in human rights abuses by the government, military, and law-enforcement agencies. In the case of drugs, such human rights violations are prevalent throughout the Americas, Russia and Eastern Europe, the Middle East, and Asia.[50] If one defines incarceration of nonviolent drug users as imposing intolerable human rights costs, the number of such countries and continents grows larger.

National Security and Violent Conflict Costs … and Human Security, Again

Finally, in the context of violent conflict, the presence of a large-scale illicit economy, such as illegal drug production and trade, can greatly exacerbate security threats to the state.[51] Sometimes organized crime involved in such illicit economies can become so violent, such as in Mexico over the past decade, that it can overwhelm a state's weak law-enforcement capacity, and its actions can amount to a national security threat, not merely a public safety threat, causing more deaths than many a civil war or insurgency.

Over many decades, numerous militant groups have tapped into the drug trade and other illicit economies. In the drug trade alone, groups as diverse as the Taliban in Afghanistan, Sendero Luminoso (Shining Path) in Peru, FARC (Fuerzas Armadas Revolucionarias de Colombia) and rightist paramilitaries in Colombia, to name just a few, have participated. Groups involved in illegal logging and mining include again the Taliban and Haqqani network in Afghanistan, FARC and other leftist guerrillas and rightist paramilitaries in Colombia, the Revolutionary United Front (RUF) in Sierra Leone, and the Khmers Rouges in Cambodia, plus many others.

From their sponsorship of or participation in illicit economies, militant groups derive a multitude of benefits. With profits in the order of tens of millions of dollars annually, belligerents can immensely improve the physical resources they have for fighting the state and their rivals: they can hire more combatants, pay them better salaries, and equip them with better weapons. In fact, the increase in the belligerents' physical resources is frequently immense.

Better procurement and logistics also enhance what I call "the freedom of action" of belligerents—that is, the range of tactical options available to them and their ability to optimize both tactics and grand strategy. Prior to penetrating illicit economies, belligerents frequently have to spend much time and energy on activities that do little to advance their cause, such as robbing banks and armories to obtain

WHY ILLEGAL WILDLIFE & DRUG ECONOMIES MATTER

money and weapons, or extorting from the local population for food supplies. Once their participation in an illicit economy solves the belligerents' logistics and procurement needs, they become free to concentrate on high-value, high-impact targets.

Violent conflict of course overlaps with natural environments. Some 80 percent of major armed conflicts have occurred within biodiversity hotspots.[52] Forests, being particularly dense, provide good hiding places. Nor is it surprising that militant and terrorist groups have also exploited the illegal wildlife trade to feed their soldiers and to generate funding. In Africa, these include the Lord's Resistance Army (LRA) poaching elephants in Uganda, South Sudan, and the Congo,[53] and perhaps the Janjaweed Arab militia of Sudan, who have been accused of butchering thousands of elephants in Cameroon, Chad, and the Central African Republic.[54] (Oftentimes, West Africans call Janjaweed any Muslim outsiders or any outsiders.)[55] The Muslim Seleka rebels in the Central African Republic, including armed fighters from the Sudan, have similarly been accused of poaching in the Dzanga-Ndoki National Park. RENAMO (Resistência Nacional Moçambicana) also traded in rhino horn and ivory during Mozambique's civil war.[56]

Such involvement by militants in poaching and wildlife trafficking is not new. During the 1970s and 1980s, militant groups such as UNITA (União Nacional para a Independência Total de Angola) in Angola as well as the militaries of African governments killed thousands elephants for bushmeat and to generate revenues from ivory. Even today, both the Democratic Forces for the Liberation of Rwanda (FDLR), linked to the Rwandan genocide, and the Coalition of Congolese Patriotic Resistance (PARECO), as well as the Congolese national army fight each other in the country's national parks, where they also poach.[57] In India, the Nationalist Socialist Council of Nagaland in the country's northeast and militant Islamist groups in Bangladesh have traded in many poached species and genera, such as birds. In Myanmar (Burma), the United Wa State Army and other militant groups traffic wildlife into Yunnan and northern Thailand, along with methamphetamines and other contraband.[58] In Afghanistan, the Taliban (as well as powerbrokers associated with the Afghan government) facilitate the hunting of houbara bustards, snow leopards, and saker falcons for wealthy Saudis and Emiratis.[59] War refugees have also on occasion turned to poaching to make ends meet when humanitarian assistance is inadequate, such as in refugee camps in Tanzania since the 1990s.[60] War may displace local populations into protected areas where their presence results in hunting, timber gathering, or cattle grazing, all of which can have detrimental environmental effects.

One of the most bizarre instances of militancy being funded by wildlife trafficking occurred in Nepal's civil war between 1996 and 2006. The collection and international trade in *yarchagumba*—a form of caterpillar fungus scientifically known as *Ophiocordyceps sinensis*—was a significant element of the Maoist insurgency's fundraising. Used in Traditional Chinese Medicine and viewed as a potent aphrodisiac and cure for a variety of ailments, including cancer, *yarchagumba* was (and continues to be) highly profitable. So large were the Maoists' revenue from the illegal trade that the Nepalese army devoted significant resources to pushing the Maoists out of the subalpine grasslands where the caterpillar fungus was found.[61] Despite the fact that collection of *yarchagumba* was legalized in Nepal in 2001, the Maoists were not excluded from the trade and continued to derive significant revenues from it. Even after the legalization of the trade and the end of the civil war, *yarchagumba* prices have continued to rise, from $400 to $800 per kilogram in 2004 to $4,600 to $5,900 in 2010, an astounding order of increase reflecting growing demand and scarcity.[62] Since the civil war ended, the *yarchagumba* trade no longer funds militancy.

Pushing militants out of sensitive ecosystems can reduce poaching. During Nepal's civil war, for example, fifteen rhinoceros were killed in the prized Bardia National Park in one year alone.[63] After the war's end, poaching declined substantially, to less than five rhinos and five tigers killed per year, and rhino populations rebounded.[64] Several reasons account for this poaching decline. First, the Maoists became a dominant political force and acquired access to the legal economy. Second, surplus Nepalese military troops, eventually augmented by integrated Maoist units, were sent to guard the national parks. Still, the military units deployed to national parks are handicapped in their anti-poaching operations by a lack of sufficient intelligence, mobility, and rapid-reaction assets. Most patrolling takes place on foot, with often only one car—for the commanding officer—available for the entire battalion deployed in a national park.[65]

WHY ILLEGAL WILDLIFE & DRUG ECONOMIES MATTER

Importantly, official militaries and paramilitary forces, not just anti-state militants, have been involved in wildlife poaching and smuggling. The South African apartheid state and its military and intelligence units traded in ivory and rhino horn in the 1970s and 1980s.[66] The militaries of Uganda, a close US defense partner in Africa, and of the Democratic Republic of the Congo, as well as Congolese militias, have been accused of poaching elephants.[67] Sudanese armed forces are believed to have massacred thousands of elephants in 2005 for ivory destined for China.[68] The Zimbabwe military likely poached more elephants in Zimbabwe's Gonarezhou National Park than Mozambican RENAMO units.[69] Myanmar's military is as much an actor in illegal logging as ethnic insurgents.[70] Militaries deployed in the name of conservation have engaged in land theft from local populations, such as in Honduras.[71] Conversely, members of anti-poaching units, prized for their military skills taught by private security companies and foreign technical advisors, have at least on one occasion joined an armed anti-state rebellion and insurgency, notably in the Central African Republic.[72]

Sadly and paradoxically, however, peace can produce even more detrimental results for protected species and natural ecosystems than war if it enables greater access to valuable natural resources, and hence habitat destruction. Ceasefires involving ethnic insurgents in Myanmar have been bought by allowing them and other actors to engage in poaching and free-for-all resource extraction.[73]

This "peace detriment" is only one example of the complexities of the linkages between militancy/terrorism and poaching. Combating terrorism also gets much better funding than preserving biodiversity. Exaggerated and one-sided portrayals of the nexus skew policy responses toward tough law-enforcement approaches, often at the expense of protecting the rights of local communities and compensating them for displacement and food loss. Other times, local poachers and local communities are unjustly labeled as militant sympathizers to allow for their expulsion from national parks and other protected areas and for the dispossession of their land.[74] Yet most poachers are not terrorists, and most militants and terrorists are not poachers.

Conservation scholars critical of such conflations label this "green militarization."[75] They criticize the violence exacted on local communities in the name of suppressing the alleged terrorism–poaching nexus.

THE EXTINCTION MARKET

Some question outright whether any linkages between militancy and poaching exist. They are also wary of the use of private security companies and the deployment of drones and modern technologies to combat poaching.

At other times, the state's exaggeration of national security threats is used to prevent environmental conservation measures, such as the demarcation of protected areas. In Laos's borderlands, powerful businesses and state-linked industries, including the Laotian military, engage in profitable extractive industries such as logging, including the felling of rosewood, a desirable but endangered and hence prohibited species. Manufacturing security threats becomes a politically convenient way to prevent environmental protection and perpetuate profitable unrestrained legal and illegal economies.[76]

Violent conflict not only overlaps with poaching: often it also permeates conservation efforts, even today. Sometimes forced displacement is masked by counterinsurgency and counterterrorism narratives. In Colombia's Tayrona National Park between roughly 2005 and 2010, counterinsurgency efforts enabled one ecotourism company to displace people in order to generate economic revenues from ecotourism. There are many valid and important national security reasons to pursue militants into protected areas, but in the Tayrona case in Colombia, the counterinsurgency effort served as a cover and an enabler of dispossessing marginalized local communities of their land.[77] Elites and powerful businesses profiting from ecotourism have similarly sponsored the forced displacement of local communities in the name of conservation in Honduras.[78]

The eradication of illegal drug crops often also leads to the forced displacement of people. Sometimes this is simply the consequence of people losing their illegal livelihoods. At other times, displacement can be orchestrated by vested economic interests seeking to acquire land. As post-insurgency land in Colombia has become valuable, including for the cultivation of legal crops such as coffee, cocoa and oil palms, or for logging or mining, the forced displacement of people is taking place well beyond Tayrona.[79]

Yet it is the terrorism–poaching nexus that is frequently exaggerated by many countries that feed the supply chain of the illegal traffic in wildlife. In Kenya, for example, al-Shabaab and Somalis more broadly are pervasively blamed by government officials and the public for poaching and all kinds of criminality.[80] Indeed, the narrative that Somalia's terrorist group al-Shabaab is behind elephant poaching in Kenya received wide coverage in the press.[81] Yet the claim turned out to be flimsy at best, with some reports casting strong doubt on much of the evidence.[82]

It is possible that al-Shabaab does tax ivory smuggled from Kenya into the Somali port of Kismayo. Al-Shabaab controls important parts of the surrounding Juba region, including key corridors to Kenya. Many terrorist, militant, and criminal groups tax legal goods, illegal contraband, and all kinds of economic activity within their sphere of influence. So does al-Shabaab.[83]

Oftentimes, militants are pushed into or come to control territories with illicit economies new to them. Eventually, they often come to tax the economies, and sometimes even seek to displace other traders from a particular market to make more money. That has been the case with many militant groups with respect to drugs as well—from Sendero Luminoso in Peru and the FARC in Colombia to the Taliban in Afghanistan to Abu Sayyaf in the Philippines.[84] As income streams, militant groups might see little difference between drugs, timber, and wildlife. That does not mean, however, that such nexuses emerge every single time.

Moreover, even militants make choices of what and what not to get involved in. Sometimes they refrain from participating in a particular illegal economy, including the drug trade; at other times they prohibit even local populations from participating in it.[85] When Sendero Luminoso, the FARC and another leftist guerrilla group in Colombia, the ELN (Ejército de Liberación Nacional, National Liberation Army), as well as the Taliban, first encountered drugs, they tried to ban illegal drug production. However, they all learned that such a policy prevented their ability to consolidate control, and undermined their capacity to develop support and gain acceptance and legitimacy in the eyes of local populations. Thus they rescinded the prohibitions, and came to tax and trade the contraband. Many subsequent militant groups did not repeat their mistakes and embraced drugs and other illicit economies right away. Some groups, however, do not learn from the experience of other militants, and are tone-deaf to the pushback from local populations who are dependent on illegal economies. Unlike the Taliban, Islamic State in Afghanistan, for example, still imposes a ban on poppy cultivation in the province of Nangarhar, and chooses to use brutality to rule rather than legitimacy.[86]

Moreover, to the extent that ivory is exported through the port of Kismayo, the odds are that the Kenyan Defense Forces present in southern Somalia, including Kismayo, and Ahmed Madobe, president of Juba State and a former al-Shabaab commander who defected, get a substantial cut. Both allegedly tax smuggled sugar and charcoal, produced from acacia illegally logged throughout East Africa and transported to the Middle East.[87]

Yet for many countries that are the source of wildlife products, a preoccupation with poaching terrorists is a convenient distraction from addressing the issue of corruption among rangers and anti-poaching militias and military units, ecolodges, and high-level government officials. Without rooting out this pervasive corruption, and ending the economic dependence of local communities on participating in or tolerating poaching, many conservation efforts will fail, no matter how sophisticated the rangers' equipment against poachers becomes.

To the extent that it exists at all, the participation of militant groups in poaching is only a fraction of the illegal trade that goes on. This is not merely the case simply in terms of market share; they also play only a small role in global smuggling chains, particularly beyond their locus of operation. For many reasons, including attempts to reduce militants' income sources, it makes sense to push them out of protected areas when they are involved in poaching. But as long as there is strong demand and limited enforcement capacity, someone else will take their place as poachers. With the rise of poachers' firepower, militants do not have a comparative advantage in poaching. As Chapter 6 explains, a lot of what law enforcement does is not to determine whether an illicit economy exists or not, but rather who runs it and has access to it.

Local communities can often (though not always) choose whether to participate in poaching carried out by militants, whether to yield to it if the community is unable to mount effective resistance, or whether to mobilize against it. Neither the various ethnic militant groups in Myanmar nor the Nationalist Socialist Council of Nagaland in India, for example, experienced any popular pushback from their supporters or local communities against their participation in hunting and wildlife trafficking since the local population also participated in the illegal economy and hunting was a centuries-old tradition.

But while also operating in northeast India like the Naga militants, the Bodo insurgents had a very different experience with participating in poaching and wildlife trafficking. Building on a local culture of animal protection, the Bodo Security Force (BdSF), which has been seeking to liberate "Bodoland" from Assam, took it upon itself to enforce anti-poaching laws. For example, they singled out known rhinoceros poachers and told them to desist from their activities. If the poachers failed to comply, the Bodo militants killed them. The BdSF's location in the Manas National Park and the need to fund their insurgency later encouraged the group to violate its own edicts and dabble in wildlife trafficking. But the subsequent public outcry from the local population that constituted its base was so strong that the BdSF aborted its participation in the illicit trade and went back to enforcing environmental protection.[88]

Similarly, although the *dacoits* (robbers and bandits) in Uttar Pradesh, India, in the 1980s so undermined public safety that their activities amounted to a insurgency, the *dacoits* sought to protect the unique wildlife in their area by such acts as burying poachers alive.[89] In the Central African Republic, local communities which embraced conservation were so determined that in one instance they resorted to using violence against outside herders. During one altercation, they encouraged officially sanctioned anti-poaching guards to use their firearms to prevent the herders from coming back, even suggesting that the guards kill the herders' cattle. They also mobilized volunteers to help the existing anti-poaching militias take on the intruders.[90]

The Political Capital of the Illegal Drug Trade and Illegal Wildlife Trade

However, local communities often do not mobilize against poaching, illegal grazing, or logging, but rather are willing participants in them. Economies and behavior labeled as illegal by authorities are often considered legitimate by local communities. Suffering intense hardships as a result of being forced to obey the law, they rebel against it.

Under such circumstances, the sponsorship of illicit economies greatly increases not just the physical resources and freedom of operation of the sponsors, but also crucially, their political capital—that is, the extent to which the population welcomes and tolerates the presence of the sponsors.[91] This is true regardless of whether the sponsors are belligerents, criminal gangs, individual powerbrokers, or powerful local poachers.[92]

Beyond providing marginalized populations with livelihoods, sponsors of illicit economies sometimes use the proceeds of their activities to distribute real-time economic benefits to local populations, such as otherwise absent social services. Sendero Luminoso and the FARC, for example, provided clinics, roads, sewage, trash collection, and schools.[93] Sponsors of illicit economies can also provide protection and regulation services to the illicit economy itself, including protecting producers from, for instance, brutal and unreliable traffickers or military forces who want to eradicate drug crops or stop poaching by local communities.

Examples of such political capital accruing to sponsors of illicit economies also exist in wildlife trafficking. For example, Annette Michaela Hübschle shows, in her excellent study of rhino horn trafficking, how local heads of poaching groups in Mozambique and South Africa, whom she refers to as local "kingpins," claim to fulfill important social welfare functions.[94] Some promise to provide poachers with legal support if they are arrested or provide "life insurance" to their families if they are killed. They justify their poaching as a necessary activity forced on them by the neglect of the state and its failure to provide jobs and livelihoods. Other poachers are drawn to areas of successful poaching, sometimes even from other countries, but they usually have connections such as kinship ties with people in that area. Economic spillover from poaching revenues and activities also creates other jobs within communities, such as that of traditional healers who prepare good-luck charms to enhance the success of poaching expeditions and who foretell when a poaching expedition will be successful. Many of the younger poachers are buying modern houses in their villages while others are buying them hours away on the coast of Mozambique. Others are purchasing cars, even luxury ones. Poachers from the older generation, on the other hand, continue to buy cattle, a sign of prestige and affluence. One of the local poaching kingpins Hübschle interviewed bought himself a hotel.[95]

The local poaching kingpins thus develop recognition, prestige, and legitimacy within their communities. Poaching is portrayed not merely as a matter of economic necessity but as a means of claiming reparations for the loss of land now designated as an environmental area and which they can no longer access for economic opportunities, such as grazing, logging, hunting, and agricultural production. Hübschle describes the local kingpins as "self-styled Robin Hoods." "We are using rhino horn to free ourselves," they claim.[96] Similarly, abalone poaching gangs in the Western Cape province of South Africa justify their illegal wildlife trade, and gain political capital with local communities, by labeling commercial fishing quotas as unjust and unfair to struggling grassroots communities.[97] Moreover, while the community knows who the poachers and perhaps middlemen are, it is not willing to provide that information to the state.

Four factors critically influence the extent to which illicit economies bring political capital to their sponsors.[98] First, the state of the overall economy in the country/region; second, the labor-intensive character of the illicit economy; third, the presence or absence of independent traffickers; and fourth, the government's response to the illicit economy. The state of the overall economy determines the size of the local population who cannot obtain legal livelihoods. The labor-intensive character of the illicit economy determines the extent to which the illicit economy provides livelihoods to marginalized populations. The presence or absence of thuggish traffickers separate from other sponsors of the illicit economy, such as militants, determines the extent to which those sponsors can bargain of behalf of the population for better prices and otherwise protect them from the traffickers. Finally, government responses to an illicit economy determine the extent to which local populations become economically and politically impoverished or empowered through the illicit economy and dependent on sponsors for the preservation of their illegal livelihoods. In a nutshell, if the state of the overall economy is poor, the illicit economy is labor intensive, thuggish traffickers are present, and the government tries to suppress the illicit economy without providing for alternative livelihoods or compensation, sponsors of illicit economies obtain large political capital.

Political capital can be further augmented by other historic or normative grievances that make locals see laws as unjust, such as when poaching becomes a form of anticolonial resistance and a form of self-empowerment, or when the killing of an animal is seen as a rite of passage, taking a boy into manhood.

In sum, threats from poaching and wildlife trafficking, like those from the illegal drug trade, range from environmental and economic threats, to justice and political threats, to national security threats. But conservation efforts, like counter-narcotics efforts, can also severely threaten the human security and human rights of marginalized communities. By and large, mitigating the threats posed by the drug trade is easier than assuring the key objective of suppressing poaching and wildlife trafficking, namely, preserving species and ecosystems.

The one structural advantage that poaching and wildlife trafficking have in terms of policy mitigation is that even when poaching involves the large-scale willing participation of local communities, it still tends to be less labor intensive than poppy or coca cultivation. The more labor intensive an illegal economy is, the more difficult it is to generate adequate alternative livelihoods and provide sufficient compensation for those severely hurt by efforts to suppress the illegal economy, and the more political capital sponsors of the illicit economy have.

A failure to generate political support among local communities for measures to suppress an illicit economy is politically very costly. Widespread opposition to law enforcement and widespread rule violations severely complicate law enforcement in terms of resource demands and ethical burdens. Laws are far easier to enforce if they are internalized by most.

5

THE ACTORS

The structure of an illicit economy critically determines the effectiveness of various policy interventions. The design of law-enforcement strategies should be in response to how illegal production and smuggling networks are organized. Who and how many the various participants and stakeholders in illegal (and also legal) wildlife economies are influences the effectiveness and sustainability of the various policy alternatives. For example, if wildlife smuggling in a particular region is dominated by a narrow hierarchically structured organized crime ring that does not employ many people and that services a narrow niche market, law enforcement and interdiction have a high chance of being effective. Conversely, if ground-level poaching is conducted by thousands of poor people, enforcement without compensation may become politically and ethically very costly, and alternative livelihoods strategies or community-based natural resource management approaches should be explored.

Policies to address the illegal wildlife trade, or any illegal economies, need to be informed by answers to two essential questions. First, who are the current poachers, traffickers, and consumers? Second, how will they adapt in response to business competition and policy responses? The second question draws on experience from the drug trade, where the structure of smuggling networks and producers' and traffickers' behavior are to a large extent determined by the policies adopted. Both structure and behavior also reflect business competition from other traffickers and producers. They are not an immutable constant.

This chapter examines the structure of wildlife poaching and trafficking, and challenges and problematizes aspects of current policy mantras. Dominant narratives today often overemphasize organized crime as an aspect of wildlife trafficking and underemphasize the corruption of government institutions and the wildlife industry in many wildlife-supply countries. Equally, inadequate attention is given to the involvement of local communities in poaching, and at times the intersection between these communities and global organized wildlife trafficking.

The most important characteristic of poaching and smuggling networks is their diversity. Some have become highly organized and vertically integrated. Other wildlife trafficking supplying global demand is organized but dispersed, with no kingpins or top-level traffickers. Other illegal wildlife trade involves the extensive participation of local communities. Some of the time, communities poach merely for their own subsistence; at others, they also sell illegally obtained wildlife products to local, regional, and, via middlemen, global markets. Sometimes, local communities who interact with organized global poaching networks may join them to generate greater revenue or because they are physically unable to resist them. At other times they may try to oppose them.

While recognizing these many variations, three basic types of interdependent actors can nonetheless be identified: consumers, suppliers, and middlemen. Each of them crucially structures the way that wildlife trade networks function. The dispersion or concentration of participants along the nodes of demand, supply, and transshipment privileges different types of interdiction, alternative livelihoods, or community-based resource management approaches, as well as demand-reduction strategies.

Consumers

Although China's consumers dominate the global wildlife market, demand for wild plants and animals is increasing throughout Southeast and East Asia, exacerbated by the region's growing population and its

THE ACTORS

increasing affluence. What had previously been mainly source and transshipments locales, such as Thailand and Vietnam, have rapidly become important consumer markets. East Asian diaspora communities, including in the United States, are also important consumers of wildlife products.

Demand for wildlife products is present and increasing in other parts of the world as well, including Latin America and Africa. Some types of demand, such as for bushmeat in Africa, are traditional and go back centuries or millennia. Even the long presence of these markets does not necessarily mean that they are sustainable: high population growth in some regions, such as West Africa, can lead to such an expansion of demand that it produces unsustainable rates of hunting.

Other types of demand for wildlife, including in Africa and Latin America, are newer, emerging, and expanding as a result of the greater affluence and disposable income in those regions. Beyond meat for subsistence and sale, pelts and antelope horns are widely utilized in Africa and sometimes fed into global fur supply chains. In Latin America, a robust market for pets persists, whether they are local parrots and parakeets or small monkeys, such as tamarins and marmosets. The trade in reptile skins for shoes and clothes also seems to be expanding in Latin America.

Far more detailed studies are needed to enable us to understand the state and evolution of the demand for wildlife products and the markets that meet it around the world. It is crucial to understand who the current users of wildlife products are, which groups might emerge as new sources of demand, how and why customers exit the market for illegal wildlife products, as well as what precisely motivates their consumption.

Suppliers

The primary motivating factor for wildlife hunters and traders is economic, ranging from small-scale subsistence needs for some to major high-profit business for others. At the start of the smuggling chain are the hunters of animals and collectors of plants and minerals. This group consists of both poor (often subsistence-level) hunters and professional hunters. Beyond need and greed, other motivations include the rejection of colonial or outside international values and a form of political rebellion against the imposition of norms that are seen as alien, discriminatory, and against the basic interests of the community.[1]

Hunters who are poor include those for whom hunting and the utilization of natural resources, including pastoralism, are part of a long and deeply established tradition. One example is the forest-dwelling communities of Central Africa, who are dependent on bushmeat for subsistence. But traditions of hunting and grazing are present more broadly across Africa, as well as in Latin America and Asia.

In Asia, examples include the Nagas of northeast India and the Pardhis of Gujarat and Maharashtra. The Nagas have lived and exploited the forests of northeast India for centuries, and animal hunting was a matter of both subsistence and prestige (though eliciting less admiration than human headhunting, which they also used to practice). The Nagas have long traded both skins and wild animals, and the tradition has been slow to decline despite the efforts of India's government and conservation NGOs.

The Pardhis represent an example of how social and economic marginalization perpetuates the participation of particular groups in illegal economies. Often poor, illiterate, and mostly nomadic, Pardhis were designated as "criminal" tribes in 1871 under the British Raj Criminal Tribes Act. The fact that the police (and Indian society more broadly) often assumed that these people participated in assorted kinds of crime made it difficult for many to obtain legal employment, and indeed drove some to crime, including poaching. Today, some Pardhis are expert poachers both because they have acquired the necessary skills over generations and because they have meager legal, economic, and social opportunities. Just as in Kenya, where the default attitude is to blame poaching on the Somalis, in India a frequent reaction is to claim that the Pardhis are behind poaching.[2]

Looking beyond Pardhi traditions, there is a more basic cause: at least 50 million people living in and around India's forests depend on non-timber forest products (NTFPs) directly or indirectly for subsistence.[3] In dryland India, wildlife products constitute between 14 and 23 percent of household income, rising to 57 percent during periods of drought.[4] For some marginalized communities in Laos, Cambodia, and Myanmar (Burma), the dependence sometimes tops 70 percent of people's income. What they eat and what other products they depend on crucially influences the environmental impact of forest-dwelling communities: if they eat an abundant species with high reproductive rates, the impact on wildlife may be minimal (analogously, if more people in the United States ate venison, the negative environmental impacts of overpopulated deer destroying forest habitats might be mitigated). On the other hand, if forest-dependent people, as in Central Africa, extensively hunt critically endangered species, such as gorillas and chimpanzees, their environmental impact can be severely detrimental.

Moreover, the illegal wildlife trade can reshape traditional hunting and other forms of forest exploitation. Not all indigenous communities nowadays hunt purely for food. Many traditional hunters have replaced their bows and arrows with firearms.[5] Such a technological evolution is not always bad: hunting with firearms can be discriminatory, and the hunter has the capacity to avoid endangered species. Traps, even traditional ones, often injure or kill a much wider array of species. Sometimes outside wildlife traffickers recruit local populations because local hunters have superior knowledge of the locations and habits of animals and their labor is cheap. Indeed, the increasingly commercialized trade in bushmeat is believed to be one of the main causes of unsustainable hunting.[6]

Sometimes marginalized communities can be pushed into hunting as a negative side effect of other policies, without their necessarily having previously hunted on such a problematic scale. For example, marginalized ethnic communities along Myanmar's borders have participated for decades in various illegal economies, from the production of and traffic in drugs to illegal logging and illegal mining of gems. Although these activities have brought large revenues to militant separatist groups and their leaders (as well as to the Burmese military, logging companies, and Chinese and Thai traders), many of the primary producers, miners, and loggers continue to be desperately poor. Suppression of one illegal economy—poppy cultivation and opium and heroin production—drove some to switch to the illegal wildlife trade.

The situation in Special Region No. 4, in Myanmar's Shan state, which borders the Chinese province of Yunnan and is controlled by the National Democratic Alliance Army (NDAA), illustrates this phenomenon. With its capital Mong La sitting on the border with China,

THE EXTINCTION MARKET

Special Region No. 4 used to be a major area of opium poppy cultivation. The poppy eradication drives of the late 1990s left many farmers impoverished, often with food for only eight months a year.[7] Farmers coped by resorting to logging, with the timber destined for China, and catching any animals they could find in the forest, both for consumption and for sale in China.[8] With the collapse of Mong La's gambling and prostitution enterprises following the Chinese government's decision to restrict access of Chinese tourists to the city, the wildlife trade intensified, and Mong La became one of the five biggest wildlife smuggling hubs in Myanmar, with Myanmar's biodiversity-rich forests swept empty of animals.[9]

Professional Hunters and Middlemen

The arrival of regional or international wildlife traders often triggers a community's participation in wildlife trafficking. The role of Vietnamese traders in poaching and wildlife trafficking markets in Southeast Asia provides an illustration. Once wildlife was depleted in their home areas, Vietnamese traders orchestrated extensive wildlife hunting in the Nakai-Nam Theun National Protected Area in Laos, the country's largest protected area. The subsequent hunting of pangolins, civets, turtles, and monitor lizards in Nakai-Nam Theun resulted in a significant decline in these species.[10] With the arrival of middlemen who facilitate marketing, prices for wildlife increase. In a village in Laos, for example, before the arrival of Vietnamese traders, a golden turtle would sell for about $100. After the middlemen arrived and connected the local market to Chinese markets (where Chinese traders marketed the turtles as a cancer cure), the turtles would sell for about $1000. Overharvesting led to scarcity, which in turn contributed to the price increase.[11] Middlemen also stimulate the diversification of poaching. Thus illegal collection expands from orchids to insects, hunting from civets and bears to pangolins, and from langurs to salamanders, with an emptied forest left behind.

The low-tech hunters sometimes evolve into, and sometimes are joined or altogether displaced by, professional hunters. As hunting empties forests, wildlife scarcity makes trapping more time consuming and requires greater skills facilitated by sophisticated equipment. Thus many less-skilled hunters drop out, and the remaining ones become professionalized.[12] Highly skilled professional hunters are sought after by trafficking networks, who frequently facilitate their mobility within a country and at times even between countries. Such industrial-scale hunters, mostly of European extraction, dominated ivory hunting in the Central African Republic during the 1960s and 1970s, for example.[13] This second group of high-tech hunters also includes recreational hunters who violate hunting regulations. Both types are supported by local trackers, guides, and carriers.

The more hierarchically structured and vertically integrated trafficking networks do not necessarily seek to eliminate unattached poachers. The latter are convenient for distracting law enforcement. Thus in Southern Africa, for example, wildlife poaching bosses prefer that multiple rhino hunts take place concurrently and tolerate hunting groups which operate independently of them.[14] Similarly, drug-trafficking organizations often allow local criminal gangs to operate and merely tax them. Often they organize more than one smuggling run at once, coupling it with decoy runs. In addition, many of those at the top of the drug trade prefer to have large organizational layers underneath them to hide their role and complicate law enforcement's efforts to prove their culpability.

Middlemen are also crucial nodes in the international dimensions of the illegal wildlife trade. In addition to being able to organize local poaching, they are connected to global markets and top-level traffickers. Both middlemen and top traffickers can also establish networks of corruption, such as among customs officials, and cultivate political patrons. When wildlife goods are not smuggled by boat or truck and instead human couriers are employed, middlemen organize them. Convenient couriers include diplomats, whose diplomatic pouch is not searched by customs officials, business people who travel between supply and destination countries, and even students and tourists. These modalities of organizing illegal supply and transshipment networks involving wildlife are no different from those used in the drugs trade.

At the apex of the smuggling chain are sometimes big traders who facilitate wildlife traffic across the globe. One notorious trafficker with a global reach is Wong Keng Liang, better known as Anson Wong. A Malaysian, Wong first established himself in the illegal (and legal) trade in reptiles, selling anything from legal geckos to illegal Komodo dragons, Chinese alligators, and Madagascar ploughshare tortoises, a critically endangered species with less than 100 remaining in the wild. That early start earned him the nickname the Lizard King. Later, he diversified into any wildlife of high value, such as rhino horn, Spix's macaw (believed to be extinct in the wild), and panda skins. He used a private zoo that he owned in order to launder endangered species and as a cover for his global illegal trade. After imprisonment in the United States for wildlife trafficking, Wong went back to Malaysia to run breeding farms and private zoos, but allegedly continued trafficking in wildlife.[15]

Another wildlife kingpin, Sansar Chand, gained notoriety for allegedly organizing the large-scale poaching of India's tigers and sales of their products throughout Asia.[16] He was charged with wildlife trafficking by Indian authorities but died while on trial in 2014 as a result of a terminal illness. The Poon family from Hong Kong has traded in ivory and shark fins for generations. Along with several other long-established ivory traders and craftsmen, the Poons moved from Hong Kong to Singapore in the 1980s to take advantage of loopholes in Singapore's wildlife laws. The Environmental Investigative Agency, an NGO that specializes in exposing the illegal wildlife trade, alleged that the Poon family also traded in illegal ivory.[17] Perhaps Asia's largest known wildlife trafficker has been the Laotian Vixay Keosavang, often dubbed the "Pablo Escobar of wildlife trafficking." Vixay's trading company, Xaysavang Trading, was implicated in the smuggling of ivory from Kenya and rhino horn from South Africa, and a myriad of other animals, including lizards, turtles, and snakes. They were shipped to Vixay's breeding facility, which was licensed to breed endangered animals and sell them in Laos and internationally, where they were issued with fake documents and resold. The trading involved tens of thousands of snakes, turtles, and monitor lizards per single order, indicating a vast laundering scheme as the breeding facility could not possibly produce such numbers. Trucks carrying tiger cubs and carcasses, pangolins, and reptiles to Vixay's businesses were intercepted in Thailand. To supply the global market with rhino horn, Vixay's presumed deputy Chumlong Lemtongthai hired Thai prostitutes in South Africa to pose as licensed trophy hunters. In fact, the rhinos were killed by crooked operators of a private ranch, with the horns shipped

THE ACTORS

to Vixay's addresses in Laos. While his deputy Chumlong is to serve thirteen years in prison in South Africa, the politically influential Vixay remains at large.[18] Nonetheless, evidence obtained from Chumlong's computer and made public during his trial in South Africa hit Vixay hard. In November 2013, the US government announced a $1 million reward for information leading to the dismantling of Vixay's trafficking operation and specifically named his company, Xaysavang Network, in its announcement.[19] After the announcement, Vixay is believed to have removed himself from the wildlife trade and switched to smuggling cars into Vietnam.[20]

Many of these traffickers did not diversify into the wildlife trade from other illegal markets, such as drugs. Many started their criminal careers in wildlife. Before entering wildlife trafficking, Chumlong used to sell fruit in a Bangkok street market. The Bach brothers, mentioned in the previous chapter, purported global wildlife traffickers, also run legitimate businesses in wholesale agriculture and forest products, construction materials, electrical equipment, hotels, and food services.[21]

That does not mean that contagion effects will not take place and that criminal groups do not learn from each other about business opportunities in other domains, such as wildlife trafficking. This appears to have already taken place in the case of at least some Mexican criminal groups, which now also seem to participate in totoaba bladder smuggling from the Gulf of California to China.[22] But even if such learning takes place, it is unlikely that a single predominant type of network organization will emerge. Rather, just as demand will remain diverse, so will the types of networks that supply it, as is also the case in drugs.

Thus it is necessary to question the current fad for arguing that the dominant shape of wildlife smuggling networks is a highly organized hierarchical organization, mimicking drug cartels, with kingpins at the top, and sometimes even intermeshed with drug cartels. Sometimes global drug traffickers do sit at the top of the smuggling chain. Indeed, professionalization and intensification of poaching have taken place in the case of some smuggling networks, particularly in commodities such as ivory and rhino horn.[23] One widely cited report argues that one smuggling network was responsible for trafficking over 14 tons of ivory products from up to 35,000 elephants poached over two years.[24]

But not every drug or wildlife smuggling organization is a tight-knit, hierarchical network with a big trader at its apex. In Myanmar, one of the world's poaching and smuggling hotspots, much of the poaching and smuggling is carried out by poor, low-level poachers and smugglers who sell poached animals in Mong La and Tachilek to both low-level traders and middlemen, and often directly to consumers. Similarly in Indonesia, many poachers in Kalimantan, Sulawesi, and Ambon are unorganized, low-level opportunists, not wildlife kingpins with a global reach.

Rhino horn smuggling into Yemen, for decades one of the most important markets for the product, also turned out to be decentralized. The horns mostly originated from South Sudan, South Africa, Kenya, and the Democratic Republic of the Congo. But instead of one or two, or even several, vertically integrated groups smuggling them into Yemen, the smugglers were an assorted group of Sudanese nationals and Yemeni sheikhs and diplomats whose luggage was not checked by Yemenia Airways. These smugglers became particularly important when foreign navies' antipiracy operations in the Gulf of Aden made rhino horn smuggling by boat difficult.[25] But even during the years of boat smuggling in bulk, the horn was carried by fairly low-level traders who arrived in Djibouti in small dhows also transporting livestock, other contraband, and smuggled migrants from Somalia, Ethiopia, and Eritrea. Although these traders were smugglers of various kinds of contraband, they were not highly organized or part of sophisticated and vertically integrated criminal organizations.

In Nepal, Chinese and Indian high-level brokers often sit at the apex of the country's wildlife smuggling networks and work via middlemen in Kathmandu and district-level towns. Nonetheless, outside brokers cannot operate without the cooperation of locals, whom they hire to work as spotters and hunters. Middlemen also often hire women and children to carry body parts of slaughtered animals, on the assumption that they will receive less scrutiny from law-enforcement officials and lesser penalties if caught. Wildlife law-enforcement officials frequently point to members of the Chepang and Tharu community as poachers, but poaching and trafficking in Nepal rarely entail the cooptation of an entire community. Nor are local-level poachers politically organized or aligned with a particular political party. Instead, poaching operates at an individual level with middlemen cultivating political protection simply on the basis of preexisting patronage networks. Prominent political leaders exert pressure on park management and the courts to release apprehended poachers and traffickers.[26]

Indeed, drug-trafficking networks are equally highly diverse, with some fairly hierarchically organized, such as the Medellín cartel in Colombia and the Sinaloa cartel in Mexico, but other smuggling networks, such as in the case of drug smuggling in China, are loose and atomized.[27] But even the Medellín cartel was tightly organized only in comparison with other drug-trafficking groups. In fact, it was far more decentralized than is often assumed. Even the archetypal cartels, such as the Medellín and Sinaloa ones, never had the capacity to set the market price for drugs and relied on layers of subcontractors, franchises, and loose affiliates.[28] As Peter Reuter pointed out in his *Disorganized Crime: Economics of the Visible Hand*, it is often erroneously assumed that professional, "organized" crime necessarily implies hierarchical structures. In fact, many organized-crime and drug-trade structures are distributed networks.[29] Nor does the top person control everything. The role of the criminal boss, the market share of the criminal group, and the vertical structure of the illegal industry are often greatly exaggerated. This erroneous assumption results in promoting supposedly knock-out blows against key players, vastly overestimating the extent to which such effects can be delivered and their effectiveness, whether in combatting drug trafficking, wildlife trafficking, or other illegal economies.

Moreover, different structures have different built-in mechanisms for post-arrest adaptation and replacement. The term "kingpin," used to describe top bosses in the wildlife and drug trades, only reinforces this tendency to overestimate the effect of removing any single individual (or a single trafficking network). Very rarely does a criminal organization fold when the top boss is out. Unlike in the case of some, and only some, terrorist groups, removing kingpins has not yet dismantled or substantially reduced any illegal economy. Many have the ability to replenish their leadership, and have an interest in doing so. Trafficking organizations can most of the time survive the removal of individuals; networks can survive the elimination of entire nodes and reestablish them.

This is not to imply that the Anson Wongs and Vixays of the world should not be arrested and prosecuted. There are ethical and political reasons to do so, as well those pertaining to the rule of law. But one should not promote such interdiction strategies focused on removing top traffickers as the way to end wildlife trafficking. Others will replace them. After what appears to be Vixay's withdrawal from wildlife trafficking in 2014, the Bach brothers are alleged to have expanded their own wildlife trafficking operations to fill the gap.[30] Similar allegations have also been made against two Lao companies, Vinasakhone and Vannaseng. Officially licensed for legal trade in wildlife products and tiger farming in Laos, they are accused of having become major wildlife smuggling outfits.[31] (At the 2016 Conference of the Parties to CITES in Johannesburg, the government of Laos promised to look for ways to phase out tiger farms.)[32]

Diaspora communities often serve as important connecting links in the global illegal wildlife trade. As with all social mobilization, personal connections and networks play a critical role. Typically retaining the cultural traditions and predilections of their home, such as *ye wei* ("wild taste"), they may fail to become well integrated into a new home or temporary work locale abroad. The resulting sense of isolation and marginalization breeds susceptibility to recruitment by criminal rings. Or they may be merely opportunists. In Nepal, important wildlife smuggling networks operate within the Tibetan refugee community. Lhasa is a major wildlife smuggling hub, and big-cat furs and ivory are highly prized among the Tibetan, including lama, community.[33] Nepal's border officials tend not to diligently search "men in red clothes" (Buddhist monks), who as a result can easily hide illegal wildlife body parts in their robes.[34] The Dalai Lama has himself recognized the Tibetan community's role in the demand for poached animals and wildlife smuggling, and actively campaigned against them.[35]

As with other illegal economies, profit mark-ups grow significantly the further downstream the smuggling chain the product has moved and the more law enforcement it has had to overcome. Such mark-ups are not small even within a country. While a poor hunter in Tam Dao National Park, Vietnam, can earn perhaps a few hundred dollars a year, an owner of a restaurant in Tam Dao will be able to make $1,000 to $1,500 selling wildlife meat to tourists, while a medium-sized trader in Vinh Yen will earn more than $15,000 a year.[36] In Hanoi, the trader's income will be greater yet. Similarly, a pangolin caught in Myanmar traded there in the late 2000s for $3 per kilogram. When smuggled into Kunming, China, the price rose to $57. Upon arrival in Guangdong province, the price further went up to $86 and could reach $171 during special occasions, such as the Chinese New Year.[37] Reflecting growing scarcity and growing demand, a decade later, in 2015, live pangolins sold for $120 per kilogram in Vietnam, another country with a huge appetite for them. In Kenya, a kilogram of rhino horn may fetch $9,000, while in China it will bring upward of $70,000.[38] A top-level Mozambican poacher was paid $2,700 per kilogram of rhino horn in 2009 and over $8,400 in 2011. With the average weight of a horn being 6 kilograms, the payoff would come to between $16,200 and $50,400.[39] However, that does not necessarily mean that a top poacher can keep the entire sum to himself; he will have to divide the profits with lower-level poachers, spotters, trackers, and other support-team members. Finally, in the destination market of Vietnam, a kilogram of rhino horn will sell for between $25,000 and $65,000 per kilogram, depending on its quality and the purpose of use, whether it be medicinal or as a symbol of prestige.[40]

Most of the time, profits are low in the initial stages of smuggling. While many poor participants, such as hunters, processors, and even small traders, are better off economically than they would be in the absence of participating in the illegal economy, they rarely manage to escape poverty. In Myanmar's Mong La wildlife market in 2006, I was offered a female accipiter hawk for $4. Unfed for several days, the bird was barely alive, had a broken leg, and without better treatment would likely would die within days. When I asked the seller, a middle-age local woman, whether she knew that she was doing something illegal, she reduced the price to $3. Throughout the market, terrapins, monkeys, rodents, and birds were crammed into tiny cages, all for sale for a few dollars. So were bear claws and bile, and the dried genitals from all kinds of animals.[41]

Such small profits also characterize vast portions of production and smuggling chains in drugs. Most farmers of illegal crops will not escape poverty. Although their participation in the illegal drug trade assures their human security and gives them a possibility of some socioeconomic improvements, they will remain poor and marginalized. Such relative and absolute poverty often also characterizes drug retail markets, including in the United States and Western Europe: most street dealers do not make much money, and that is the reason why so many of them deal only part-time and seek other employment to make ends meet. There simply is not enough work relative to the number of willing participants to make retail drug dealing full-time work and a source of riches.

Many key transshipment centers, such as Vientiane in Laos and Linxia in China, remain areas of low living standards and underdeveloped legal economies. The same basic logic explains why. If everyone in a wildlife-smuggling hub could make $100,000 per year from wildlife trafficking, then tens of thousands of people from other parts of the country would move there, bringing wages down even with the large volume of wildlife smuggled. So even when earnings from the illegal wildlife trade can be large compared to the earnings from available legal jobs for low-level poachers and middle-level traders, being a low-level poacher or middle-level trader still does not garner big bucks compared to the pool of available participants. Thus wildlife trafficking may destroy nature but is highly unlikely to lift a country out of poverty.

Other Actors and Stakeholders

As with drugs, timber, and gems, "laundering" within the wildlife trade is not only of profits, but also of actual animals and plants. Since captive-bred animals are exempted from CITES prohibitions on trade, breeding farms are used to launder poached animals, as in Vixay's case. Public and private zoos also provide good cover for smugglers since a zoo can claim to have a breeding program for endangered animals, and thus explain the arrival of new animals. As detailed in Chapter 7, the laundering of animals and falsification of certificates have plagued controls on the ivory trade in China, Hong Kong, Japan, and Thailand, where supply, trade, and retail sales have been legally permitted to a degree. In Thailand, ivory from domesticated elephants can be legally sold, and smugglers from Myanmar or Africa often claim that their ivory has come from Thailand's domestic animals.[42] Such laundering problems also occur with tiger products in China, where sellers claim that their tiger products come from animals raised on tiger farms, not from poached animals in India and Indonesia.[43]

Beyond zoos, corrupt wildlife industry officials tasked with issuing licenses and setting hunting quotas, corrupt veterinarians, or taxider-

THE ACTORS

mists who can fake a rhino horn trophy to mask illegal hunting for Asian markets are important players in the illegal trade. Knowing regulation loopholes allows legal wildlife industry actors to be effective in evading controls. For example, strict regulation of the Russian fur trade created significant barriers to entry for criminal groups and other outsiders. Instead, knowledgeable insiders, such as registered hunters and law-enforcement officials, dominated the illegal trade hidden by the legal one. Only wealthy Russians could hire them to procure rare furs.[44]

Ecolodges, private reserves and parks, and trophy-hunting outfits play critical roles not just in influencing whether the legal wildlife economy can generate enough income to suppress the temptation to poach, but also in disrupting or enabling actual poaching, as analyzed in Chapter 7.

Other stakeholders in the regulation of wildlife trade and conservation—who are thereby at least indirectly stakeholders in the development and implementation of responses to poaching and wildlife trafficking—include logging companies, agribusinesses, the fishing industry, local police and enforcement forces, private security forces, and governments.

The logging industry, for example, frequently demands access to new areas where timber is abundant. New logging routes open access routes for poachers and traffickers. In remote areas, loggers themselves frequently hunt local wildlife for food. Around the world, the legal fishing industry, frequently subsidized by governments, often fishes illegally by harvesting outside of designated areas, in excess of legal quotas, catching immature fish or endangered species.

As detailed in the following chapter, local rangers, police, and wildlife law-enforcement officers frequently earn only small salaries, little prestige, and limited chance of promotion by enforcing regulations against those involved in the wildlife trade. The pressures of corruption that they, as well as top government officials, face are high.

Other private actors are increasingly hired to supplement wildlife law-enforcement authorities, such as private security companies, former soldiers, foreign trainers, intelligence units, technical operators, and anti-poaching militias. Standing militaries, both domestic and foreign, can also be mobilized to supplement wildlife law-enforcement forces, as discussed in the previous chapter. During the 1980s and 1990s, French soldiers carried out "border control operations" against conservation foes in the Central African Republic, for example.[45] At times, such actors crucially assist and enable desirable environmental outcomes, such as in Nepal. However, they have also been implicated in severe human rights abuses.[46] Regardless of their actual effectiveness and the side-effects of their use, like other legal actors in the wildlife trade, they become advocates of particular anti-poaching policies, such as beefed-up law-enforcement efforts. That may well be the right response, but it also serves their budgets and institutional interests well. Environmental NGOs also critically influence which policies are adopted and funded. Over the past decade, many of them have advocated interdiction, bans, and law enforcement.

The broader governance of the environment and wildlife trade, and the role of various actors in legal natural-resource-based economies, cannot be divorced from the illegal wildlife economies. The legal side and its players impact and shape the illegal side, and vice versa.

In sum, policies to curb the illegal trade in wildlife and to ensure the conservation and preservation of biodiversity need to address the diverse and actor-specific drivers of the illegal wildlife trade. Neither the drug trade nor the global illegal wildlife economy is uniformly structured. There is great variation in how hierarchical, vertically integrated, or disorganized the illegal trade is in particular places—whether in wildlife or in drugs. And just as with the drug trade, much of the illegal trade in wildlife is actually rather disorganized, built around loose networks, with only occasional vertical integration. Thus policy designs need to be very specific to particular places and particular times, and they may have to be revised as illegal markets and networks adapt or evolve. But the fact that an illegal economy is disorganized does not mean that it does not cause serious and severe threats and harm to states, societies, and nature.

Interdiction patterns as well as socioeconomic interventions need to be tailored to local patterns and networks. That is the case at both tactical and strategic levels: specific constellations of structures embedded within a particular social contexts could lead one to give up on an entire class of policy interventions, not just on particular interdiction patterns or particular design of alternative livelihoods strategies or CBNRM approaches. But conservation policies will also be ineffective if they ignore corruption among the presumed "good guys"—government institutions, the legal wildlife industry, and law enforcement.

6

POLICY RESPONSE I

BANS AND LAW ENFORCEMENT

The global poaching crisis has induced large segments of the conservation community, including many international conservation NGOs, to call for far tougher law enforcement and for outlawing some of the existing legal markets for wildlife products, particularly in ivory. These actors have also opposed allowing any additional international legal sales of rhino horn. In calling for bans, they thus contradict the desire of some African countries, such as South Africa, that are asking for a legal means to sell their stocks.

Indeed, some legal markets in wildlife have crucially facilitated poaching. Wildlife trade and anti-poaching efforts have long been accorded the least priority by many law-enforcement authorities around the world. Given this low baseline of enforcement efforts, boosting enforcement of wildlife regulations is overdue.

But better and tougher law enforcement is not a silver bullet. There are inherent limitations to what even determined law enforcement can accomplish. The parallel existence of legal trade further complicates and weakens law enforcement. At the same time, as Chapter 7 shows, there are very sound and powerful reasons to make as much wildlife trade as possible legal, even at the cost of complicating law enforcement.

Even when all trade in a commodity is illegal, law enforcement, particularly interdiction, is enormously difficult to implement effectively. It often fails in its critical task of radically curtailing supply, and it can have highly problematic side effects and, depending on the design of interdiction efforts, outright counterproductive effects. Moreover, decades of experience with drug policies show that prohibition and the enforcement of prohibition have complex effects on prices and demand.

If large segments of the populations have not internalized laws and consider them illegitimate, enforcement becomes more difficult yet. But law enforcement can also motivate compliance and even norm internalization. Law enforcement thus needs to be designed to minimize problematic social side effects and counterproductive environmental outcomes.

The Regulatory Theory

In theory, the prohibition of trade in and consumption of a commodity is supposed to constrict supply and discourage demand. Interdiction is aimed at reducing supply further, by seizing illegal goods and reducing the capacity of traffickers and suppliers to supply the product, whether they are *cocaleros* in Colombia or poachers in Tanzania. Law enforcement also creates inefficiencies in supply, raising trade barriers and boosting prices. The more effective law enforcement is, the greater this "tax" and the price of the illegal product become. Higher prices make the product less affordable, thus discouraging use. In the case of drugs, price-driven reduction in demand applies not only to casual users but also to hardcore addicts.[1] Casual users often spend so little on drugs that price is not the main factor in their decision to use them, whereas addicts may have limited income and spend so much on drugs that they are surprisingly sensitive to price fluctuations. Similarly, discouraging use via higher prices is the central purpose of high taxes on cigarettes and alcohol.

Prohibition and enforcement also introduce additional physical barriers for both users and suppliers. Not every trader has the brains and stomach to evade law enforcement. Consumers may not know how to go about, nor bother to invest the time in and absorb the risk of finding an underground store. Searching the darknet for tiger bone takes more time and skill than searching a legal internet market for a product. Making something illegal also limits product innovation.

BANS AND LAW ENFORCEMENT

Thus the form of sale affects the level of consumption of a prohibited product. Open, visible, easily accessible markets with plentiful advertising will result in greater consumption. If customers just go to a restaurant which has endangered species on its regular menu, they will consume them in greater quantities than if they have to request a special secret menu with the rare exotic endangered species on it, or if they have to invest time in locating a special underground restaurant with such products. That is why driving retail and consumption markets in illegal wildlife products underground and minimizing advertising make good sense even if those markets are not eliminated altogether. Particularly if consumption is driven by a desire to ostentatiously display status, power, and wealth, such as by wearing ivory bangles or coats from endangered species, a ban that drives such visible displays down may well shrink demand significantly.

In Uruguay, marijuana legalization reduction has been designed to minimize commercialisation, by permitting state monopolies, self-growing, and small co-ops, but prohibiting for-profit business. In contrast, the design of marijuana legalization in the US state of Colorado permits undesirable heavy commercialization and marketing.[2]

Prohibition also has normative effects on both suppliers and users. Among both there are subgroups that have internalized norms and laws and will not engage in behavior that is illegal and harmful. The mere proposal of placing devil's claw (a plant from southern Africa used as a natural remedy for arthritis and traded globally with a principal market in the European Union) on the CITES Appendix II list so diminished the market in Europe that it generated very substantial income losses for poor African subsistence harvesters who depended on it for income.[3] Similarly, the 1989 ban on ivory significantly contributed to a decline in sales in Europe and the United States, and, along with an extensive NGO campaign about the horrors of elephant poaching, apparently also helped to drive down demand for ivory in those markets. A ban on wild bird trade in United States and Europe registered similar positive outcomes. Official exports of wild birds from four of the five leading bird-exporting countries fell by more than two-thirds between the late 1980s and 1990s, a decline often attributed to the US import ban.[4] In both cases, the bans were implemented in societies with high degrees of environmental consciousness and strong law-enforcement capacities.

Yet prohibition does not generate the same normative outcomes and drop in demand among all sectors of a population. Some young people do drugs precisely because violating authority is cool. During my fieldwork on crime markets in Medellín, Colombia, I was repeatedly told of the so-called *paisa* culture that celebrates law evasion. This is practiced not just in terms of drug use, but more broadly in terms of making money by evading the law. In East Asia, in contrast, much consumption of wildlife takes place because it is considered hip and normal. Peer pressure plays a strong role, but not because consumption is a form of rebellion against the dominant norm. Other considerations than legality and normative preference influence decisions. Many Afghan opium poppy farmers admit that they engage in illegal and undesirable economic activity but can see no other economic survival opportunities. But just because not every user and supplier is deterred by prohibition does not mean that the removal of prohibition will take place without supply and consumption increasing.

The Realities and Challenges of Interdiction and Law Enforcement

Precisely because the normative effects of prohibition are not sufficient, law enforcement is crucial. There are many reasons to enforce rules and regulations. Nonetheless, drugs provide a sobering lesson as to the limits of law enforcement. No commodity market has seen as much law enforcement as the illegal drug economy. Tens of billions of dollars a year have been spent around the globe for decades on reducing the global drug supply. The level of resources dedicated to wildlife regulation enforcement, on the other hand, is several orders of magnitude lower—at most in the hundreds of millions of dollars per year. Even when one thinks of the extent of law enforcement dedicated to combating drugs in terms of intensity per dollar—that is, the amount of dollars dedicated to law enforcement compared to drug profits—the intensity still significantly surpasses those of wildlife law enforcement.

When the dedication of large resources to counter-narcotics efforts results in an effectiveness rate of 40 percent, it is deemed a great but rare success. However, a 40 percent interdiction success rate may still be insufficient to halt the devastation of a species, particularly one of low numbers and slow reproductive rates.

BANS AND LAW ENFORCEMENT

Moreover, at which point in the smuggling chain interdiction takes place matters enormously. In the case of drugs—a nondepletable resource than can be produced in very large volumes indefinitely—seizing drugs close to production, such as in Colombia, or Peru, or Burma, is not very effective in terms of the cost of their replacement and knock-on effects in terms of retail prices. Seizing drugs close to consumption and retail markets, such in the United States or Europe, boosts prices much more.

However, for political, social, and justice reasons, interdiction in Colombia, even though not very effective in terms of boosting retail prices, is still preferable to eradicating the crops of hundreds of thousands of poor farmers who depend on drug cultivation for their livelihoods and will mobilize to oppose eradication. Interdiction in Colombia can have other desirable effects, such as reducing smugglers' proclivity toward violence and reducing their capacity to corrupt institutions or penetrate political systems. Well-designed interdiction can also limit the access of militant groups to drug resources.

There are analogous effects in poaching and the illegal wildlife trade; there are also big differences. Like eradication, preventing poor marginalized communities from subsistence hunting is ethically problematic and can become politically unsustainable. However, focusing interdiction closer to retail markets instead of supply areas can also be highly counterproductive. For example, drug traffickers assume they will lose a certain percentage of drugs and simply pay for the production of larger volumes to cover their predicted losses. They even sometimes welcome interdiction and eradication efforts since these boost prices and make stockpiles more profitable. The traffickers' ability to increase and adjust supply to offset losses is one of the reasons why prices of drugs have not gone up high enough to reduce the capacity and motivation of consumers to purchase them.

Increasing the volume of animals poached to maintain supply despite law enforcement is a most undesirable outcome in the wildlife trade. Traffickers of rare parrots from Indonesia, for example, fully expected a 90 to 95 percent mortality rate as a result of their smuggling methods. To evade law-enforcement agencies, they stuffed the parrots into plastic bottles and threw them into the sea so as to retrieve them on open waters outside of the reach of the authorities.[5] The fact that less than 10 percent of the parrots survived was not a deterrent, as profits on the remaining specimens were more than sufficient. In fact, prices can be boosted by scarcity so much that absorbing huge losses and driving a species to extinction is highly profitable and attractive for traffickers. The rarer the species, the greater its value.

Law enforcement must avoid creating those transshipment inefficiencies that motivate smugglers to organize the poaching of many animals in order to deliver a few to the market. Avoiding these effects is particularly imperative in the case of markets in rare wildlife, where prices and profits increase the rarer the animal is. Enforcement *in situ* to prevent animals being killed in the first place is far more effective in minimizing species loss. Shutting down retailers of illegal wildlife commodities is important. But transshipment seizures create counterproductive effects, resulting in a greater number of animals poached.

Similarly, placing a species on the endangered species list—and thereby announcing its scarcity—may result in its habitat conservation in the United States, but in Asia or Africa it may stimulate greater poaching.[6] Similarly, designating an area as a newly protected wildlife zone may attract poachers to that space.

This does not imply that it is a good idea to forego the protection of threatened wildlife species, but rather it means that it is necessary to have at the ready the necessary law-enforcement assets to protect the park or species before the new rules are announced. Time-lags between the announcement and the implementation and effective enforcement of prohibitions are particularly problematic as they are likely to boost the rate of poaching. Indeed, precisely to avoid attracting poachers, scientists these days typically do not disclose specific locations or publish photographs of rare or newly described species.

Overall, bans on trade in wildlife have had highly varied and complex effects on species conservation. Before elephant poaching escalated in Africa in the late 1990s, it had shifted to Asia, where remaining wild elephants became targets of intense illegal hunting.[7] Despite an initial decline in the international trade in parrots with Europe and the United States, Brazil's domestic trade in parrots, especially involving popular mimics such as the blue-headed parrot, has continued unabated and is having significant negative consequences for particular species.[8] A generalized prohibition on hunting went into effect in Kenya, in the

BANS AND LAW ENFORCEMENT

1970s. Yet poaching is prevalent and since the ban was issued, Kenya has lost some 70 percent of its wildlife.[9] One can always argue that without the ban, the species collapse would have been even greater and more acute (although of course there is no way to prove such a counterfactual claim).

The 1977 CITES ban on rhino horn shows the complexities and unevenness of outcomes. For almost two decades after the ban was promulgated, intense poaching continued. By 1994, the population of the black rhino (one of the rhino species) had declined by 97.6 percent, while the northern white rhino had become extinct in the wild. The ban alone was not enough. It was only in the early 1990s that the ban seemed to work when, as a result of intense international pressure, including the threat of US sanctions, countries where there was a large demand for rhino horn—Japan, Taiwan, and South Korea—shut down domestic demand. However, without a comparative case it is very difficult to know whether continued rhino-horn trade would have caused the species to decline even more.

Crucially, during the 1990s and until the early 2000s, *in situ* law enforcement against rhino poaching in protected areas, combined with thriving wildlife tourism and legal hunting on private reserves (such as in South Africa) or in community-owned lands (such as in Namibia), seemed to work very well, so much so that South Africa became the model for allowing legal hunting, as did Namibia for community-based resource management strategies. As a result, the southern white rhino population in both countries stabilized and grew. In South Africa, poaching losses were only about 14 a year.[10] Removing horns from live animals also negated the poachers' incentive for killing rhinos. (Innovative strategies, like removing horns, are of course not always possible in other species; removing sharks' fins in order to reduce their attractiveness to poachers will kill the sharks.)

But in the early 2000s, demand for rhino horn reemerged in China and exploded in Vietnam. Poaching rapidly escalated in South Africa, shattering anti-poaching enforcement. Corruption among South African rangers spread, and despite new bush wars against poachers, the country has lost some 10 percent of its rhino population per year, and is now at the epicenter of a rhino-killing vortex.

So was the ban effective or not? It is invoked to demonstrate both the effectiveness of bans and their failures.[11] Clearly, the ban was not effective on its own; it was only effective when demand also dropped. But demand dropped only as a result of intense campaigning and international pressure enabled by the ban. Yet campaigning did not extend to suppressing new demand in new markets. And the new rise in demand overwhelmed both *in situ* law enforcement and legal regulations to prevent poaching.

Effective Interdiction Is Hard

By and large, wildlife law-enforcement efforts are severely underfunded, often well below what attempts at boosting interdiction effectiveness rates require. There is a strong positive correlation between the number of guards and the success of protecting conservation areas.[12] Sustaining that correlation, of course, depends on the guards being motivated to perform their duties and not to poach themselves.

Even in the United States, where the Obama administration made combating wildlife trafficking an important policy focus, in 2014 there were still only 205 federal investigators and an additional 130 federal wildlife inspectors nationwide (the latter number a decline from the previous 140). Only 38 out of the country's 328 ports of entry had full-time inspectors on-site. In 2014, those officers at ports of entry were tasked with examining 180,000 wildlife products. Just three wildlife detector dogs were on the force as of November 2015. Subsequent budget requests in 2015 and 2016 sought to increase those numbers. Because of the reduced staffing, investigations fell from more than 13,000 in 2012 to 10,000 in 2014.[13]

Nonetheless, dedicated efforts can produce important results and be a source of inspiration. Operation Crash was one of the largest US and worldwide interdiction operations against wildlife trafficking. It targeted US and international rhino poaching networks as well as the illegal trade in ivory. A handful of US Fish and Wildlife special agents, eventually partnering up with other law enforcement officials and agencies in the United States and abroad, spent several years painstakingly uncovering and tracking criminal networks. By spring 2015, Operation Crash had issued indictments against thirty-three individuals and businesses and achieved twenty-two convictions for wildlife trafficking. Another large coordinated international interdiction operation

BANS AND LAW ENFORCEMENT

against wildlife trafficking—Operation Cobra III—brought down smuggling networks in May 2015. Organized by the Association of Southeast Asian Nations Wildlife Enforcement Network (ASEAN-WEN) and the Lusaka Agreement Task Force established to combat wildlife trafficking in 1999, the operation involved law enforcement units from 62 countries across Africa, Asia, the United States, and Europe, and resulted in 139 arrests and more than 240 seizures of ivory, rhino horn, pangolins, rosewood, tortoises and other species.

New technologies can help. Spatial Monitoring and Reporting Tool (SMART) software, adopted by over 147 conservation projects in 31 countries, allows rangers, NGOs, and communities to report incidents of poaching and suspicious behavior to a common database and facilitates mapping and analysis.[14] Given the vast spaces that need to patrolled, any greater visibility of a space is enormously useful. So too is the capacity to predict where and what time poaching is most likely to occur, and to move forces there. The deployment of drones in the Serengeti National Park, lamented by scholars who decry green militarization, is at the same time being credited with a drop in poaching in the park. But even with night-vision goggles, drones, signals intelligence, and high-power weapons, patrolling vast territories is not easy.

That is not to say that interdiction can never be effective. Intensified law enforcement resulted in the reduced poaching of elephants in Zambia in the 1990s,[15] though this achievement was not sustained when a new wave of global demand hit the country and region. A crackdown in Bali, Indonesia, on illegal catches of sea turtles during the first decade of this century was also a success. Used in traditional Balinese ceremonies, marine turtles had been caught at a rate many times surpassing the 1,000 per year allowed under local regulations. In 1999, 27,000 turtles, for example, were killed. Profauna, an Indonesian environmental NGO, encouraged zero-catch quotas and pushed for greater law enforcement by the police and other law-enforcement agencies, such as the Forestry Ministry. The confiscation of turtles by the police increased significantly, and illegal catches decreased by 80 percent.[16]

Monitoring beaches on a small island is, of course, much easier than monitoring vast tracts of forest or the open ocean. The fact that police units on Bali are less corrupt than elsewhere in Indonesia was a significant cause of the campaign's success. Greater international presence in Bali than in other parts of Indonesia, such as Ambon and Kalimantan, helped in monitoring the enforcement effort and maintaining public pressure for the campaign. The localized nature of demand was also a crucial factor.

In contrast, freshwater and marine turtles and land tortoises captured in Indonesia for the international market in China, Thailand, and Vietnam have not fared well. At the beginning of the 2000s, for example, Malayan box turtles were widespread across Indonesia. But within ten years, they, as well as two endemic Sulawesi land tortoises, fell victim to the Traditional Chinese Medicine (TCM) craze. So that they could be eventually shredded in blenders into jelly and paste, villagers in Sulawesi would collect the creatures and sell them for 5,000 Indonesian rupiahs (about 50 cents) per turtle or tortoise. According to a biologist from the Pacific Institute in northern Sulawesi, a subsequent three-month field research project in the area in 2007 found only two specimens of what used to be several plentiful species, including some found nowhere else.[17] The turtles and tortoises were literally eaten off the island. A law-enforcement effort was mounted, but with neither with the vigor nor with the successful outcome of the Bali effort.

In fact, the point to take from many of the above examples is that what crucially facilitates and oftentimes underpins the effectiveness of interdiction, law enforcement, and bans is a reduction in demand for a commodity—whether as a result of a ban, purposeful demand reduction strategies, or exogenous factors.[18] But a reduction in demand and compliance with a rule are often dependent on and interlinked with law enforcement. Bans, law enforcement, and interdiction do more than constrict supply; they are one set of the factors that drive normative changes and the internalization of rules.

The Paradoxes of Effective Interdiction

Prohibition even with determined law enforcement often produces paradoxical effects. First, if seizures stimulate greater killing of animals to supply demand, they are outright counterproductive, as discussed above.

Second, the more effective law enforcement becomes in stopping the supply of a commodity, the greater the price of that commodity that remains in circulation. Thus those who are not deterred or incapacitated by law enforcement will have the incentive of higher profits to stay in the business (as well as facing high risks in doing so).[19] These are not one-to-one relationships. A higher price does not necessarily mean that profits (in an economic sense) are also greater—particularly if evading law enforcement becomes very complex and costly. But merely imposing higher costs on producers and traffickers through law enforcement does not mean that they will go bankrupt or that the illegal trade will become "low profit" (as advocates of tougher wildlife enforcement like to argue) and traffickers and poachers will give up the illegal wildlife trade. Instead of making poaching and trafficking high-risk and low-profit ventures, law enforcement can make them high-risk and higher-profit.

Third, no market, least of all a criminal one, is static. In response to competition from other groups and pressure from law enforcement, wildlife traffickers and poachers learn to be innovative with their evasion, corruption, and smuggling methods, just as drug traffickers and farmers of legal crops, factory managers, and fast-food franchises do. The stratagem of Vietnamese rhino-horn traffickers to hire Thai prostitutes working in South Africa or Czech nationals traveling there to pose as licensed trophy hunters is one example.[20] Similarly, wildlife traffickers have adapted to the nondisclosure procedures of scientists to protect rare endangered species by scouring scientific journals for information on their location.[21]

The fourth problematic consequence of effective enforcement is an undesirable displacement of illegal production or trafficking to a new area—what in counter-narcotics efforts is known as "balloon effects." The displacement of elephant poaching from Africa to Asia following increased enforcement efforts in Africa in the 1990s serves as an example.

These balloon effects may also inadvertently push poachers and traffickers to expand the menu of poached products, such as from civets to hornbills, in the same way that drug traffickers expand their products from cocaine to methamphetamines and heroin in response to law enforcement.

Sometimes displacing an illegal economy from one area to another can be a useful and desirable objective of law enforcement. Not all environmental areas are equally sensitive to disturbances. Interdiction can also seek to drive hunting to areas where a species may be abundant and its population can take sustainable hunting—the operative and difficult thing, of course, being to keep hunting sustainable.

Again, there are analogies in anti-drugs efforts. The illegal drug economy, just like other illicit economies, does not have the same and equally intense problematic effects around the world. In Afghanistan, illegal drugs underpin the livelihoods of millions people and account for a large portion of the country's GDP. They also fuel criminality, power abuse, corruption, and are a crucial source of financing for the Taliban and other militant groups, as well as for pro-government militias and powerbrokers. But with all the many threats and challenges in Afghanistan, the problem of drug cultivation there is still less severe from the perspective of global security than if drug cultivation was displaced to Pakistan. Profits and political capital from drugs there could be captured by far more dangerous militant groups seeking to destabilize a country with nuclear weapons, which could leak to militants. Such destabliziation could even provoke a nuclear war between Pakistan and India, with catastrophic repurcussions for the entire region.[22]

Fifth, the more effective law enforcement becomes, the more it can give rise to tougher, meaner, more powerful, and more dangerous traffickers. Often law enforcement takes out the traffickers who have the least capacity to evade, corrupt, and intimidate law-enforcement officials. The traffickers with competence and political power survive and are needed by lower-level traders as patrons. Such survival-of-the-fittest vertical integration in a criminal market happens when law-enforcement officials act on tactical intelligence without a prior strategic analysis of how interdiction will reshape the criminal groups and criminal market. A prime example of such an inadvertent outcome is the vertical integration of the Afghanistan drug market between 2003 and 2008 as a result of interdiction.[23] Before interdiction was undertaken, the Afghan drug market was far more decentralized and populated by many small traders, and hence less harmful.

Limitations of Going after the Kingpins and Advantages of Middle-Layer Targeting

Whom interdiction efforts target makes a large difference to the effectiveness of interdiction. Arresting low-level poachers often has a limited effect. Since 2012, more than 800 suspected rhino poachers have been arrested in South Africa, and yet rhino poaching levels increased at least until 2015, when they dipped but still remained at a very high level.[24] In Tanzania, an investigation into wildlife crime arrested more than 1,000 people in just over a year, including Yang Feng Glan, a Chinese businesswoman and a prominent ivory trafficker dubbed the Queen of Ivory.[25] Yet elephant poaching has not lessened. (Arrest numbers are of course easy to inflate.)

Environmental NGOs and technical advisors sometimes argue that interdiction aimed at disabling smuggling networks is easy and should be the thrust of law enforcement since trafficking seems to be concentrated among a few groups. No doubt prosecuting top wildlife traffickers rather than poor trackers and poachers is normatively desirable. Moreover, if powerful actors systematically escape prosecution by intimidating law-enforcement agents and justice officials so as to escape prosecution and punishment, the rule of law becomes weakened, and large segments of the population may stop internalizing laws and norms.

However, even when interdiction succeeds in capturing a particular kingpin or takes down a network, new traffickers step into the market opening. Far more so than terrorist group leaders, crime kingpins are replaceable.[26] It may still be desirable to capture top traffickers and systematically target smuggling networks as they rise up, reconstitute, and reemerge, but it would be much too optimistic to assume that arresting kingpins will end poaching and wildlife trafficking.

Targeting the middle operational layer of a smuggling group can be far more effective in disrupting its operational capacity than the so-called high-value targeting of bosses. But going after the middle layer will only significantly disrupt the operational capacity of a smuggling group if much of the operational layer is arrested at once. If capture of middle-layer operatives, as of kingpins, is done piecemeal, one at a time, the criminal group will find it rather easy to replenish its lost functional capacity.

As much as possible, the goal should be to produce system-wide effects. Thus targeting all ivory traders at once in one large block (such as in Fuzhou, China) will have far greater deterrent effects than taking them down one by one and allowing them time to adapt to the new regulations and law-enforcement strategies.[27]

However, middle-layer targeting requires more sophisticated skills, greater resources, less corruption within law-enforcement agencies, and lengthier investigation periods than piecemeal opportunistic targeting, even of kingpins. Pulling the entire middle layer of a criminal group down at once requires far greater capacity and time to identify its members and develop adequate intelligence and evidence. Unlike in the case of drugs, time matters acutely, especially when animals are being poached at extinction rates. Moreover, sitting on intelligence for a long time risks it leaking out. Both operational security and assuring effective intelligence flows become more complicated if local park rangers need to pass information to distant headquarters or liaise with multiple law-enforcement agencies.

But as long as demand for an illegal commodity persists, even entire organizations and structures can be replaced. For example, as discussed in the previous chapter, when Vixay Keosavang appeared to remove himself from wildlife trafficking in 2014, other wildlife trafficking outfits—the Bach brothers and the Lao companies Vinasakhone and Vannaseng—stepped into the gap in the market and the illegal wildlife trade has continued.[28]

Capacities and Corruption

Elementary skill sets continue to be deficient among rangers and wildlife-enforcement officers in many parts of the world. Much of Western assistance now focuses on transfers of high-power firearms, night-vision goggles, and other technical assets to wildlife-enforcement officials in Africa and Asia. This militarization is in part a response to the increasingly lethal firepower of poaching networks, often with deadly consequences for park rangers, many of whom are courageous, selfless, and committed individuals. For example, in the Virunga National Park in the Democratic Republic of the Congo, a crucial habitat for forest and savannah elephants and critically endangered mountain gorillas, perhaps over 150 rangers have been killed by poachers since 2004, and the park's warden was seriously wounded in a 2014 ambush.[29] The Thin Green Line Foundation estimates that 595 rangers were killed worldwide between 2009 and 2016, with 189 in Africa.[30] Meanwhile, poachers' violence against rangers encourages the militarization of wildlife enforcement. In South Africa, many former soldiers of the apartheid-era South African Defence Force joined various private military companies during the 1990s, with many such units used to guard national parks and protected areas. In Nepal, the deployment of the country's military to guard national parks has improved environmental conservation. The Nepalese military colonel Babu Krishna Karki, director of national parks and wildlife reserves in the Nepal army, has been recognized worldwide for his environmental efforts. The use of Botswana's military to patrol protected areas is sometimes credited with the low incidence of poaching in the country compared to its neighbors.

Yet despite recent efforts to boost the capacity and status of wildlife-enforcement officials, in much of Africa, Latin America, and Asia rangers continue to be poorly paid and resourced. A career in wildlife enforcement will produce less recognition and material benefits, and lead to slower promotion, than a career in combating other crimes. Many wildlife enforcers lack investigative skills and capacities, including evidence-collecting, maintaining secure custody chains, conducting forensic examination, developing informants, working undercover, and obtaining intelligence from controlled delivery. Moreover, being a wildlife-crime investigator is in fact a very different job than being a regular beat ranger. Training sufficient numbers of wildlife-crime investigators is imperative.

Insufficient and missing wildlife-enforcement capacities also characterize developed countries efforts. A European Parliament study of wildlife crime enforcement in the European Union found a "varying and often low level of sanctions, a lack of resources, technical skills, awareness and capacity among police forces, prosecutors and judicial authorities, [a] low priority given to wildlife crime by enforcement institutions and a lack of cooperation between agencies."[31]

These missing capacities might seem to have fairly straightforward solutions. New wildlife enforcement networks and units are being set up, such as ASEAN-WEN and similar wildlife enforcement networks in Central America, Central Africa, the Horn of Africa, and southern Africa. Other international law-enforcement organizations, such as INTERPOL, are dedicating increased resources to combating wildlife trafficking.[32] Increasing numbers of NGOs and for-profit companies are rushing to train rangers around the world.

Yet a crucial part of the problem is corruption, which is pervasive throughout many wildlife supply chains and particularly prevalent among many African countries as well as in India, Nepal, and Indonesia.[33] Corruption is of course hardly unique to the wildlife realm. In much of Africa, most governance is characterized by resource scarcity, institutional weakness, pernicious patronage systems, and corruption—all exacerbated by deficient oversight and accountability mechanisms.[34] Where political will is lacking to implement conservation rules, vested interests easily mobilize to oppose and subvert wildlife law enforcement and perpetuate poaching by establishing corrupt networks throughout the political system. In all too many countries, the purpose of the state and the core of governance have long been to accumulate rents for particular powerbrokers and their cliques, not to deliver governance and needed public goods.

Even the militarization of park protection does not escape these problems. It is often rangers themselves who provide crucial intelligence to poachers about where animals and law-enforcement assets are present. Former members of Kenya's police and army have also been caught poaching.[35] In much of Africa, rangers as well as park wardens are selected not on merit but on the basis of ethnic quotas and political patronage, and they are often notoriously unmotivated. For example, two patrol rangers I interviewed in the Tsavo West National Park in Kenya (one of whom claimed to be a "big boss" without giving me his specific rank) first denied that any poaching was taking place in Tsavo. Then they admitted that poaching was going on, alleging it was all by Somalis or Somali-Kenyans (the favorite scapegoat in Kenya for all manner of crime), and that nothing could be done about it because they lacked vehicles and, most importantly, gasoline, which higher-up bosses siphoned off for themselves. As the interview concluded they asked me for a hefty bribe, saying that without it they would not be able to protect the animals. I refused to pay.[36] Another ranger at the Tsavo West rhino sanctuary was more forthcoming, disclosing that poaching was taking place, not only of rhinos and elephants, but also of lions to stop them from attacking cattle, and of zebras, with the meat sold on the local market.[37] Although he had been posted to Tsavo for more than ten years, he did not like his job. His family lived nearly 800 kilometers away and he got only thirty days off per year, thus seeing them only every two months. He found his $170 monthly salary vastly inadequate for his own and his family's needs, and he felt bored. He really wished that he had some kind of diploma so he could get a job with the Ministry of Energy.

In her studies of poaching and conservation in the Central African Republic, Louisa Lombard reveals how anti-poaching militias and rangers fluidly switch between being protectors of wildlife and poachers themselves since they do not make enough money from their official job. This crossing from protection to poaching and back is also practiced by the local community in the area, even though community-based conservation has been formally delegated to it.[38]

Throughout Asia and Africa, former poachers have been hired as rangers—not by mistake, but by design to keep them from poaching. Unsurprisingly, outcomes have varied. Sometimes poaching declined radically because the new rangers stopped poaching and had an intimate knowledge not only of the forest and animals, but also of poachers' methods and identities.[39] When the income they make from being rangers at least approximates the money they made from poaching and satisfies their family's needs, they can make extraordinarily effective guards. But hiring the fox to guard the chicken coop is a risky strategy, for it can spread corruption throughout the ranger force. It thus requires careful and diligent monitoring, and immediate action is needed if corruption and poaching by rangers persist. Moreover, a local community living in an area where poaching occurs may not easily accept poachers going "clean and green" if new rangers have to enforce anti-poaching regulations against a community that participates in poaching itself. Threats to the lives of rangers and their families and tensions within the community can arise.

A system of hiring based on patronage not only means inadequate candidates are recruited but also that mechanisms to fire incompetent and corrupt ones are lacking. Yet even within a patronage network, appointed rangers and wardens often have to buy their appointments, paying bribes to get licenses, school certificates, and actual appointments, and they expect to recover those investments by taking a cut from poaching.[40] Seizures of wildlife products, pointed to as an indicator of effective enforcement, often mask deep corruption problems. What happens with the seized ivory? Is it simply resold by rangers on the criminal market? Moreover, corruption is linked to politicians. Thus, poaching regularly increases at the time of elections because resources, including those earmarked for park protection, get diverted to support electoral campaigns and distributed among potential supporters, and because poaching too can fund campaigns.[41] This phenomenon is not new: in Kenya, for example, the linkages between political dynasties and poaching networks are thick and deep.[42]

There is increasing recognition of corruption and patronage problems among rangers. New efforts are being tried to recruit rangers locally in order to bring conservation money to local communities, sometimes even selecting individuals who might otherwise be recruited as trackers or poachers. Rangers need to be carefully vetted and given a boost in status and resources. Increases in salaries are often merited. It may well be that traffickers can always outbid a ranger's salary, but as long as rangers can obtain a sufficient livelihood and have a sense of social advancement, status, and pride for carrying out honest work, many will resist the temptation of corruption.

Corrupt wildlife-enforcement officials are also of course present in Europe and the North America. In 2011, a former member of the Royal Canadian Mounted Police was arrested for smuggling narwhal tusks (whale ivory).[43] Nor are major African ports such as Mombasa or Kismayo the only transportation hubs through which illegal commodities are trafficked. The challenge of checking large numbers of containers applies to all ports and airports, and illegal wildlife contraband is smuggled through Heathrow, Los Angeles, Rotterdam, and Dubai.

Individual officers can be corrupted in any country. But there is a difference between corrupt individuals and systems of corruption and patronage where whole institutions are corrupt. Many countries that are key actors in wildlife trafficking are wracked by pervasive and systemic corruption. South Africa, for example, ranks sixty-first on the Global Corruption Index of Transparency International, while the least corrupt country, ranked first, is Denmark. Other key states include Thailand (76), China (83), Vietnam (112), Tanzania (117), and Kenya (139); among Western countries are the United Kingdom (10) and the United States (16).[44] Crucially, anticorruption mechanisms, such as vetting, oversight, and internal affairs units, tend to be much weaker or nonexistent outside of Europe and North America.

BANS AND LAW ENFORCEMENT

Rotating rangers to new areas might reduce their susceptibility to corruption, but it may also spread poaching to new areas. Moreover, effective interdiction requires getting to know the local terrain and developing positive relations with local communities—both of which require time. Human intelligence is still crucial and not fully replaceable by signal intelligence in prosecuting and preventing poaching. The vetting of rangers, just like the vetting of all law-enforcement agents, needs to take place continually, not just at the time of recruitment.

Merely firing corrupt rangers might mean that they will join poaching and smuggling rings outright, given their knowledge and skills. Imprisoning them may remove them from poaching for a while, but prisons often serve primarily as schools for criminals. Depending on the severity of the offenses and the degree of corruption, some non-prison community-based supervision of fired rangers may at times be preferable to imprisonment. But that assumes that the local community is diligent in the monitoring and not itself implicated in poaching.

Yet progress can be and has been achieved. The leadership of the Mara Conservancy by Brian Heath provides an inspiring example. In 1994, the famous Maasai Mara National Reserve was divided into two different administrative areas. The eastern side, located in what used to be the Narok district,[45] is managed by the Narok County Council. The northwestern sector, known as the Mara Triangle and located in the then Transmara district of the reserve, comprising about one-third of the Maasai Mara protected area, is administered by the Mara Conservancy. The Mara Conservancy was established in 2001 as a rescue operation to improve the reserve's management. When Heath took over the protected area more than a decade ago under contract from the local Kinguri Council to run the newly established Mara Conservancy, the reserve's roads were in a terrible condition and the reserve even lacked patrol cars, with rangers regularly calling ecotourism lodges to ask for transport. Its staff had not been paid for four months, and its headquarters was totally dilapidated. Serious poaching as well as cattle rustling was going on throughout the reserve.[46] Many of the rangers themselves were actively involved in poaching.

Heath's first step was to pay all his staff and buy them uniforms and cars. Although he was not allowed to fire rangers, some of whom have been there for over two decades, diligent oversight and management made a difference. He created incentives for rangers to actually catch poachers and established night patrolling operations—something that had not been done before in the reserve—with Heath going along with them on the patrols. Every day would start with a security and management briefing, followed later in the day by a planning meeting, new practices introduced when he took over. Eventually he established a canine unit, greatly feared by local poachers.

Developing a decent working relationship with his law-enforcement counterparts on the Narok side of Maasai Mara and in neighboring Tanzania was difficult, but he got permission to cross into the Narok side and into Tanzania when in hot pursuit of poachers.

Establishing trust and support within the local community, which was heavily engaged in poaching and illegal grazing, was the greatest challenge. This region of the protected area, both on the Kenyan and the Tanzanian sides of the border, has a long history of forced ecological conservation driven by white colonial powers, who brutally evicted and resettled local Maasai and other communities without adequate compensation. Prohibitions against entering parks for grazing, logging, and hunting reduced community access to subsistence resources, and generated resentment against the parks and conservation efforts that still persists. Conservation continues to be seen as a project of rich, white, ex-colonial lords catering to the luxury desires of white tourists, while the black local community sees itself as prohibited from exploiting local resources for its economic advancement.

What ultimately changed the community's perception of Heath was his success in stopping the high amount of cattle rustling around and within the reserve, relieving the Maasai community of the need to devote a large amount of resources to guarding their livestock. His innovative plan also allowed the community to regularly take cattle through the reserve to a salt lick, and occasionally, under guard, permit some grazing within the reserve to create a grassland mosaic, as shorter grass is highly beneficial to some animals, such as cheetahs and Thompson's gazelles. The community still got access to some of the reserve's crucial resources and did not have to suffer total exclusion, while environmentally detrimental effects of encroachment were reduced.

What also helped Heath gain support from the local community was that many of the poachers were armed bandits, holding up vehicles, robbing people, and sometimes kidnapping them and absconding with them to Tanzania. Heath successfully exploited the overlap between conservation imperatives and the local community's outrage over other forms of criminality. The result was a significant increase in intelligence flows from the local community. Catching some 250 poachers annually over several years, Heath managed to turn the reserve around—despite warnings that as a white man he would not last in the Mara Conservancy for three days.[47]

Still, the poaching of elephants and rhinos continued on the edges of the park, with animals killed both by automatic weapons, suggesting industrial-scale ivory poaching, as well as spears, showing persistent poaching by local communities. In 2012, some 138 elephants were killed around the reserve, and one inside the protected area. Heath responded by cultivating a network of informants within the local community. He organized sting operations, with undercover informants posing as ivory buyers. When Heath and I talked in May 2013, on the very day when a new sting operation was under way, he still feared more poaching pressure was coming.

Intensified illegal grazing has compounded the new poaching wave. With the top of the Mara escarpment (the boundary of the protected area) now heavily cultivated, more and more people have been trying to graze their cattle inside the Conservancy and the broader Mara reserve. The acreage of grazing land outside of the park has declined because of non-Maasai people moving into the area, who have expanded the cultivation of maize, beans, and millet. Population growth and persisting corruption in Kenya's Forestry Service[48] have also intensified deforestation, particularly in privately owned forests around the park, which have been cleared for more agricultural land and charcoal. The park's elephants have come into conflict with people more frequently, and as a result have sometimes been killed. But Heath's biggest challenge was to institutionalize the procedures and improvements he brought to the park so that they would last after he leaves, creating resilience in the Conservancy against corruption.[49]

Heath's story, while exemplary, is unfortunately atypical. Instead, in the context of highly corrupt and inadequate law-enforcement capacities, a favorite law-enforcement response is to set up special interdiction units (SIUs). Sometimes, SIUs can score spectacular successes,[50] and they can have a deep impact on rooting out corruption and weakening organized crime. One of the most effective special law-enforcement and prosecution units was South Africa's Directorate of Special Operations (or the Scorpions), active between 2001 and 2009. A multi-agency task force within South Africa's National Prosecuting Authority, comprising many of the country's best prosecutors, police, and financial, forensic, and intelligence experts, the Scorpions investigated and prosecuted organized crime and corruption. An international version of SIUs and tribunals is the International Commission Against Impunity in Guatemala (La Comisión Internacional Contra La Impunidad en Guatemala, CICIG), which has managed to effectively prosecute entrenched corruption and organized crime in a region deeply pervaded by crime and entrenched connections between politicians, law-enforcement forces, and criminals.[51]

Yet even though elite SIUs can be effective in dismantling organized crime and its political backers, the political will to maintain them can dissipate quickly. After the Scorpions began investigating top-level politicians (such as then Deputy President Jacob Zuma) for corrupt deals, they were rapidly disbanded, absorbed into other law-enforcement departments, gutted of power, and withered.[52]

Conversely, SIUs can end up going rogue. Within intensely corrupt political systems with powerful criminal organizations, the political pressures on and the coercion of SIUs can be irresistible, particularly if foreign assistance and monitoring cease. After all, the best way to become a country's top drug trafficker is to head its top counter-narcotics unit, just as the best way to be the top *capo* is to be the minister of counter-narcotics or justice—as Du Yuesheng in China in the 1930s, Peru's intelligence czar Vladimiro Montesinos in the 1990s, Mexico's drug czar José de Jesús Gutiérrez Rebollo, and the Zetas (Mexico's elite counter-narcotics unit in the 1990s, and one of its most violent drug-trafficking groups after 2007) all learned. With superior training, they can prevail against their criminal rivals and capture criminal markets, while hiding their rogue activities under the cloak of law enforcement. Similarly, the best way to be an ivory trafficker is to be minister for the environment or the head of the primary wildlife law-enforcement agency.

Penalties and Deterrence

Effective incapacitation and deterrence require not only sufficient arrests at the appropriate level of a smuggling network, but also effective prosecution and punishment. Good cooperation with prosecutors is rare in much of Africa and Latin America, not only in wildlife-crime cases, but also in cases of drug smuggling and homicide. In dealing with wildlife crime, prosecutors (sometimes because they are corrupt, sometimes because they do not know any better) often charge poachers only with wildlife offenses and ignore opportunities to add charges—of illegal firearms possession, for example, or of attempted murder if rangers were fired at—to secure higher penalties.

Penalties tend to be low. In the Mara Conservancy, Kenya, for example, a poaching offense would often be prosecuted only as a trespassing offense (for which the prosecutors would not have to bother gathering evidence) and result in a fine equivalent to about $25, while illegal grazing would carry a $2.50 fine. Such penalties have been insufficient to deter violations. Moreover, given extensive corruption in the judicial system and community resentment against conservation, many offenders escape prosecution. The reserve's wildlife-enforcement officials thus found themselves capturing the same poachers over and over.

Globally too, wildlife-crime penalties tend to be fines and small prison terms. In Singapore, for example, a person found guilty of smuggling a CITES-protected species can be fined $50,000 for each species and up to $500,000, and face a two-year jail term.[53] In Malaysia and Thailand, custodial penalties range between three and five years.[54] In India, the penalties for tiger, snow leopard, or leopard poaching and trafficking are a are a minimum fine of $440 (rising for subsequent offenses) and three to seven years imprisonment, while killing an elephant carries a penalty of approximately $100 to $500.[55] In Vietnam, one of the world's hubs of wildlife trafficking, the typical maximum penalty is no more than three years imprisonment.[56] In China, though illegal hunting or devastation of a rare species can result in life imprisonment, such extreme penalties have not been issued, and the more standard penalties are between five and ten years imprisonment. Recently, however, tough penalties have been handed out. In November 2013, the supreme court of China's Guangdong Province upheld twelve- and fourteen-year prison sentences for two ivory traffickers convicted of smuggling 1.04 tons of ivory.[57]

Low penalties for wildlife crime also are common within the European Union. In most EU countries, penalties can be up to five years imprisonment, but usually only month-long sentences are handed out, while wildlife crime prosecutions are only rarely undertaken.[58] In the United States, the largest penalties issued for wildlife trafficking recently range only between 2.5 and 5.8 years.[59]

In Africa, as a result of active lobbying by environmental NGOs and the visibility of international wildlife trafficking, the trend has been to toughen penalties for international traffickers. Instead of a Chinese national being charged a fine of $1 per ivory trinket (a 2013 case in Kenya),[60] Asian traffickers have lately been receiving very lengthy prison sentences. For example, in March 2016, two Chinese citizens received a twenty-year sentence.[61] Chumlong Lemthongthai, identified as the key Africa-based operator of one of the most notorious wildlife traffickers, Vixay Keosavang, was originally sentenced to forty years in prison in South Africa, but after a round of appeals the sentence was reduced to thirteen years and a fine of about $70,000. Unfortunately, African prosecutors and justice officials have not shown similar resolve when it comes to charging and prosecuting high-level politically linked African traffickers.

While increasing penalties makes sense in many parts of the world, penalties need to be commensurate with the income and socioeconomic condition of the offender. A several-hundred-dollar penalty may be laughable as punishment to a middle-level trafficker, but it can wipe out the family assets of a subsistence-level poacher. Of course, with differentiated penalties, trafficking networks have every incentive to pretend the poaching was subsistence driven, just as they hire children to carry poached animal parts. Still, there is a good reason to apply tougher penalties with some judicial discretion, instead of mandating high minimum-level penalties, even though judicial discretion can easily be warped within highly corrupt systems.

Moreover, deterrent effects of increased penalties level off. Punishment that deters but also facilitates rehabilitation is what is needed. Flooding prisons with low-level subsistence poachers is not a desirable strategy. They and their families will suffer great hardship and likely will be unable to cease subsistence poaching anyway.

Robust evidence from criminology shows that most criminals' careers end by the time they reach their forties.[62] Much of this evidence is derived from studies of homicides, gangs, and drug smuggling.

There could be differences with wildlife crimes since subsistence poaching or illegal grazing and logging may occur later in a person's life. In my research on jaguar poaching in Pantanal, Brazil, I came across a hunter in his seventies who claimed to have poached hundreds of jaguars over decades while on the payroll of cattle ranchers. Though he complained he was no longer physically able to go out and hunt, he was full of enthusiasm about grooming his sons to continue this profitable activity.[63] Similarly, retired rangers in Africa in their fifties and sixties can pose significant poaching threats. If they see substantial drops in their income and standard of living after retirement, they will be a primary target for recruitment by wildlife trafficking rings since they have superior knowledge of anti-poaching methods and wildlife movement patterns.

However, extreme approaches, such as shooting poachers on sight and, in particular, shoot-to-kill policies, are ineffective deterrents and undesirable. Several countries with intense poaching rates, including Botswana, Tanzania, and India, have at least temporarily implemented a shoot-to-kill policy at various times. Botswana's President Ian Khama, for example, endorsed such an unwritten approach. Often celebrated in the West as a determined conservationist, he is a controversial person in Botswana, where local communities often resent his bans on hunting and where he is accused of catering to the interests of rich and powerful ecotourism companies. Indeed, despite Botswana's reputation for ensuring better training of its enforcement units and as one of Africa's least corrupt countries, the policy led to the highly questionable killing of thirty poor Namibians accused of being poachers during an anti-poaching operation into Namibia by Batswana soldiers.[64]

In more corrupt Tanzania, an October 2013 law-enforcement action to crack down on the poaching of elephants and rhinos also embraced a shoot-to-kill policy. Involving Tanzania's military, local police, and special anti-poaching militias, Operation Terminate produced widespread allegations of extrajudicial killings, rape, torture, and extortion from innocent people, and was itself terminated.[65] Although South Africa has not formally adopted a shoot-to-kill policy, encounters between poachers and rangers have been lethal. Between 2008 and 2013, at least 300 suspected poachers were killed in the Kruger National Park alone.[66] Trained by South African private military companies, the park staff of Malawi were implicated in 300 murders, 325 disappearances, 250 rapes, and numerous instances of torture in Liwonde National Park between 1998 and 2000.[67]

Sometimes rangers' attempts at self-defense might require shooting to kill. But making shoot-to-kill the first and primary response is unlikely to create a deterrent to poaching, and it entails serious human rights violations. Indeed, wherever shoot-to-kill and other maximum-force policies are in place, law-enforcement officials have been tempted to present dead bodies as criminals or insurgents, regardless of the reason the people were shot. In one of the most notorious cases in Colombia, hundreds of innocent people were killed by Colombian soldiers who claimed their victims were FARC guerrillas. These were not mistakes due to the fog of counterinsurgency. With promises of jobs, victims were purposefully lured by soldiers to a remote place, murdered, and dressed up as guerrillas. The greater the numbers of killed guerrillas, the longer vacation, and larger bonuses the soldiers received.[68] Many superior officers encouraged the murderous deception.

Policies of indiscriminate force beyond what it strictly necessary for self-protection easily turn law-enforcement units into gangs of brutal thugs. Policing remote protected areas is particularly susceptible to the carrying out of violations since little institutionalized or outside oversight exists. Such accountability problems are compounded when private security companies or irregular militias are deployed as antipoaching units.

Crucially, a policy of killing poachers on sight severely compromises the ability to develop official as well as community intelligence on poaching and trafficking networks. It exacerbates schisms between the local community and law enforcement, and thus limits intelligence flows. Moreover, a dead trafficker will provide much less useful information than a live one. Middle-level and top-level traffickers often welcome shoot-to-kill policies: they can easily hire new poachers, while the dead ones cannot provide information to implicate them. (When Brian Heath took over the management of the Mara Conservancy, he stopped the existing practice of shooting to kill on sight. It was not deterring poaching, or providing intelligence on smuggling and trafficking networks. Rangers working for Heath now have to make every effort to arrest and interrogate the poachers they come into contact with, later handing them over to Kenyan or Tanzanian authorities).[69]

Overall, the extreme proposals voiced among the conservation community calling for very lengthy prison sentences and death penalties for poachers and wildlife traffickers, and a widespread adoption of shoot-to-kill policies, are inappropriate. They involve major human rights problems, can overburden justice and penal systems, and will not deter poaching—creating instead more hardened criminals out of poachers. The wildlife conservation community would be wise to avoid the inappropriate excesses of the war on drugs—especially the dysfunctional effects of severe punishment.

Crucially, effective deterrence is not just about severity of punishment, but critically about consistency and certainty of punishment. If one low-level poacher is killed during an encounter with rangers or a corrupt customs official in Mombasa is sentenced to death for ivory smuggling once every few years, the overall deterrence effects will be negligible. For law enforcement to have pronounced deterrence effects in reducing homicides, for example, prosecution rates are often estimated to require at least 40 percent effectiveness. Indeed, environmental studies themselves have shown that the probability of detection significantly influences decisions to poach.[70]

7

LEGAL TRADE

In Chapter 6 I examined bans on hunting and trade and law enforcement because these solutions to the wildlife trade are currently the most strongly advocated by many environmental NGOs and some conservation biologists. Not all conservation biologists agree, and many economists consider legal trade the best guarantee for species survival. People refrain from extirpating animals and transforming their habitat into agricultural or industrial uses only if they can make money from the preservation of the habitat and the species: "wildlife stays, if wildlife pays." The premise of CITES itself is that trade in wildlife is allowed and is only restricted if a species becomes threatened or endangered.

There are strong arguments for mobilizing tangible and robust economic incentives to reinforce the ethical and emotional motivations for species conservation, and to allow considerable scope for the legal supply and marketing of wildlife. But like with bans and law enforcement, allowing a legal supply is fraught with difficult dilemmas and less-than-perfect outcomes. The controversies over permitting the legal supply of threatened wildlife resemble some of the arguments over permitting a legal supply of drugs and the expected positive and negative outcomes of drug legalization. However, the basic motivations for legalizing drugs are quite different from the basic objectives of permitting legal trade in wildlife.

THE EXTINCTION MARKET

The Regulatory Theory for Making Drugs Legal and the Counterarguments

Advocates of drug legalization and critics of existing law-enforcement drug approaches, many of whom do not support drug legalization, make the following arguments. Drugs cannot be eliminated from the world, and interdiction and other law-enforcement measures cannot be made more effective than they are now. Criminalizing production and use overburdens law-enforcement and justice systems; it empowers criminal groups and augments their resources; it severely undermines the human rights of users, low-level non-violent dealers, and their families; and it compromises the health of users and, via communicable diseases, the health of wider society.[1] The suppression of drugs can also compromise and undermine counterterrorism and counterinsurgency efforts, and increase political instability as poppy and coca-producing populations see their livelihoods disappear.[2] Some also contend that if drugs were legal, organized crime would be disempowered and poorer; law enforcement could become more strategic and focused on violent crimes, such as murders; many fewer people would be incarcerated; and public health and human rights would be enhanced. Instead of the state having to spend resources on expensive law-enforcement and correction facilities, the state would make money from taxing drugs.

Legalization advocates point to economic data from the many US states with laws allowing the sale of medical marijuana products. In place for over a decade and supplying over two million patients as well as users who use medical marijuana for recreational highs, sales of medical marijuana products are estimated to generate billions of dollars per year.[3] In Colorado, the first US state to fully legalize marijuana for nonmedical recreational use in 2012, and where the first recreational pot store opened in 2014, marijuana revenues in 2015 were $88,239,322. Taxes collected from alcohol were considerably less ($48,949,617), but from tobacco considerably more ($197,037,085).[4] Nonetheless, since Colorado adopted an intense commercialization and aggressive marketing approach to its legal marijuana, it is likely that the Colorado marijuana market will expand further, and that tax revenues will increase. Nationwide, in 2015 Americans spent $5.4 billion on legal medical and recreational marijuana.[5] If the United States also legalized marijuana at the federal level, the market might amount to $36.8 billion in retail sales annually,[6] that is, considerably more than the total global estimated value of the illegal wildlife trade ($23.5 billion per year).[7] That does not necessarily amount to an expansion of the market overall, but rather to the absorption of the existing illegal market into the legal one based on current prices. When legal marijuana prices fall, the total revenue picture may look quite different, with an additional factor being how much demand for cheaper marijuana will rise among casual users and abusers.

Those critics of the war on drugs who do not advocate comprehensive legalization—particularly not of highly addictive drugs such as heroin, cocaine, and methamphetamines—posit that many of the promised effects are highly contingent on local institutional and cultural settings, and may not materialize. They also contend that proponents of legalization underestimate the effects of drug addiction and its post-legalization increase. Finally, they maintain that many of the highly problematic outcomes of the war on drugs can be mitigated by decriminalizing drug use, adopting harm-reduction approaches, and changing the prioritization and timing of supply-side suppression measures, including eradication, interdiction, and alternative livelihoods even without outright legalization.

The effects of legalization on the power and violence of organized crime groups are also disputed. Some challenge the notion that legalization alone will reduce either the proclivity to violence or the power of organized crime groups.[8] Moreover, legal markets also have to be monitored, controlled, and enforced; thus legalization will hardly eliminate all of the costs of enforcement. Nor would poor drug farmers necessarily benefit from legalization; their political and economic marginalization may cut them out of the newly legal market, just as they are cut out from other legal agricultural markets.[9]

Another controversy regarding banned drugs concerns the extent to which their legalization would increase consumption and the addiction that is devastating individuals and communities. Quite likely, legalization will mean greater drug use and addiction levels than when a drug is prohibited; but the critical questions are how large such increases would be and whether drug addicts would be able to receive better treatment, medical care, and social support if drugs were legal.[10] The problematic abuse and expanded-use effects are expected to be much weaker for the far less addictive and dangerous drug of marijuana than for methamphetamines, for example.[11]

Harm-reduction measures in the Netherlands, a country with well-established social services, have helped reduce some of these problems, and violent criminality associated with drug distribution also remains low.[12] Drugs are still illegal in the Netherlands. Even marijuana is only decriminalized for use, and is sold in specialized commercial establishments known as "coffeeshops," with strict limits on how much an individual can buy. The production of and wholesale trade in marijuana remain illegal.

On the other hand, in countries where the state is weak and medical care provision underprovided, such as in West Africa, it is difficult to imagine how harm reduction measures such as those carried out in the Netherlands could be delivered. Other mechanisms, such as traditional health-care networks, could perhaps be tapped for the provision of drug-abuse assistance. The broader point is that local institutional and cultural settings and different modes of legalization and commercialization will impact the results of legalization, including levels of demand and use and their harmful consequences, and the effectiveness of policies to mitigate those consequences.

Moreover, the legalization of drugs is not simply one uniform policy. The many policy choices and trade-offs involved will affect the kind of regulation model one adopts. Will legalization be done under a restrictive, state-controlled system? Will it involve only small cooperatives of growers, or will big-business commercialization be allowed? Will commercialization produce undesirable regulatory capture that prevents the fine tuning of policy designs? What kind of advertising will be permitted? How will drug abuse and dependence as well as accidental overdose be prevented? How high should taxes on legal recreational drugs be? High taxes might discourage wider use and bring more money to the state coffers, but they sustain or encourage a parallel black market, and vice versa. Will drugs be taxed on the basis of their potency? Will there be limits on the levels of potency of marijuana that can be sold legally in stores? Will casual drug users be charged very high taxes so that they are discouraged from use while addicts are sold drugs with very low taxes so that they do not need to commit petty crime to feed their habit? Or shall addicts be given drugs free under supervised conditions?

In short, even when it is decided that permitting legal trade is the best way to mitigate the threats, costs, and harm of drug use, the drug trade, and drug policies themselves, there are a myriad of difficult policy questions and regulatory dilemmas that need to be answered, and these all influence ultimate policy effectiveness and its side effects. Of course, making such judgments is not only a matter of scientific parsing of the evidence for and against. Policy lobbies, commercial businesses, and NGOs all play a role in the policy process and seek to shape regulations to benefit their interests and beliefs.

Regulated Legal Trade in Wildlife: Theory and Practice

The regulatory arguments for permitting legal trade in wildlife center on different sets of objectives and dynamics. At its core, the current debate about legal wildlife trade is not over whether legalization will reduce the harm associated with wildlife trafficking and consumption or whether it will increase total demand, but whether production from illegal and unsustainable sources will be displaced by a legal supply.

Some wildlife advocates decry and oppose any hunting and consumption of (wild) animals. But most of the conservation debate is about how to reduce the numbers of wild animals illegally hunted, and about how best to preserve wild populations of animals and their habitat. Licensed supply of wildlife promises to accomplish the environmental goals by, first, reducing pressure on wild resources, and second, giving those close to wildlife an economic stake in the conservation of species and their habitats instead of converting them into agricultural or industrial land. Licensed supply of wildlife can come from different sources. It can be based on managed and well-regulated hunting, whether by local communities or by professional and trophy hunters. Licensed supply can also come from farms and captive-breeding facilities, which sometimes may be permitted to acquire their initial stock from the regulated capture of wildlife.

The first argument for legal trade is to take pressure off wild resources: instead of obtaining animals caught in the wild, consumers will obtain them from farms.¹³ For example, the US ban on the import of wild birds was mostly deemed successful since the demand could be satisfied by parrots bred in captivity in the United States (each bird is accompanied by a certificate of origin), thereby reducing nest-poaching in the neotropics.¹⁴ The ban similarly helped suppress demand for wild-caught birds in Europe. Farmers and ranchers of captive-bred species could even become proponents of bans on hunting of wild animals in order to increase their market share.

Another important success of licensed farming has been the reduced hunting of crocodilians (crocodile, caiman, and alligator species). By the 1970s, over two million crocodilian skins supplied the trade, and there was strong evidence that many wild crocodilian populations had drastically fallen. In 1975, CITES prohibited trade in all crocodilian species sourced from the wild, but licensed commercial trade in them if they were bred in captivity, as both industry actors and many conservation economists and biologists promoted.¹⁵ Farming of crocodilians was crucially facilitated by the fact that breeding them in captivity is not difficult.

Thus, after 1980, skins from captive breeding came to dominate the trade, while the number of skins taken from wild animals massively declined and wild populations rebounded in many countries. Incidences of the laundering of skins from poached animals were mostly addressed effectively, largely because of the requirement that all crocodilian skins and skin parts be recorded and tagged.¹⁶ The eleven most commercially valuable species of crocodilians are the species least threatened with extinction.¹⁷ What is perhaps most striking about the outcome is that unlike in the case of other species and taxa, the success of legal trade in and conservation of crocodilians took place regardless of the country in which they occurred, despite variations in their level of economic development, quality of governance, and environmental consciousness. The trade in crocodilians came to be widely considered as a model of the effectiveness of market-led conservation.

This success is all the more impressive given that crocodilians are not cute and cuddly. At the same time, precisely because crocodilians are not seen as cute as pandas or as close to humans as primates, there were many fewer objections to farming them for their skins. Lately, however, some animal rights groups have begun mobilizing to prohibit the legal farming of crocodilians, arguing that keeping any wild animal in captivity is cruel.[18]

Yet allowing the legal trade in crocodilians has hardly eliminated all poaching and trafficking. Skins from crocodiles are considered highly desirable for boots and clothing among Latin Americans and Latino communities in the United States, and with rises in disposable income among these groups there appears to be a new rise in poaching.[19] But little is understood about this new poaching increase. Is it driven by laxer law enforcement of protected areas, such as in Latin America, and less intensive protection provided to crocodilians than at the height of the crisis? Are new customers less environmentally conscious or more driven by the lower price of illegally sourced animals compared to the price of farmed ones? Is the renewed poaching of crocodilians merely riding the coattails of generalized poaching around the world, or is it somehow specific to certain countries which supply the skins, or to countries in which there is demand for them? Such questions need to be answered in order to devise effective policies. One answer might be to expand the number of crocodilian farms. Another would be to beef up law enforcement in protected areas so that the most easily accessible and cheapest supplies still come from farms. A further solution might be a combination of the two.

The second argument in favor of legal supply is that it gives hunters, ranchers, businesses, large-scale landowners, local communities, and others close to wild resources a stake in sustainably preserving species and entire ecosystems. Farmers, herders, and ranchers often see wild animals as a source of competition for grass and grazing, or as vermin carrying diseases that can decimate domestic herds. Wild animals eat and damage crops, and can hunt and kill domesticated animals. In turn, locals seek to extirpate them. Managed hunting puts money in people's pockets legally and, in the case of forest-dependent communities, assures their food security and perhaps even raises their standard of living.[20] Managed hunting also reduces the pressure for converting land from its natural state to pasture or farmland.

One of the most celebrated success stories is the recovery of the Andean vicuña, a relative of the llama. Local peasant communities in Peru came to support conservation after being allowed to obtain income from the sale of vicuña products as well as being given a new school. Eventually, they also obtained property rights to land on which the vicuñas graze.[21] In Kenya, one of the best-regarded wildlife reserves, the Laikipia Conservancy, which used to be managed by Brian Heath (manager of the of the Mara Conservancy, described in the previous chapter), allowed the managed hunting of zebras. Permitted to sell the zebra skins and meat, the local community developed a material stake in conservation, and bonded together to prevent outside poachers from hunting in the reserve. Before Heath licensed the hunting of zebras, both zebras and antelopes were seen by the local community as a nuisance as they destroyed crops, and their poaching had been prevalent. Later, when managed hunting was again prohibited (the default policy in Kenya since the 1970s), intense poaching escalated again.[22]

Allowing the managed hunting of species that breed easily, such as antelopes, also makes sense in peripheral areas where there is little potential for ecotourism, but where conserving buffer-zone land may be crucial for conservation success.

Such economic incentives for the preservation of species and habitats apply not only to poor communities but also to a variety of other landowners who do not necessarily have an altruistic attachment to nature conservation. If bans and other restrictions on land use and requirements for conservation impose significant costs on local landowners, they may even want to extirpate the species from their lands to avoid conservation costs. Giving people economic incentives to conserve species and land, such as through tax breaks if natural ecosystems are not converted to other uses and robust wildlife populations are kept, can be crucial conservation motivators.

Similarly, the frequent argument for encouraging the consumption of wild meat and allowing the trophy hunting of big game on private properties in Africa is that it motivates landowners to preserve the ecosystem instead of burning down the bush for cattle ranching. In South Africa, for example, some 11,600 registered game parks in 2015 covered 21 million hectares of land,[23] and contained between 16 and 20 million animals, including 27 percent of the country's white rhinos and 20 percent of the black rhinos.[24] Rhino-stocking preserves numbered 380 and covered 2 million hectares, roughly the size of the Kruger National Park. Many rhinos were sold through an auction system to private reserves after the success of breeding programs exceeded the carrying capacity of national parks in South Africa in the 1960s and 1970s.[25] The substantial funds generated through such auctions can be used for a variety of conservation measures, whether acquiring more protected land, as has often been the case,[26] or providing compensation to marginalized communities excluded from land, a less usual case. Indeed, in South Africa, where controlled hunting of the white rhinoceros is permitted, the species increased spectacularly for many years, growing by 130 percent between 1997 and 2007,[27] before the current poaching and trafficking crisis struck.

Paradoxically, intensified poaching is not only wiping out desirable environmental outcomes, it is also threatening the economic incentives for conservation. The costs of keeping poachers out of reserves may surpass ecotourism and hunting revenues. Some South African private reserves that used to buy rhinos from state-breeding facilities no longer want to do so because of the cost of protection enforement.[28] This is a replay of the 1970s, when a blanket ban on trade in live rhinos and their products undermined the economic incentives for conservation. Since private game-park owners could not derive enough money from tourists viewing rhinos, and the costs of rhino protection were high, they were reluctant to stock them.

If a trade ban eliminates the economic value of a species, or if charismatic species favored by ecotourists and licensed hunters are wiped out and tourists no longer come, the pressure to convert land to agricultural use also grows. The entire ecosystem and many species, not just the protected ones, can be destroyed.

Finally, regulated trade can also raise money for conservation. Trophy hunting is a big business that can raise vast sums of money—for those who manage to obtain the revenues, such as state agencies and wildlife businesses. For example, between 1,000 and 1,600 trophy hunters visited Tanzania each year between 2000 and 2010, according to Tanzanian national wildlife authorities.[29] With international hunters paying thousands and even tens of thousands of dollars for a hunting license, the generated revenues can be very high.

Moreover, many African countries end up with stocks of ivory from elephants that have died naturally, and they want to be able to sell it to pay for the costs of conservation, such as park rangers' salaries and habitat preservation. Countries with well-managed elephant populations that were successful in preventing poaching, such as Namibia, Botswana, and South Africa, have been given a license to sell some of their ivory stocks. The 2008 sell-off raised $15 million.[30] Moreover, sometimes, such as in the Kruger National Park in the 1990s, elephant populations were so large that they were damaging habitats. Park managers had to resort to elephant culling regardless of whether or not they would be able to sell the ivory. At the time, licensed culling did not lead to the increase in poaching that plagued countries without a license to sell ivory, such as Tanzania, Sudan, Zambia, and the then Zaire (currently the Democratic Republic of the Congo). So at least before the latest poaching crisis, poor environmental management, problematic property rights, and other governance problems in countries such as Zambia, Tanzania, and Sudan seem to have had a far greater impact than the international regime.[31] But as poaching has greatly expanded throughout Africa, even South Africa's elephants are now killed at levels greater than when the country was issued the 2008 license. Nonetheless, as of 2016, elephant poaching in South Africa is still nowhere as extensive as in countries that did not apply for a license, such as Tanzania, Mozambique, and Cameroon, and the country's elephant population remains stable. On the other hand, rhino poaching in South Africa has been very intense.

Other outcomes of licensed trade schemes have not been uniformly positive. To deal with unregulated and rampant trade in blue-fronted parrots (also known as blue-fronted amazons) taking place throughout the area of the bird's range, the government of Argentina instituted a regulated trade scheme in the Chaco region in the 1990s. The scheme was aimed at reducing the levels of unlicensed trade to a fraction, while also raising money for conservation and providing almost 20 percent of income for poor peasant landowners who were otherwise clearing their land of trees and converting it to soy cultivation.[32] The scheme appeared to be a great success. Other experts, however, have questioned the robustness of the Chaco data. Moreover, studies of licensed trade in several parrot species in Nicaragua showed their collapse by 80 percent over a ten-year period.[33]

LEGAL TRADE

Where Does the Money Go?

Park revenues and hunting fees do not necessarily go to conservation, for corruption can divert resources to private pockets. Nor are the revenues necessarily collected effectively. For example, several years ago Kenya implemented an electronic smart-card system for national park entry tickets after it was discovered that many employees systematically underreported entry-fee revenues and pocketed the difference, and that tour companies were printing fake tickets. But the first version of the electronic smart-card system was hacked and large amounts of revenue were stolen, though the system seems to have become more secure since 2010.[34]

The Maasai Mara National Reserve system receives some of the largest numbers of ecotourists in Africa and should have one of the highest revenues for a national park in Africa, around $50 million a year, a dream income for many national parks in other parts of the world. For example, when a national park in Ethiopia raises $20,000 annually, it is considered a large amount.[35] Yet in the Narok part of the Mara reserve, only meager revenues have long been reported, and the quality of roads and park conservation remains poor as most of the money generated does not go back to the park or into improving the socioeconomic life of the local community. Extensive illegal grazing by cattle owned by park rangers and councilors as well as illegal tourist camps were present in the reserve at the time of my research in May 2013, and have been present for years.[36] Political interference by local authorities has actively undermined management of the reserve. But when Kenya's national government and the Kenya Wildlife Society sought to take over the park, the Maasai community mobilized against them and effectively blocked the move.

The mismanagement present in the Narok part of the park is in stark contrast to the abutting Mara Conservancy managed by Heath, where revenue collections are high, financial reports are regularly posted on the internet to increase transparency, and an outside company has been subcontracted for financial oversight to reduce opportunities of corruption.[37]

Similar problems occur in the national parks of Ethiopia. The Lake Abiata-Shala National Park has one of the highest revenues from park entry fees among Ethiopia's protected areas. Close to Addis Ababa and featuring flamingos, it receives many visitors and collects some $1,500 a month in entry fees, a sizeable amount by Ethiopian standards. The park was established in the 1970s to protect a key stopover area for migrating shorebirds, such as ruff and barn swallows, flamingos, and resident big mammals. However, massive encroachment in terms of illegal logging, farming, grazing, and outright settlement in the park—and even the presence of an industrial plant—resulted in massive degradation. The park's income has not been able to stop the destruction nor radical species decline. As the park's warden lamented: "Were it not for the flamingos still staying around, why would people come? In the 1970s, this was a beautiful pristine area. Now, despite the income we get, what's the difference between here and a village outside the park?"[38]

In Tanzania, Zambia, Zimbabwe, and many other African countries, and of course elsewhere in the world where corruption is pervasive, ecotourism and trophy-hunting revenues meant to support conservation are often stolen for personal profit. This is not merely incidental. If legal supply alone could wipe out all poaching, then revenues needed for law enforcement might be eliminated. But legal supply alone has not done so—even in the celebrated case of the legal supply of crocodilians, some poaching persists and law enforcement is necessary. It also has to be paid for and often is very expensive.[39]

Moreover, those who benefit economically from a farmed supply of animal products may not be those who need to be compensated for foregoing hunting or land conversion. Farming may take place far away from protected or ecologically valuable areas. Without receiving adequate compensation for not transforming their land, local landowners may want to level the forest for soy or cattle, or local communities might feel economically compelled to poach. Who benefits from conservation as well as how much they benefit matters critically.

There are often many competing claims to wildlife revenues, all the more so if other sources of revenue are lacking in the country. Among those seeking a piece of the action are national governments, wildlife enforcement authorities, central state development and infrastructure agencies, local government authorities, local state conservation authorities, wildlife tourism businesses, and local communities.[40] At the same time, most protected areas typically generate only low revenues compared to alternative uses and fail to cover their operating costs.[41]

LEGAL TRADE

The Industry Eats Its Tail

Legal supply does not always take pressure off the wild. The fishing industry is the most potent example. Although fishing is a multibillion dollar industry and billions of people and most countries in the world have a stake in preserving it, more often than not fishing is done at an unsustainable rate, devastating species after species to such an extent that scientists now worry about the possibility of empty oceans in a few decades unless radical regulation is instituted. Because of the vested interests of fishing fleets and the governments of the countries in which they harbor, such as Japan, the needed radical regulation of fishing remains elusive. Many animals and plants are placed on the CITES Appendix I list precisely because their exploitation fails to encourage stakeholders to manage their use sustainably.

Similar overexploitation dynamics abound in other markets with depletable resources. The Thai and Indonesian logging industries, for example, learned that after they unsustainably logged out their forests, they could move their operations to other parts of the world or to other extraction activities, such as mining, or African oil palm cultivation.[42]

Breeding wild animals on farms can also reduce incentives to preserve the species in the wild, thereby increasing pressures to convert land to agricultural use and eliminate habitats.

Objectives of private wildlife reserves may not always be aligned with environmental goals, even in the absence of corruption. For example, private ranches in South Africa are increasingly keen to stock non-indigenous species for hunting, such as fallow deer and wild boar from Europe, even though they may have a highly detrimental impact on local species. The private reserves and ranches are also increasingly interested in stocking and breeding animals with certain desirable trophy hunting traits, such as specific colors, thus interfering with natural selection and negatively influencing the genetic stock of species.

A similarly questionable practice is the so-called "put and take" in which owners of game ranches buy endangered species such as rhinos from the state only to have them shot during trophy hunts soon after. While this practice may mean that natural habitats and other native species are preserved on private lands, it does not contribute to increasing the size of population of the endangered species itself in the wild. It may still be valuable if it reduces poaching pressure on other wild specimens of that species as long as put-and-take trophy hunting satisfies all demand and does not increase demand and facilitate laundering.

Another telling case concerns bushmeat. In northern Sulawesi, Indonesia, the Christian community has a strong taste for bushmeat, with anything that can be hunted often being highly craved for dinner (and very pricey in the Langowan and Tomohon bushmeat markets). One of the greatest delicacies—its consumption being a symbol of status and affluence—is the black-crested macaque, a primate endemic to Sulawesi. Over the past three to four decades, the species has experienced an 80 percent decline. Although deforestation in Sulawesi has eliminated much of the macaque's habitat, hunting these days actually poses a far greater threat to the species. In addition to its highly prized meat, its fur is worn during traditional dances to signify bravery, and its skull decorates masks and costumes. Protecting the threatened primate has become an environmental priority for conservationists in northern Sulawesi. One NGO leading the effort to save the macaques near the Tangkoko Reserve—the Selamatkan Yaki project—has emphasized the importance of environmental education. It explains to consumers that if they do not reduce hunting to sustainable levels, all the macaques will be gone; there will be no more pricy meat, and there will be an end to the fun of hunting the primates, a factor which many hunters identified as an important motivation. But persuading local hunters to hunt sustainably has been difficult. Moreover, many of the wildlife traders I interviewed across the Indonesian archipelago about the critical depletion of the species they were selling, and about the negative impact on their business if the animals were extirpated in the wild, were shockingly unaware and indifferent. They would insist that the birds and animals would always be in the forest.[43]

If Legal Supply is Expensive and Difficult, Poaching Will Persist

If poached animals or plants remain cheaper than farmed ones, the legal supply will be undercut. The more governments try to discourage the use of particular wildlife products by imposing heavy taxes on them, or the more complex and costly it is to raise an animal on a farm, the more likely poaching and the illegal trade in wildlife will persist. This is the reason why an extensive and highly profitable illegal market in cigarettes exists alongside the legal one. In China, the price of illegal ivory products is often one-third of the price of legal ones.[44] Moreover, many consumers in Asia prefer wild animals to farm-raised ones, believing them to have greater curative potency.[45]

Furthermore, allowing legal supply may not necessarily satisfy demand for two reasons. Since raising some species in captivity is difficult, many farms resort to replenishing their stocks with wild animals, often quite extensively. And if the availability of supposedly farmed-raised wildlife products increases consumers' desire for them, and distinguishing between legally sourced and illegally sourced ones is difficult for both consumers and law enforcement officials, pressure on the species in the wild may even increase. As Susan Wells and Jonathan Barzdo aptly point out, for a farm product to displace a wild product, it must be available in large quantities and cheaply (ideally cheaper than the wild product), or be of better quality than the wild product. But all of these factors are often difficult to achieve in breeding programs, especially in their early stages.[46] Examples of problems abound. A 2015 TRAFFIC study of the farming of Tokay geckos in Indonesia, for example, shows that the cost of legal breeding is high and that farms also extensively deal in poached animals.[47]

Laundering through the Legal Supply and Legal Hunting

This problem directly connects to a fundamental problem with legal supply, namely, that captive-breeding and licensing schemes often allow the leakage of illegally caught wildlife into the supposedly clean legal supply chain. Since monitoring a wildlife product from farm to retail outlet through its various processing stages, especially when it crosses international borders, requires a large amount of law enforcement resources, it is often relatively easy to add illegally obtained products to the legal supply.[48] Frequently, neither customers nor law-enforcement officials have the capacity to determine whether a wildlife product was obtained from the wild, from a captive-breeding facility, or from a legal supplier. Effective monitoring and interdiction would require intense patrolling of farm facilities, frequent random inspections, and tight monitoring of distribution chains, along with robust prosecution and punishment. Very rarely are these measures applied to wildlife obtained from legal farms in East Asia.

China's and Hong Kong's legal trade in ivory is marked by many regulatory and enforcement loopholes, enabling the laundering of vast amounts of illegal ivory, which in turn drives poaching. Fake and sloppy certification is common, and ivory registration is systematically manipulated to hide illegal ivory. Consumers and law-enforcement authorities are frequently misinformed by storekeepers that the ivory products they sell come from mammoths and from pre-1989 sources, antiques, or trophies, the sales of which are all legal. Gifting and private sales outside of licensed stores are another mechanism used to avoid regulation.[49] Although Japan nominally had extensive laws governing the sale of ivory, for a long time it also failed to stop illegal ivory leaking into its markets. Under pressure from environmental NGOs, Japan ultimately took some steps to tighten control. With laws permitting the selling of ivory from domesticated animals, Thailand serves as a major center of illegal ivory laundering. Permitting the antique ivory trade provides another loophole: traders in illegal ivory simply make it look like old ivory.

Despite the commitment of the Chinese government to monitor them closely, tiger farms in China, whose stock of 5,000 tigers now exceeds by almost five times the numbers of tigers in the wild, are often suspected of serving as facilities to launder poached tigers while encouraging consumer demand for tiger products.[50] Although the Chinese government banned all trade in tiger products (as well as in rhino horn) in 1993, by the end of 2016 the ban had not been properly implemented and products, such as tiger-bone wine, have repeatedly been advertised and sold in China. In fact, despite its public commitments to end the domestic tiger-product market, Chinese authorities have been quietly issuing licenses to farms to sell such products.[51] When a captive-bred tiger dies, its bones and teeth can be legally sold with a license. Quite apart from the often awful condition of tigers on farms, this legal supply also serves to launder body parts from poached tigers. Illegal imports of tiger products into China from abroad also continue, from tigers poached in the wild and apparently also from tiger farms licensed in Laos, Vietnam, and Thailand. (In September 2016, the government of Laos promised to explore how to close down its tiger farms.) Since the existing illegal and legal supply of tiger products has not satisfied the existing and likely growing demand for tiger products, increasingly lion bones (both legally traded and from poached lions) are increasingly falsely labeled as tiger bones in East Asian markets. And although tiger farms have been present in China for several decades, the number of tigers in the wild in China remains in single digits and has not recovered. The key and very difficult question is whether, if the farms were truly closed down, demand for tiger products would fall rapidly and *in situ* law enforcement would be able to protect wild tigers at least as well as it currently does, or whether demand would persist and poaching would intensify.

Chinese farms that raise the giant salamander, praised for its meat, also launder poached animals.[52] Moreover, licensed facilities which illegally sell one particular wildlife product often sell other illegal wildlife products, fronting as covers for the broader illegal wildlife trade.

Another example of laundering is the legal trade in wild birds in Europe. The majority of legally exported birds reportedly come from captive-breeding facilities in China, Vietnam, and Malaysia, with the EU and Japan their principal customers.[53] After the avian influenza outbreak in 2005, the EU severely restricted the import of birds, and the legal export of birds from Southeast Asia almost completely stopped. The EU ban apparently contributed to the reduction of poaching of wild birds.[54] This indicates that wild-caught birds were leaking into a supposedly captive-bred supply.

The many enforcement challenges for ensuring that legal supply chains are not used to launder illegal products include the ability to detect falsified licenses and prevent corrupt government officials from handing out inappropriate licenses. If and when drug legalization spreads to countries with pervasive corruption, corruption in licensing will be difficult to avoid. Among officials tasked with supervising the legal wildlife trade, corruption can be massive.

In large parts of Africa, licensed trophy-hunting is riddled with corruption that eviscerates the promised conservation benefits of legal hunting. Licenses are issued in violation of rules and without regard for their environmental impact. Collusion between unscrupulous hunters, licensing officers, veterinarians, breeding facilities, and park rangers often result in hunters killing far greater numbers of animals than the license permits.[55] Pregnant or breeding females are often shot during hunts even though they should be off limits. During the 1980s, investigations in South Africa uncovered private ranches owning unregistered rhinos in violation of regulations.[56]

Local authorities may have few incentives to properly monitor licensing. The resulting lack of oversight is especially pernicious if one blanket license is issued for a long time.[57] Trophy hunts can be staged to mask hunting for illegal wildlife products. Such pseudo-hunts also push up the price of legal hunts. At the height of the pseudo-hunts feeding the Vietnamese and East Asian markets for rhino horn, the cost of a hunting license doubled to between $90,000 and $110,000.[58] Thus, after allowing their rhinos to be poached, park owners would have enough money to buy a new one.

New technologies, such as DNA testing or microchipping, both of which are being implemented for some species, can reduce the law enforcement challenge of sorting the legal from the illegal to some extent, but they are unlikely to eliminate it. For DNA testing to be effective beyond random catches and for it to produce deterrent effects, the database of known genetic samples needs to be large, since DNA will vary for each specimen. Testing also needs to take place any time the product changes hands. Neither requirement is likely to be achieved any time soon in Southeast Asia and in many other parts of the world. Not only are such portable technologies unavailable to the beat ranger, in many countries they are missing at the national level (where often even forensic capacities for homicide are lacking). Indeed, much of the DNA testing of seized ivory to determine its origin is conducted out of one lab in the United States. Even in the visible case of ivory, where countries want to demonstrate their resolve to stop illegal flows, countries are often reluctant to give scientists seizure samples since DNA testing may reveal very uncomfortable truths and jeopardize overseas funding.

Having detailed, complete, and continuously updated surveys and databases of legal stocks and incoming legal supply is critical, but rare. Thus in Japan, with all its vaunted regulations, it is not mandatory to register whole tusks held in private collections if there is no intention to trade them, registration does not cover the movement of ivory onward from processing, and much information can be falsified. A

LEGAL TRADE

systematic DNA database, even in a country that should have the institutional capacity to implement it, does not exist.[59]

DNA testing challenges grow enormously if testing is supposed to ascertain that every single reptile sold by a breeding facility is in fact the genetic offspring of parents owned by that facility. Moreover, not all samples have a long shelf life. Many timber samples, for example, degrade rapidly, and testing is no longer possible within a matter of days. New technologies are being developed to overcome such problems, but again they are of very limited availability, and technological fixes by themselves are rarely sufficient if massive corruption persists.

Allowing subsistence hunting in protected areas by poor communities can also complicate and confuse law enforcement. If a shot is heard, how do law enforcement officers know whether to rush to the scene and round up the poachers or whether the hunting going on is permitted? Many indigenous communities no longer hunt with bows and arrows but now use firearms. In fact, environmentally it is preferable that communities hunt with firearms, targeting only designated species, rather than with traps in which many other species, including endangered ones, can get caught. In theory, hunters can be issued with not just a license, but also some sort of tracking device. But whether that would be effective in allowing law-enforcement officers to determine where and where not to act would depend on how diligently the licensed hunters were in carrying the device. Many members of marginalized communities might fear such tracking devices, regarding them as further intolerable violations of their freedoms and capable of being abused for nefarious purposes by those in power. There is also the risk that such devices could find their way into the hands of industrial poachers and unlicensed hunters through unauthorized sales, bribery, and theft.

Under the best of circumstances, authorized hunters will cooperate with law-enforcement agencies and provide intelligence on unauthorized ones or those who violate quotas. But such cooperation cannot be counted on: authorized hunters may not have this information themselves, they may fear retaliation by industrial poaching outfits and corrupt government officials, or become corrupted in the process; the poachers may be relatives of the authorized hunters; or the authorized hunters may simply resent and refuse to interact with wildlife-enforcement authorities.

But while total bans on wildlife trade simplify the intelligence requirement of distinguishing between the legal and illegal, they do not necessarily reduce, but can augment, the overall resource requirement for effective interdiction. Law enforcement of course does not need to eliminate all poaching to achieve substantial environmental gains. Eliminating the most dangerous and detrimental kinds of poaching would be valuable.

One way to deal with the limited capacity of law enforcement to differentiate between the legal and the laundered is to encourage self-regulation of the wildlife industry through education programs and government or NGO certificates of good corporate behavior. But self-regulation sometimes has poor results: the fishing, logging, and mining industries have often failed to effectively self-regulate, and external oversight is needed.

Experience with certifying sustainable timber is illustrative of the challenges. Ideally, such certification covers the entire supply chain. The traded timber should be certified from the moment it is carefully, legally, and sustainably selected for cutting in the forest to the moment a customer buys a piece of furniture or lumber in a Western store. Any gap in controls in the chain increases the chance that illegal timber enters the trade and is effectively laundered.

The Forest Stewardship Council (FSC), an independent international NGO, employs a certification system that tracks timber from forest to shelf and is often considered the current gold standard of certification for timber. However, by the end of the 2000s, the FSC still certified only approximately 220 million acres of forest, of which 110 million (or one half) are in North America, while there are 10 billion acres of forested land on Earth.[60] Less than 2 percent of tropical timber was covered by FSC certification.[61] Certifying forested land is expensive, costing about $50,000 per concession, and customers are not always eager to absorb the higher cost that this entails. Tests by Home Depot, one of the largest purveyors of FSC-stamped products in the United States, suggest that less than one-third of customers would pay a 2 percent cost premium for certified products.[62]

Given the size of the trade and the complexity of certification (timber changes hands many times along a supply chain and is processed into numerous, often minute pieces, over extensive periods of time), the reliability of the process is frequently problematic, with many opportunities existing for fake certificates, falsification, or timber laundering along the way. The more timber is subject to certification, the more challenging it is to maintain the quality and reliability of certification. When I asked a logging company representative in Samarinda, in Kalimantan, Indonesia, about whether not obtaining green certification meant the company feared losing access to Western markets, he just laughed: "For us, it's just another bribery item. We pay for the inspectors. And anyway, they go out for two days out of a year—how much can they see?"[63]

Beyond the problems created by the sheer volume of products traded and the time and resources needed, as well as problems of fake licenses, certification schemes are plagued by other problems. A product may come from a certified site but may not be harvested sustainably. Some of the certification is very limited, confirming only that the timber originates from a particular concession and that the company had the necessary permits. Other certification can require more rigorous evidence of compliance with harvesting regulations and other operational matters.[64] Even then, sustainability may not necessarily be part of the certification system. Most legislation mandating the certification of wood, including the US Lacey Act and the EU's due-diligence reporting requirements, centers on its legality, not sustainability.

Certification problems often start with forest management plans. The design and implementation of forest management plans are both frequently pervaded by serious problems, even though the existence of such a plan can qualify logged timber for certification. Forest engineers can be incompetent and corrupt. Since natural forest regeneration often takes upward of fifty years in the tropics, it may be difficult to determine whether management programs assure species and ecosystem sustainability. Rarely do customers actually understand what is being certified, whether it is mere legality, sustainability, or biodiversity protection.

Some certification labels are of poor quality or fake, "greenwashing" unsustainable wildlife products as legal and sustainable. In some cases, major retailers, even in the United States and Western Europe, have appropriated eco-labels, including that of the FSC, without really being certified.[65] At other times, timber and wood-product suppliers have obtained the FSC's chain-of-custody certification, indicating that they have adequate capacity to check their supply chains without actually handling any FSC certified timber.[66] Extensive unreliability of certification can encourage greater, and undesirable, consumer demand. Monitoring the watchdogs who issue certificates is necessary.

The kinds of laundering and certification problem discussed here also exist in drug markets. Marijuana traffickers from as far away as Cuba have relocated to legal-marijuana Colorado. They cultivate marijuana there without licenses and illegally export it to other US states, though the scale of the problem is not yet clear.[67] The bulk of law-enforcement violations so far have to do with cannabis and cannabis products, such as marijuana candy, being illegally sold in other US states.[68] Such claims of illegal diversion are easy to exaggerate. But the Rocky Mountain High Intensity Drug Trafficking task force has documented a substantial rise in marijuana smuggling from Colorado to other states, from an annual average of 52 seizures between 2005 and 2008 to 316 seizures in 2014, following Colorado's legalization of marijuana.[69] However, this increase in seizures does not necessarily mean that marijuana smuggling has increased dramatically; it could mean that law enforcement agencies now prioritize going after marijuana smuggling more than they used to.

Substantial diversions of opium for legal medicinal use into the illegal production of heroin exists in some countries where the cultivation of the opium poppy and harvesting of opium resin are allowed, such as India. A producer of opium for both recreational use and medical purposes for several centuries, India after independence was given an international permit to cultivate the poppy for morphine and other medicinal opiates. Over time, India's opium poppy cultivation has declined, largely as other competitors, such as Australia and France, supply the international pharmaceutical market with a superior product—opiates based on the alkaloid thebaine, which have a much smaller chance of being diverted for illegal use. Nonetheless, the legal cultivation of the opium poppy in India still employs several hundred thousand farmers. Yet at least 25 percent of the opium resin collected is diverted from the legal industry to the illegal production of heroin.[70] Control mechanisms on cultivation and processing remain poor, and corruption in licensing is widespread.[71] In order to maintain large employment opportunities in legal opium poppy farming, India has not adopted the so-called poppy straw method, in which morphine and other medical alkaloids are derived from the whole plant (excepting the seeds), which greatly minimizes the chance of producing heroin from the plant. Instead, opium in India is obtained by scraping the resin from the lanced poppy capsule, the same way it has been harvested for centuries, which enables conversion into heroin. This harvesting process also prevails in Afghanistan, the country with the largest amount of illegal poppy cultivation in the world.[72] Moreover, illegal cultivation of the opium poppy, perhaps in increasing amounts, also takes place in India.[73] The diversion of licensed opium for the illegal production of recreational drugs stands in stark contrast to other licensed suppliers, including Turkey, which adopted the poppy straw method. Overall controls on poppy cultivation in Turkey are robust, diversion into the illegal trade does not take place, and although Turkey was traditionally, until the 1970s, a large producer of opium, no illegal cultivation takes place there today.[74] Illegal drug use and extensive drug smuggling through Turkey are nonetheless robust. The illegal diversion of opium in India would likely be even greater if not for the so-called 80:20 rule, by which the United States guarantees to buy 80 percent of opium containing morphine from India and Turkey, in order to minimize the illegal drug trade there.

In fact, political will as much as technical capacity determines the success of insulating the legal trade from the illegal one. There is no guarantee that a legal supply will weaken organized crime. Arguments for legalizing marijuana so as to bankrupt Mexican criminal groups and reduce violence in Mexico have been made by many drug-legalization advocates, including the former Mexican president Vicente Fox.[75] Proponents also argue that legalization would free Mexican law-enforcement agencies to concentrate on other crimes, including murders, kidnappings, and extortion.

Reducing the number of people arrested and imprisoned for nonviolent drug offenses is a crucial and worthy goal; and that is a very important reason to at least decriminalize drug use. Imprisoning users does not for the most part reduce drug use, and under some circumstances can even exacerbate it. Imprisoning people can violate human rights and can destroy people's lives and social productivity.

Crowded prisons are financially costly and often, particularly in Latin America, schools for criminals. Stigmatizing and punishing users undermines efforts to stem the spread of HIV/AIDS and other communicable diseases.

However, whether and when legalization reduces crime and weakens criminal groups is a very different matter. Such goals are unlikely to be achieved in a country with weak and corrupt law enforcement and strong and violent criminal groups, such as Mexico, for several reasons.

First, smuggled marijuana likely constitutes not much more than a fifth of the revenues generated by the drug-trafficking organizations or about $1.5 billion a year, as a 2010 study argued.[76] These are not numbers that will cause bankruptcy.

Second, without robust state presence and effective law enforcement, both elusive in significant parts of Mexico, organized crime can remain in the legal drug trade, and have numerous advantages over legal companies, such as preexisting knowhow and intimidation capacities. Nor does legalization imply increases in state effectiveness: much illegal logging persists alongside legal logging. In Mexico itself, legal and illegal logging and mining concessions as well as avocado growers are objected to extortion by drug-trafficking groups in states such as Michoacán, Guerrero, and Chiapas.[77] If the state does not physically control the territory where marijuana is cultivated—which is often the case in Mexico—drug-trafficking groups can dominate the newly legal marijuana plantations, still charge taxes, still structure the life of growers, and even find it easier to integrate into the formal political system. Many US oil and rubber barons started with shady practices and eventually became influential (and sometimes responsible) members of the legal political sphere. But there are good reasons not to want very violent Mexican crime syndicates to become legitimized.

In Italy, gambling, including slot machines, was legalized to take gambling away from the Mafia. Even as the gambling lobby and gambling itself have rapidly expanded, causing socially ruinous gambling addiction among some, the Mafia has still managed to dominate the legal gambling business. It has been able to increase its profits, use gambling to enter new regions of Italy and set up loan-sharking and extortion rackets there, and exploit legal gambling as a means of laundering illicit drug money, just as it has previously used agriculture, trucking, and restaurants.[78]

LEGAL TRADE

Third, a gray drug market would likely emerge. If drugs became legal, the state would want to tax them to generate revenue and to discourage greater use. The narcos could then set up their own plantations avoiding taxation, corner the market and reap the profits, and the state would be driven back to combating them and eradicating their fields.

Fourth, and worse yet, Mexican drug-trafficking organizations might intensify the violent power struggle that already exists in relation to remaining hard-drug smuggling and distribution. Notably, the shrinkage of the US cocaine market is one of the factors that precipitated the current wars among Mexican drug-trafficking groups. To compensate for revenue losses, they could intensify their extortion practices in other illegal, informal, and legal economies in Mexico, increasing their political power. Nor would law-enforcement agencies necessarily become liberated to focus on other issues or become less corrupt. All legal markets, whether in wildlife or drugs, need enforcement.

Boosting Demand: Reduced Moral Opprobrium, Marketing, and Innovation

Crucially, legal supply can also reduce the moral opprobrium of trading in a particular species of wildlife. The availability of artificial fur coats helped reduce demand for fur from wild animals. But over time, it also undermined the shaming campaign of the anti-fur community in the United States and Europe, since wearers of real fur could easily assuage criticism by saying it was merely artificial fur; so overall, the wearing of fur became once again socially accepted in the West.

Indeed, one of the most undesirable effects of licensing may be that it does precisely the opposite what it is set up for, namely, to eliminate demand for illegal products in threatened species. If licensed trading chains are extensively pervaded by illegally obtained wildlife, licensing may serve only to whitewash consumer consciousness. Instead of feeling any guilt at all and moderating their demand, consumers may in fact increase their consumption of the supposedly harmlessly obtained product, and the impact on wildlife may be even worse than if the trade were fully illegal and the moral consequences unambiguously clear. Greater availability of a species may also reduce its price, thus in turn stimulating demand (even though perhaps reducing poaching rents).

How intense these negative effects are depends on numerous factors, such as how many consumers and potential consumers care about consuming only legally and sustainably obtained wildlife products, and how many are indifferent to the legal and normative dimensions. Even if consumers are green-conscious, do they have the capacity to reliably distinguish between a legally sourced and illegally sourced product, or are there other watchdogs in the system who can reliably do the job for them? What are the price differentials between legally sourced and poached animal products or animals? If illegally sourced products are much cheaper, will morality trump prices? Comprehensive and extensive surveys of the extent to which customers are willing to put up with higher prices in the name of environmental conservation are often lacking. But in the case of legal marijuana, at least one survey found that about 60 percent of surveyed marijuana users would be willing to pay $2.50 or more per gram extra to access legal marijuana.[79] With illegal marijuana costing $10 per gram in the Unites States, that shows a willingness to absorb a 25 percent price hike.

Whether these negative side effects should discourage licensing also depends on how much demand there would be regardless of whether a product is legal or not. Ascertaining such outcomes depends on difficult empirical work pervaded by numerous problems of measuring illegal economies, and soft variables, such as consumer preferences that consumers may be loath to disclose in interviews. From a policy perspective, which is easier: to design demand reduction campaigns to discourage all use of a particular wildlife product, or to build in enough nuance to persuade customers to buy only from certified sources? The answer partially depends on how malleable demand is to start with—that is, can consumption be simply eliminated or will some consumption, and how much, remain despite demand-reduction campaigns?

Compounding this problem is the fact that suppliers also often experiment in new and exotic species, and legal supply in one species may thus create a new and undesirable demand in other species through the same established supply networks. How strong the marketing and commercialization pressures are will vary from one regulatory scheme and one institutional and cultural context to another. In the case of wildlife smuggling, the promotion of new products and constant innovations in the marketing of Traditional Chinese Medicine already go on, regardless of whether or not the species is defined as endangered and thus whether trade in it is prohibited or not. For example, Chinese traders only recently decided to promote the supposed curative qualities of manta ray gills, switching to promoting this product aggressively after the supply of large shark species in Indonesia dramatically fell due to overfishing.[80] Also in Indonesia, Chinese traders in Kalimantan have helped drive intensive new hunting of hornbills, labeling the bird's bill as the poor man's equivalent of the rhino horn and inventing claims about its curative qualities.[81]

Crashing the Price by Flooding the Market with One-Off Sales or Artificial Products

Because interdiction and law enforcement are costly and difficult, and also because of commercial interests, some governments have repeatedly suggested a one-off sale of a wildlife product, such as rhino horn, to flood the market and send the price plummeting, thus undercutting illegal traders.

There are at least two ways to implement such a scheme: first, from a genuine legal supply; or second, from an artificial legal supply. Currently, if Swaziland had succeeded in getting a license to sell its stock of rhino horn at the September 2016 meeting of CITES signatories, it is almost certain that it would not have been able to crash the market, and nor would any subsequent sale of South Africa's much larger stock. In addition, a sale could well set off many counterproductive effects. To start with, South Africa may be able to sell rhino horn for a lower price than poachers. But, since it is unlikely that South Africa would be able to maintain its legal supply at that same level indefinitely, the illegal market would persist and could increase. The difficult question is: would demand increase as a result of legal rhino-horn sales, or would it remain stable?

Part of the problem is that traffickers have a variety of adaptive mechanisms at their disposal, and not much of their capital is tied up in the trade, with sunk costs low. Well aware of the effects of scarcity and depletion, (illegal) traders would have every incentive to hoard stocks and await a price hike, just as they did during previous bans on ivory and whale parts.[82] Even if a one-off sale wiped out the illegal market in a particular species for a number of years, traffickers would maintain revenue flows by poaching other species. If that took pressure away from highly endangered species, it would be very valuable, but it would be unlikely to bankrupt and wipe out the poaching industry.

Achieving and maintaining a lower price than that of poached rhino horn would be far easier if a nondepletable artificial rhino horn could be supplied to the market indefinitely—the equivalent of artificial fur. Supplying artificial rhino horn is technically possible as the horn is made of keratin, the chief component of nails, hair, and hooves. Efforts are underway to clone rhino horn (and rhino), breed them in vitro, or bioengineer synthetic horn.[83] Taxidermists already produce replicas that look very much like real rhino horn for museums, galleries, and zoos to prevent theft. As the price of rhino horn on the black market escalated, Europe experienced a spate of such thefts. In 2011 alone, seventy-two rhino horns were stolen from museums in fifteen European countries.[84] Behind some of them was an Irish criminal gang—the Rathkeale Rovers—that specialized in the theft of high-value objects.[85] This is one of the instances in which a criminal group diversified into the illegal wildlife trade.

But whether the rhino horn supplied from non-wild sources would be fully effective in limiting the illegal market in rhino horn would depend on whether the industry, law-enforcement agencies, and customers could distinguish between the artificial and the legal supply, and whether customers would be equally enthusiastic about using the artificial product or strongly prefer the "natural" one. If customers were only interested in price and the artificial horn were cheaper, most, if not all, would likely switch to buying the artificial one, and thus reduce demand for poached animals; but if large numbers of customers strongly preferred genuine rhino horn, poachers would still remain in business.[86] Even if customers were unable to distinguish among the two, law-enforcement agencies would have to be able to—otherwise laundering would persist, just as it exists in the current markets for ivory, geckos, and parrots. Ensuring that customers cannot tell the true from the fake, but that law-enforcement agents most of the time can, is of course unlikely. Instead, the existing rhino horn used in taxidermy has also become a problem for law enforcement. Horns killed during trophy hunts in Africa were sold to the illegal black market in Asia, while fake horns were sent home with the trophy hunter, and law-enforcement agencies were not wise to the racket. Faking the horn even includes faking the smell of burnt keratin through the use of human or animal hair.

There are already fake rhino-horn products in retail markets. In Vietnam, these could amount to between 70 and 90 percent of rhino-horn volume, with the Vietnamese government often embracing the high estimates to deflect international criticism of its lack of interest in stopping the traffic in rhino horn.[87] Research into customer preferences in Vietnam by Annette Hübschle showed that Vietnamese customers preferred real horn, particularly from trophy hunts. Thus they often desired some proof of provenance, such as a hunting permit, or a photo of the supposed trophy hunt or of a dealer accompanying the hunt.[88] This is not surprising given that customers seek rhino-horn products for health reasons and attribute powers of vitality to it. Under such circumstances, consumer acceptance of replacement alternatives is thus likely to be low. Acceptance of alternatives may be higher when other characteristics are desired, as in the case of the reduction in demand for rhino-horn daggers in Yemen described in Chapter 10.

In sum, permitting a legal supply thus greatly increases the difficulties law-enforcement authorities have in differentiating illegal from legal products. Unless governments radically increase resources for officials to enable them to monitor legal supply diligently, they are unlikely to have the capacity to stem illegal laundering. The effective law enforcement of legal supply can be highly resource-intensive. There is a clear limit to how much licensing schemes can be tightened to prevent leakage, even if modern technologies are adequately distributed to law-enforcement officials. The persistent presence of extensive corruption can undermine any licensing (or prohibition) scheme. What determines the level of effectiveness of licensing are mostly the same factors that determine the effectiveness of bans: the level and quality of law enforcement; the elasticity of demand; the ability to supply licensed products cheaply and on a large scale; the strength of non-price-driven effects on consumer preferences, such as caring that one is preserving biodiversity through one's consumer choices; the property rights regimes in place; the timing of a licensing scheme; and the value of maintaining species due to such things as ecotourism.[89]

8

LOCAL COMMUNITY INVOLVEMENT

Giving local communities a stake in conservation is crucial since law-enforcement efforts often struggle to impose conservation if the local community is opposed to wildlife and ecosystem preservation. Enforcement can prevail under such circumstances but becomes far costlier in terms of resources, can become politically unsustainable, and raises ethical issues. Local community support for policies to combat wildlife trafficking will not eliminate trafficking, but it can sever existing linkages between trafficking groups and local poachers and even prevent their emergence. In Africa, a quarter of the land is made up of communal forests and rangeland; and 90 percent of the rural population access land through communal institutions.[1]

As with bans and legal trade, there is a large variation in how successful local community involvement has been in protecting habitats and wildlife, with many disappointing results. The results also strongly vary with respect to the other objectives that proponents of focusing on local communities advocate: political and economic empowerment. Some of that variation is the consequence of highly varied designs for involving local communities and how much power to devolve to them for managing local natural resources. Variations in outcomes also come from the highly diverse interests of local communities. Some communities, for example, may want to preserve natural resources for economic or cultural reasons; others may want to make money fast and move elsewhere.

The development of local alternative livelihoods to poaching and wildlife trafficking is also challenging. While frequently prescribed as the key mechanism for conservation, ecotourism unfortunately has not turned out to be a consistent remedy against poaching and wildlife trafficking, nor a reliable tool of economic growth. Despite its promise of giving communities a stake in conservation and providing them with legal jobs, ecotourism is only sometimes successful in limiting illegal economies. More expansive versions of community-based natural resource management models that transfer rights to local communities (which might or might not be based around ecotourism) have not been uniformly successful either. In some cases, communities which became richer as a result of such efforts intensified hunting and logging to further augment their resources.

The Theory Behind Local Community Involvement

Involving local communities in the design of conservation projects is aimed at much more than providing them with economic incentives for conservation. The strategy also seeks to psychologically, legally, and politically empower local communities, and to accomplish environmental conservation through means that do not harm locals and, ideally, will benefit them.

There are two types of arguments advanced by the champions of community involvement: ethics and pragmatism. The first points to evidence showing how some of the world's poorest, most marginalized, and vulnerable communities often bear the brunt of the costs of conservation and have historically suffered great harm and injustice in its name. Community-based approaches seek to prevent and redress these morally unacceptable consequences.

The pragmatic argument holds that law enforcement can be imposed without local buy-ins. But it also holds that without community participation, conservation can become enormously costly to the state and conservation advocates in terms of material and manpower resources needed for enforcement and political sustainability, especially in situations where the community feels it is being unjustly treated.

Community-based natural resource management (CBNRM) approaches aim to deliver three positive outcomes: environmental conservation, community empowerment, and social equity, or in other words, conservation, human rights, and poverty alleviation.[2] CBNRM efforts thus aim to assure sustainable development—but whether sustainability pertains to the sustainability of biodiversity (perhaps defined as avoiding the loss of one single species) or the sustainability of economic growth can be hotly contested, as the two can stand in stark opposition to each other. Moreover, CBNRM approaches also seek to maximize local residents' decision-making roles in the governance of natural resources.[3] The devolution of power and the empowerment of those who have been poor, marginalized, and deprived of rights may be not only highly economically beneficial, it can also be enormously psychologically rewarding and enabling.

Involving local communities can range from merely consulting local populations and seeking their input to the devolution of various degrees of power and authority to local communities. In a weak formulation of community-based conservation, communities are seen as neighbors of parks and thus consulted, and compensated for the costs of conservation, in order to incentivize their support for environmental regulations and enforcement.[4] In a maximalist formulation, local communities become the ultimate authority or even the sole decision-making entity in relation to managing local natural resources.[5] Along the spectrum between the two, there can be, for example, a transfer of rights to local communities, but only conditionally—such as with the proviso that local communities do not hunt endangered species (or perhaps do not hunt at all), or that they do not cause deforestation or that only carry out logging in a sustainable manner.[6] Proponents of CBNRM models often insist that the weakest formulation does not qualify as CBNRM, and that any true efforts must involve the comanagement of wildlife by local communities and some devolution of power and rights to them.

The economic means by which CBNRM in all of its formulations attempts to achieve its objectives can include ecotourism and other alternative livelihoods, as well as the local extraction of natural resources. The first two are mostly not controversial. However, disputes over means arise regarding what kind of hunting and legal trade in wildlife local communities should be allowed to engage in, what kind of logging, mining, or grazing they should be permitted to conduct, and whether they can stay within protected areas. The key issues are thus about which specific rights should be devolved to local communities.

A prominent argument of critical scholars of conservation and proponents of the more expansive conceptualizations of CBNRM is that the negative impacts of poor people on the environment are often exaggerated, and that poor people are also interested in managing their natural resources well. Such contentions, in addition to moral principles, are used to argue against excluding people from protected areas. Critical scholars, for example, maintain that the gathering of subsistence foodstuffs in protected areas causes far less damage than is frequently alleged because the poor often eat less so as to preserve productive capital and future food production.[7] They also argue that as resources become scarcer, local people dependent on them take steps to ensure their sustainability.[8]

How does the evidence brought forward in my research confirm or challenge such theories? Despite popular portrayals of the current poaching wave as solely one of organized crime rather than poverty, local communities are indeed crucial vectors in poaching and wildlife trafficking. They can resist and oppose both, participate in one or the other, or embrace both. Even though organized-crime groups may form a part of the smuggling chain, individuals in affected local communities are often complicit in poaching, serving, like rangers, as spotters and trackers, and providing access routes. This is even the case in community reserves, such as in the Maasai areas in Kenya. Here and elsewhere in Africa, illegal grazing, logging, charcoal production, and poaching are all widespread in community reserves, which are often managed far worse than the national parks or private reserves, including those where legal hunting is permitted. As a Maasai ranger from one community reserve told me: "The white tourists can have the reserve during the day. At night it is ours to graze our cattle, log, and hunt."[9] As detailed below, such statements reflect both the elite capture of the economic benefits that wildlife conservation is supposed to bring to local communities and well as their limitations in terms of revenues that ecotourism often brings, particularly in places, such as Kenya, where almost no hunting and trophy hunting are allowed. Moreover, throughout East Africa, state agencies, such as forestry ministries in Kenya and Tanzania, and official logging companies, as well as other vested interests, often dominate illegal logging and charcoal production—little wonder that poor local communities feel little incentive to respect formal regulations not obeyed by those in power.[10] In addition, as discussed throughout this book, the history of the colonial exclusion of black communities from protected areas needs to be considered. By some accounts, between 22,000 and 100,000 Maasai were evicted from protected areas in eastern Africa in the twentieth century.[11]

Even when local poachers interact with organized wildlife-trafficking networks, many poachers remain highly marginalized and desperately poor. Providing them with legal livelihoods can be an important component of policy intervention in attempts to reduce the illegal wildlife trade, as well as an important moral imperative. In Indonesia, poachers in the Moluccas and Papua are sometimes paid as little as a bowl of noodles for a day's hunting, or a pack of cigarettes for a rare bird. But that pack of cigarettes can be enough to extirpate an endangered species. Traders can be shockingly frivolous in how many individual birds or animals they are willing to have killed to ensure the survival of a few that can bring high profits on the international market. Ambonese hunters, mostly very poor, will be paid $5 for a caught black-capped lori, a bird species favored by international wildlife traders. In Myanmar, some wildlife poachers are former poppy farmers who lost their previous illegal livelihoods due to poppy eradication campaigns, with the result that they had a secure food supply for only eight months a year.[12]

Although creating economic incentives for a community's support for conservation does not address the problem of higher-up wildlife smugglers and organized-crime groups, it can simplify and focus the practice of law enforcement. It can encourage community cooperation with law-enforcement agencies, enhance the political sustainability of restrictions on wildlife trade, and reduce political conflict. It can also promote social justice. If people face a choice between dire poverty or wildlife poaching, they will frequently choose wildlife poaching as well as illegal grazing, logging, and mining, all of which can undermine conservation. If, under conditions of poverty and marginalization, a government attempts to enforce a ban on wildlife hunting, political conflict, leading even to violent opposition, may ensue. Historic social injustice may be perpetuated and augmented. Existing organized-crime groups or militant groups can exploit such situations, positioning themselves as patrons of local poachers and communities that violate environmental regulations. Weaning a community off poaching may not be sufficient for halting it—outside traffickers may still illegally hunt, even hunting out animals from the area. Nonetheless, bringing the community on board can simplify law enforcement in terms of intelligence and resources, as well as normatively. Laws and regulations are easier to enforce if the vast majority people accept them and internalize them.

Challenges for Involving Local Communities

The first issue and challenge that arise affecting the provision of benefits to local communities, and possibly full transfer of ownership rights to them, is to define who exactly is "local." Is anyone who lives within a certain radius of a protected area local? What about people who moved there a year ago? Will they be entitled to socioeconomic compensation programs or endowed with rights to land? What if the other communities dislike those newcomers and consider them outsiders? Conversely, what about people who were displaced from the area, including for environmental conservation reasons, several decades ago and have lived elsewhere? Will they have a right to receive compensation or even move back into the protected area? Thus in Tanzania's wildlife hotspots, such as the Ngorongoro Crater, disputes over who are the rightful residents, who should be allowed to stay, and which communities should leave have persisted for a long time.[13] Similarly, in the Amazon, Brazilian rubber tappers (of mixed indigenous and non-indigenous heritage) have complained that their rights have been compromised for the sake of indigenous groups and their access to forest resources unjustly curtailed.[14]

Moreover, just like all groups of people, "local communities" are not homogeneous entities. Various members of a community may have radically different ideas as to how they want to use the land and what levels and forms of compensation they want to accept. Even when ecotourism or other CBNRM approaches benefit a "local community," they do not necessarily benefit people equally. Some may not be able to partake in the benefits at all, and thus remain disenchanted and antagonized. There is not one single regulatory scheme that can resolve these challenges.

Younger generations may have different reasons for conserving nature or destroying it than their parents. Many poor people do have strong interests and values that promote environmental conservation, but others want to maximize their current—not long-term—economic benefits and make money now. In the latter case, under maximalist formulations of CBNRM, making the local community the authority, not just one of the stakeholders, of conservation, the outcome would be potentially severe environmental degradation.

The differences in the maximalist and minimalist phrasing of local community involvement can also have profound implications for countering wildlife trafficking. If a local community has the right to utilize local natural resources any way it wants (that is, the right to local land and natural resources have been transferred unconditionally), what if outside traffickers arrive and want to poach animals for the international wildlife trade, such as the Asian market in ivory? If the community opposes such hunting, the problem is relatively simple and concerns the deployment of the necessary resources to fight off the outside traffickers. But whether the state in such circumstances will be willing to devote resources to defending the community may depend on whether the state itself derives sufficient indirect benefits from the community's ownership of natural resources, such as in the form of taxes.

The situation becomes even more difficult if the local community, holding full authority over local natural resources, is tempted by offers from wildlife traffickers of accessing greater income from allowing the hunting of certain animals, such as rhinos, than it derives from ecotourism or customary grazing in the area. Some communities, coveting the natural resources they control, may opt to refuse the traffickers, but others could decide that the short-term economic gains are more attractive than their current low income, and the uncertain long-term rewards the existing conservation regime offers them.

Ecotourism and Other Alternative Livelihoods

The weaker conceptualizations of involving local communities by providing them with economic incentives not to destroy local natural resources must find ways to generate sufficient income through means other than hunting. That is particularly the case if local communities are restricted in economically exploiting protected areas or if their exploitation of such areas is environmentally unsustainable. The answer most frequently given is ecotourism. Other alternative livelihoods efforts, such as new forms of agriculture or small-scale craft production, are also put forth.

Unlike in the case of other illegal economies, such as the drug trade, the income to be gained from poaching and other kinds of involvement in the illegal wildlife trade may not be much higher than possible profits from an alternative legal economy. Since wildlife is a finite commodity, as wild stocks decline, often so too do profits for low-level poachers. Even with drugs, price profitability is hardly always the most important driver; frequently it is not, and other structural drivers have a far greater impact on farmers' decisions to grow drugs or a legal alternative.

Ecotourism

When the conditions are right, ecotourism brings money, jobs, prosperity, and development to the state, businesses, and local people, and it also preserves nature.[15] Unfortunately, ecotourism has often proved disappointing in generating sufficient economic benefits to wean local populations off hunting and illegal grazing and motivate them to support conservation. According to some studies, about 90 percent of ecotourism revenues are lost through imports of goods and profit repatriation in most ecotourism destinations.[16] Although ecotourism is supposed to provide economic incentives to local populations, often as much as 95 percent of post-tax revenues are captured by the owners of ecolodges (in South Africa these are still mostly whites). The earnings from hunting or other forms of ecotourism have also not significantly benefited local poor communities. Again in South Africa, the licensed hunting of rhinos and the permitted sale of rhino products greatly benefited the recovery of rhinos (at least, until the recent wave of poaching), and it brought large financial benefits to owners of reserves (also mostly whites) as well as state coffers, but local communities were often unable to participate in the bonanza. The conversion of agricultural land into nature reserves often resulted in the loss of employment for black people.[17] Moreover, many white landowners saw investing in nature reserves as a form of insurance against having their land subject to restitution claims by local populations.

In eastern Africa, many ecolodge owners are prominent African political dynasties. The owners are black, often top politicians, whose policies are neither equitable nor accountable. Lodges themselves have to pay a myriad of licensing fees (and bribes) to local councils and national ministries.[18] As political patronage and tribal favoritism skew the hiring of staff, very little of the proceeds from ecotourism trickles down to local communities. Ecotourism companies often bring in their own guides to avoid having to hire more expensive or less-qualified local guides.

Under such circumstances, local communities have little incentive to support conservation and refrain from poaching. In Nepal, profits from ecotourism, such as those earned by lodges near national parks in the southern Terai region, tend to be predominantly captured by rich businessmen who often reside in Kathmandu. While the lodges do employ local community members as guides, cooks, and cleaning staff, the number of jobs thus generated is insufficient, and earnings still leave most families barely coping.[19] In the Terai, alternative livelihood initiatives have by and large failed to offset the losses individual poachers and connected communities sustain from foregoing participation in wildlife trafficking.[20]

Sometimes, of course, there are model successes. In Madagascar's Anjajavy region, a seaside area of sparkling beaches with snorkeling and diving, ecotourism lodges owned by South African businesses pride themselves on only hiring staff from neighboring communities. In addition, they have built a clinic and school for the local area, and have sought to reduce overall local poverty.[21]

In Vietnam's Cuc Phuong National Park, the results of ecotourism and alternative livelihood efforts have been mixed. The park is one of Vietnam's premier biodiversity areas—in a country which has suffered extensive deforestation and tree-cover loss over many decades from the war and subsequent logging, and where hunting has significantly reduced wildlife populations. In 1995, some villagers near the park were trained in protecting the park and in developing a "family-based economy." Investments included micro hydroelectric generators, a weaving system, and microloans to support bee-keeping, deer breeding, and planting of lychees. In exchange, villagers were required to reforest an area of land and return it to the national park. Their success allowed them to reduce their dependence on forest products from the national park.[22]

The growth of ecotourism, however, did not bring much benefit to many of the villagers, who were subsequently resettled outside the park. Local elites and village heads benefited disproportionately, renting their houses to tourists. Many of the resettled villagers thus continued entering the park to hunt and gather resources, such as timber, tubers, bark, and bamboo shoots, which resulted in violent encounters with local wildlife law-enforcement agents.

The size of the generated ecotourism resources, not only the equity of their distribution, critically influences ecotourism effectiveness in discouraging the undesirable exploitation of natural resources. As conservation biologists William Adam and Mark Infield point out, there are many claimants to the ecotourism payroll: the central state, planning commissions, central environmental agencies, local government, local environmental law-enforcement agencies, ecobusinesses, as well as local communities.[23] Local communities often end up with the slimmest cut of the generated revenues, and resent environmental conservation programs which have been responsible for thriving ecolodges and ecotourism. Overall, most protected areas, even in sub-Saharan Africa, do not generate sufficient revenues to cover their maintenance costs, let alone lift surrounding communities out of poverty.[24]

Maintaining sufficient and steady incomes is a key challenge for all alternative livelihood programs designed to reduce the dependence of the poor on illicit economies. Incomes must not only assure economic survival, but also enable economic and social advancement. Delayed income, such as when ecotourism starts bringing in resources only several years after the establishment of ecotourism zones (during which time local communities are asked to forego hunting and natural resource exploitation) may be as problematic as ecotourism income that dries up within a year or two. Finding other ways to compensate local communities during such periods of inadequate income may be essential if the community is not to become disenchanted with the project and return to poaching.

On the Indonesian island of Seram, for example, twenty poachers who hunted rare parrots were converted (through the work of Profauna, one of Indonesia's NGOs most determined to fight against the illegal wildlife trade) into rescue-center staff and wildlife guides for tourists. Poaching dramatically declined. But this success depended on a steady flow of ecotourists for the newly converted poachers to guide. For that, an international counterpart to the conservation effort helped recruit US birdwatchers to travel to Seram. When the supply of ecotourists fell off, income from guiding visitors shrank and pressure on the guides to resume illegal hunting to generate a livelihood intensified.[25]

The Seram story is a micro-example of the conditions on which successful alternative livelihoods depend. If poor poachers have an assured income from other sources, they are often willing to abandon illegal hunting, even though poaching often brings more money. But income from other sources needs to be steady and assured. The problem with many alternative livelihood efforts involving ecotourism is that income fluctuates greatly and tends to be sporadic and seasonal.

For an area to draw sufficient numbers of ecotourists to generate income, it often needs to contain large mammals that can be seen fairly easily by tourists. Thus, East Africa's savannahs tend to attract many more tourists than rainforest areas. But even here, income from ecotourism can be highly seasonal, and lodges either have to build up financial reserves to pay their staff during the low season or, as frequently happens, find it necessary to lay off staff during part of the year, thus incurring the wrath of the community.[26] Additionally, to generate sufficient revenues, parks need to have good infrastructure, such as airports, roads, and lodges, as well as good security. Banditry or the presence of militant groups scares off tourists, as do other external shocks such as political instability in the country or economic downturns in tourists' home countries.

Moreover, the number of jobs available through ecotourism may be significantly lower than the pool of people susceptible to recruitment as poachers. Even if all existing poachers get legal jobs, are there other poor people in the area, or who will move into it, who will poach, particularly if compensation for poaching rises? Determining who are local people, and who requires and qualifies for assistance, can raise very difficult and complex normative issues. Poor people who migrate into a protected area to poach can reasonably ask why they should suffer grinding poverty and economic privation, and why they are turned down for the assistance enjoyed by residents in the area. In Mount Guiting-Guiting Natural Park, on Sibuyan Island, the Philippines, WWF worked with an indigenous tribe to secure its rights to lands that overlapped the protected area and succeeded, but the rest of the island's inhabitants were enraged and felt left out and cheated.[27]

Such problems also arise out of population growth. Ecotourism may, for example, employ all of the working population of a village on the outskirts of a park. However, what happens if the size of the population doubles as a result of large birthrates and greater longevity? Also, the availability of jobs may decline as a result of exogenous shocks, such as economic decline in areas from which tourists originate. What if young people decide that they now want to hunt or otherwise exploit the protected area? Will they be entitled to do so, or to compensation, or will they be punished for encroaching on protected areas?

Trophy hunting can be a very important part of ecotourism revenues. In the Okavango Delta in Botswana, trophy hunting consistently generated more revenues than non-hunting ecotourism, such as photography.[28] The number of animals to be legally hunted was determined after annual aerial surveys of species' population sizes were conducted to assure sustainability. From those earnings, some communities were able to provide social services, such as housing for the elderly or orphans, scholarships, and funeral assistance. In some villages, poverty dropped significantly, and in areas of successful ecotourism projects poaching declined.[29]

However, neither the economic nor the environmental results have been consistent in Botswana's Okavango Delta. Some communities lack the technical, managerial, and business skills to establish effective wildlife tourism ventures. They are also unable to compete with large eco-businesses. In some areas, there was corruption and the mismanagement of funds, which eviscerated the monetary gains from ecotourism. Finally, in most villages there are no mechanisms for the equitable distribution of resources.[30] Thus only some in a village developed an interest in ecotourism.

In short, the success of using ecotourism and other alternative livelihoods to prevent poaching depends not only on the number of actual poachers, but on the number of potential poachers. It is one thing to employ twenty hunters (as in the Seram example) and quite another to bring employment to several thousand people who may reside in or near an ecologically sensitive area. Moreover, whether ecotourism takes the pressure off poaching also depends on whether ecotourism companies capture large parts of the profits or whether local communities get a sufficient cut.

In some place there is also the problem with the large influx of humans—both tourists and those employed in ecotourism—generating even greater environmental damage than traditional hunting, sometimes more profoundly disturbing an entire ecosystem rather than that of a particular species. Too many lodges, vehicles, and tourists in a park can critically threaten vulnerable habitats. If high-impact ecotourism further drives animals out of the protected areas, they become even more vulnerable as communities want to protect their crops, grazing areas, and cattle, the pushed out wild animals may end up being subjected to poaching for the international wildlife trade. But officials who license lodges are often not biology conservation experts, nor do they interact with them. They can and do have other primary interests than wildlife and ecosystem preservation—in particular, to make money.

Under ideal circumstances, high revenues could be generated from ecotourists, who are both environmentally and culturally sensitive and nonintrusive. Such ecotourists do not require extensive infrastructure and modern comforts, and are willing to stay with local families living under local conditions.[31] But the trend is the opposite: increased high-impact ecotourism, including from East Asia, into protected areas around the world.

Other Alternative Livelihoods

There is also large variation in outcomes from local community involvement in managing habitats and wildlife. One example of success in weaning poor populations off wildlife poaching took place in Pu Mat National Park, Vietnam, a key biodiversity area and a site of extensive and vastly damaging wildlife trade. The trade involved 75 percent of households in the buffer zone around the park, and the illegal extraction of forest products was the only activity available to these to generate the income necessary to buy rice. Alternative development efforts led to wildlife becoming a less important part of the income portfolio of many households, and the number of poachers shrank considerably. However, this development took place in the context of the highly increased scarcity of wildlife in the national park, and thus the reduction in hunting may have taken place irrespective of the rural development intervention, while the scale of the positive effect due to rural development is not clear.[32]

Another positive example concerns a four-year project in Guinea funded by USAID and administered by the Center for International Forestry Research (CIFOR) and the World Agroforestry Center (ICRAF), which taught farmers to plant high-yield crops, establish "living fences" of bushes to pen in their livestock and prevent attacks by wild animals, and obtain fuel, fodder, and food in less environmentally damaging ways. The outcomes are impressive: incomes have risen; wild animals, including chimpanzees, have returned to the area; forest cover has increased; and illegal encroachment and human-set forest fires have decreased by 80 percent.[33]

These examples show that for alternative livelihoods to be viable, they need to be aligned with the skills of local people.[34] However, to generate sustainable income, they need to be linked to robust markets, local and/or international. The absence of such markets and established value-added chains has not only doomed many alternative livelihood and environmental conservation efforts, but also efforts to reduce the cultivation of illegal drugs. Creating such value-added chains often requires working with the private sector.

Alternative livelihoods centering on handicrafts often generate incomes too paltry and sporadic. Sometimes there are successes, such as in Botswana's Okavango Delta, where basket weaving became one of the main sources of income.[35] Overall, however, the success of handicraft-based alternative livelihood efforts tends to be even lower than the infrequent success in converting farmers from illicit to legal crops. A 2008 comprehensive study of the wildlife trade by TRAFFIC concluded that efforts to increase income, reduce poverty, and diversify livelihoods among rural communities often had relatively low impact on the illegal wildlife trade, and often did not reduce even the target community's participation in wildlife harvesting and trade.[36]

As with alternative livelihood efforts to wean farmers from the cultivation of illegal crops, the big question is whether such efforts are structurally bound to fail even among low-level hunters, or whether the programs were not sufficiently resourced and comprehensively and adequately designed.

LOCAL COMMUNITY INVOLVEMENT

A critical component of successful alternative livelihood schemes to counteract poaching has been precisely that they were instituted when wildlife was being depleted and economic interest in poaching was declining. Of course, from a conservation perspective, it can then be too late: marginalized populations are weaned off illegality, but wildlife conservation may still have failed because species were depleted anyway. What the environmental impacts are will depend on at the level at which poaching was stopped. The preservation of a species may require fewer individuals than the preservation of the species' role in the ecosystem.

Alternative livelihoods and socioeconomic measures, however, should not be limited to generating income. Also essential are increasing agricultural yields without added pressure on the environment, improvements in soil fertility and water resource management, the delivery of medical and educational services, as well as the development of local safety nets. They should also provide for infrastructure improvements that are environmentally sound—improving communications may be beneficial, but it should be noted that building roads, for example, can stimulate an influx of people into an area. Consulting the local community and amplifying its voices will be critical for success.

Alternative Livelihoods: Lessons from Counter-Narcotics Efforts

A crucial lesson from counter-narcotics efforts at providing alternative livelihoods is that replacement income cannot merely provide enough for survival; it also needs to enable social mobility. But the level of replacement income is only one factor. Many other obstacles need to be overcome.

The drug-policy scholarly community is deeply divided over alternative livelihood efforts. Drug-policy economists tend to dismiss them, arguing that drug traffickers can always pay more than a worker will earn from a legal job. Development experts often embrace alternative livelihoods and argue that they can be designed effectively. The core dispute centres on why people cultivate illegal drugs—and why they poach. Focusing solely on economic motivations, the question is whether people commit crimes to augment their income no matter what its level, or to cope with poverty. In the former case, no level of compensation or legal income can offset the resources derived from illegal trade. In the latter, delivering legal income to offset losses from not participating in the illegal economy is feasible. Research shows that some poachers and farmers of illegal crops will forego participating in criminality once their subsistence is assured, while others will not. Coupling alternative livelihoods with eventual enforcement is necessary. However, enforcement becomes both more justifiable morally and more easily implementable if it does not come at the cost of sentencing the world's poorest and more marginalized to even more extreme poverty.

Sadly, alternative livelihood efforts to reduce drug-crop cultivation have been mostly a dismal failure because of problematic design, lack of funding and perseverance, expectations of too rapid progress in suppressing drug crops, and because many peripheral areas of drug-crop cultivation are simply too remote, with transportation costs too high, to make legal production viable.[37] While alternative livelihood efforts have sometimes succeeded in reducing drug-crop cultivation in a particular area, on a country-wide level there is only one success: Thailand.

Thailand: Alternative Livelihoods and Eliminating Drug-Crop Cultivation

Thailand is widely and deservedly recognized as the model of how to reduce illicit crop cultivation through comprehensive alternative livelihood programs. It was also among the first countries to learn in the 1960s and 1970s that in order to defeat insurgencies, punitive suppression policies toward drug cultivation needed to be halted.

Five elements enabled the success of alternative livelihood efforts in Thailand. Conditions in Thailand were uniquely auspicious. Eradication policies toward drug crops were halted during its counter-insurgency operations and during several post-conflict years, while alternative livelihood programs were initiated in poppy-growing areas. Eradication was suspended until after alternative livelihood efforts generated sufficient income for opium poppy farmers—not before or concurrently, as often prescribed in international counter-narcotics efforts. The opium poppy was thus eradicated only after several years of alternative livelihood efforts, and eradication was mostly negotiated with local communities, often via a joint government–village committee which determined whether sufficient legal income was available.[38]

LOCAL COMMUNITY INVOLVEMENT

Alternative livelihoods were well designed as comprehensive rural development, focusing not just on income replacement but also on broader human capital development and the reduction of social and political marginalization of the ethnic minorities growing opium poppy. The original simplistic approaches of looking for a replacement crop, such as onions, garlic, cabbage, or more valuable crops such as apricots, were gradually supplemented by a focus on broader socioeconomic and human-capital development, including improving infrastructure connectivity, building value-added chains, and extending health care and schools to opium-producing villages. The policies were also well funded and maintained for thirty years, receiving steadfast backing from the royal family and some key international partners, such as the German development agency.

Crucially, Thailand's overall economy grew massively during the 1980s and early 1990s, creating many new job opportunities, including off-farm jobs. Thai farmers in the lowlands moved to work in factories in the cities, liberating opportunities in legal agriculture for opium-growing minorities in the highlands. A burgeoning tourist trade in the highland areas created more off-farm opportunities, while growing population density further encouraged the highland population to explore non-farming opportunities.[39] The fact that citizenship was also extended to poppy-growing minorities was essential for their ability to take advantage of new economic opportunities.

The strategy paid off. In 2002, Thailand was declared free of drugs, becoming the only country to eliminate illicit crop cultivation through predominantly alternative livelihood approaches. But it needs to be noted that drug cultivation in Thailand was always limited in scale; in fact, it was an order of magnitude lower than in most countries involved in drug cultivation. Drug cultivation in Thailand peaked in 1965 at a mere 18,000 hectares of opium, hovering mostly at around 10,000 hectares during other years.[40]

Several relevant lessons can be derived from the successful attempt at shifting opium cultivators to alternative livelihoods in Thailand, and from many other places where such counter-narcotics strategies have struggled.[41]

For alternative livelihood programs to durably reduce undesirable economies, good security needs to be established in rural regions.

Ending military conflict or violent criminality needs to be given priority. Policing and the rule of law are indispensable. Alternative livelihood programs construed merely as crop substitution are ineffective. Even if the replacement crop is lucrative, profitability is only one factor driving the cultivation of illicit crops, with other structural economic conditions often having far more profound effects on the decision of local communities to cultivate illicit crops.[42] Alternative livelihoods really mean comprehensive rural and overall economic and social development. As such, they require a lot of time, sufficient resources, the political will to concentrate them, and lasting security. Alternative livelihoods also need to be integrated into overall development strategies, with attention paid to whether overall economic growth produces job creation or capital accumulation, and whether it exacerbates inequalities. Macroeconomic policies, such as taxing labor heavily and land lightly, might have pronounced, if indirect, negative impact on the effectiveness of alternative livelihood policies: they may produce economic growth, but also perpetuate social exclusion.

Community Presence in Protected Areas

Whether local populations—often indigenous and marginalized—are allowed to remain within areas designated for environmental protection has been one of the most contentious issues in conservation policy. Historically, population relocations throughout Asia, Africa, and Latin America have taken place forcibly, without consideration of the human rights and basic economic needs of local people. Such policies have led local communities to associate environmental conservation with colonialism, and thus to reject of conservation. Even while the decentralization policies of new states nominally transferred some rights to local communities in some countries, many local communities continue to be abused while national logging and mining companies get away with environmental destruction, and even eco-businesses get away with "green grabbing."[43] Impoverished and angry relocated villagers, as well as people from surrounding areas, continue to encroach on protected land anyway, invading national parks, as they feel alienated and have no stake in conservation. Enforcement against such encroachment is sometimes brutal and at other times sporadic as many local rangers do not necessarily buy into the regulations themselves, seeing them as Western colonial impositions, and they do not want to pick conflicts with the local community.

In Ethiopia, for example, the continuing illegal exploitation of natural resources in protected areas is widespread. Although the poaching of large animals such as elephants is rare in Ethiopia, poaching for bushmeat is common. There are several reasons for the absence of big-game poaching. Big animals are scarce in Ethiopia since habitat destruction has pushed out many species, particularly elephants, to neighboring countries. Furthermore, while killing lions or leopards brings a lot of prestige and is seen as a way of protecting livestock, international smuggling networks are not very active in the country, so local hunters rarely attempt to sell lion or leopard skins or teeth.[44]

Illegal grazing and logging for charcoal are even more widespread and affect many of Ethiopia's major protected areas, including the Awash, Lake Abiata-Shala, Nechisar, and Simien Mountains National Parks. In addition to reducing grazing for wild animals, the presence of livestock also intensifies the spread of invasive plant species that cows, goats, and camels carry into the parks. Government efforts to create awareness among nomadic and pastoral communities, such as the Afar and Kereyu, many of whom are desperately poor, along with efforts to provide alternative livelihoods, such as a sugar factory, have produced meager results.

Meanwhile, since the enforcement of laws against illegal grazing in national parks is sporadic and violations extensive, the prescribed fine, equivalent to $2.50 per domestic animal found in a park, has not deterred illegal grazing. Nor have intensified efforts against illegal logging stopped deforestation and charcoal production. The economic incentives for carrying out these practices are greater than the fear of regulations and sporadic law enforcement.[45] Recruited from the local community, park rangers often have few incentives for enforcing regulations, as many of those who graze, log, and plant in protected areas may be their close relatives. Rangers may also be reluctant to enter into conflict with their fellow villagers out of fear of retaliaton or sympathy with their economic condition. To define their job (as they did to me) as the collection of fees from tourists and the protection of tourists from wild animals (mainly the non-threatening antelopes) is far easier than picking a conflict with their community for the sake of regulations they may not believe in. Thus they mostly just watch the illegal grazing and logging go on.⁴⁶ In Awash, where rangers were sporadically active, all enforcement collapsed after two guards were shot by some Kereyu for trying to prevent some illegal grazing.

Much of the land encroachment—or from the perspective of the local community, the rightful use of land that should belong to local people, not to the wild animals or white tourists—is compounded in Ethiopia by unclear landownership. Ethiopia's current constitution of 1991 reversed much of the land redistribution that took place during the rule of the authoritarian socialist Derg regime in the 1980s. The current constitution states that land belongs to the state and the people, with the government being able to take the land away at any time. While the government may provide monetary compensation for constructing a building on nationalized land, it will not pay compensation for land used for environmental protection. Meanwhile, the government has leased extensive tracts of land for farming to other countries, such as China, India, Saudi Arabia, and the United Arab Emirates. The leased land is often settled by politically weak and economically marginalized communities, such as the Oromo and Gambella, with forced displacement and land-grabbing frequent.⁴⁷ This larger context of land conflict further discourages communities from buying into environmental conservation.

Indeed, many of Ethiopia's national parks are being eaten to the ground by livestock and converted to fields and dust. In some parks, such as Abiata-Shala, even commercial farms and industrial plants, including a soda ash factory, are located within the boundaries. The artisanal gathering of soda ash also takes place, both liquidating the habitat of and pushing out flamingos and other birds, as well as antelopes, for which the park was established. It is a desolate landscape: the long grass has gone due to illegal grazing, with big mammals absent as a consequence, water level has drastically fallen in the lake due to agricultural and industrial use, and hence there are very few birds, just desperately poor women gathering soda ash with their bare hands in the scorching sun. With polygamy frequent, and second and third wives often neglected by their husbands, many of the women depend on the illegal gathering of soda ash to obtain income and support themselves and their children amidst dire poverty, even as their activity compounds the lake's decline and disturbs the birds that still stop over to feed or breed on it.

Enforcing laws against illegal grazing, logging, and other forms of exploitation of protected areas is also physically and ethically complicated in other parts of the world, and hence often sporadic and inconsistent. During my research in Nepal in May 2012, I repeatedly observed members of local communities collecting various forest resources at times rather deep in the Chitwan and Bardia National Parks, with forest officials reacting only with an oral warning, while the punishment for illegal tree cutting could nominally (and in the case of many poor villagers, inappropriately) be a prison sentence of seven years. One of the reasons for the lax enforcement of regulations was the political and economic context. In Chitwan, for example, problematic resettlements of Tharu people took place in the latter 1990s. Most of the 2,000 relocated people looked forward to the economic improvements that relocation would bring them, but they ended up on lands with poor soil, three hours away from water, and without access to any forest resources.[48] Their lingering anger makes enforcement difficult.

Moreover, in Nepal, forestry officials have been accused of colluding with traders and logging companies, issuing logging permits while engaging in minimal monitoring. The forestry sector lacks transparency, and political parties regularly protect illegal loggers and timber traders.[49] Political instability, the lack of a land use policy, and poor capacity on the part of stakeholders all limit the effectiveness of efforts to combat illegal logging and timber trafficking.[50] Combating illegal logging is also of low priority for most of Nepal's law-enforcement agencies.

Political parties further compound the problem of deforestation and illegal squatting in national parks by luring landless communities into protected areas with promises of future land titles in order to expand party support bases. Months or years later, the parties will of course fail to deliver the titles, and the communities, having meanwhile disturbed and damaged protected ecosystems, will be forcibly evicted without being granted land elsewhere. Local administrations frequently engage in similar violations of environmental regulations by moving those affected by floods and other natural disasters into protected areas, subjecting them to future evictions without compensation. Unsurprisingly, such evictions trigger *bandh*s (armed strikes that block roads).

Similarly in Indonesia, law enforcement and military forces are not only inadequate and underresourced but deeply complicit in various illicit economies, including illegal logging and mining. The corruption problem extends well beyond individual officers being in on the take. During the Suharto era, Indonesia's military had investments in large parts of Indonesia's economy. Although the military were forced to give up many of these, they continue to rely on revenues separate from the military budget for large parts of their income. A decade ago, as much as a third of the armed forces' revenue was off-budget, and that problem has been poorly tackled since and has not fundamentally changed.[51]

Local police officials and military officers not only close their eyes to illegal resource extraction, but at times actively encourage it in order to promote their family businesses. Some representatives of the mill concessions I interviewed in Samarinda, eastern Kalimantan's business hub, even claimed that local law-enforcement officials would make them accept illegally cut timber for processing or the mills would face raids. As one of the logging company executives told me:

> Look, realistically, we have few incentives to comply with regulations. Getting all the permits and licenses takes a lot of time. You have to pay bribes to local officials and to those in Jakarta. And these days, bribes are complicated and unreliable. If we don't pay bribes, it will take two years to get a license. And then what? The police or the military will hold up the logs on the river, sometimes for weeks on end, until the timber starts rotting. It's far simpler just to pay off everyone right away.

He went on to bemoan how corruption was far simpler during the Suharto era, with a 10 percent standard rate for everything. "But these days, the military are angry that the police are getting a cut too, and they're both jealous of who gets to be paid more. And yes, the coast guard and the navy make money off the coal exports." Such complicity and impunity debilitate regulatory policies.[52] This is particularly so in deeply corrupt systems, like Indonesia's, where big violators often hold great political power, sometimes by being members of Indonesia's parliament or local administrations. Such persons are rarely arrested, and even then they can avoid punishment by paying bribes.

Efforts to develop effective and equitable regulatory frameworks have also been complicated by overlapping and competing bureaucracies, unclear regulations, poor local management and administrative capacity, and lack of clear land titles. Short-term assignments meant to discourage corruption guarantee that officials find themselves having to perpetually learn about local issues. Paradoxically they often have an incentive to make as much money as fast as possible before they are sent to a less lucrative posting.

Yet sometimes successes come in the most surprising places. In the otherwise desperate Lake Abiata-Shala National Park in Ethiopia, an official's intervention managed to reduce the scale of illegal logging for charcoal, perhaps as much as by 80 to 90 percent from its peak several years before. The park's warden invested a lot of time educating the local population about the desertification caused by deforestation, emphasizing concepts from Oromo culture, which frowns upon the cutting of trees. This cultural inhibition, plus the lack of established charcoal trading networks, resulted in a suppression of logging by local Oromo. When members of the Kambaata group who worked as tree cutters and charcoal makers started arriving in a nearby city which specializes in charcoal production, local Oromo cooperated with the warden in keeping the Kambaata cutters out of their area, though not without some conflict.[53]

In addition to a local community's economic horizons and its consumption preferences, a factor that critically influences whether satisfactory environmental protection can be achieved, is population growth among those permitted to live in the protected area. In Abiata-Shala, for example, only thirty families originally resided within the park when it was established. Extensive population growth among the original community, as well as new arrivals, has created unsustainable human-to-wildlife population ratios, with several hundred families now living in the park.

Similarly, local people were allowed to stay in the Bale Mountains and Simien Mountains National Parks in Ethiopia when they were being established in order to protect native rights and avoid forced relocation. Yet in both parks, environmentally unsustainable problems have arisen as a result of high birth rates. The parks are now being grazed and logged to the ground, with land being turned into agricultural fields inhospitable to the most vulnerable endemic species, such as Ethiopian wolves and gelada baboons. Many of the people involved are often excruciatingly poor and marginalized. Ecotourism does not generate sufficient employment. In Abiata-Shala, the park warden I interviewed regarded redesignating the park's boundaries to make it smaller and relocating people as the only way to save the reserve. However, the local Oromo community mobilized against the move.[54]

Many critical conservationists who oppose resettling communities outside protected reserves argue that notions of areas free of human presence are a deeply misguided Western bias. They contend that "European, colonial, and NGO visions" of African forests and savannahs as lands without people are historically inaccurate, with many current "pristine" environments shaped by human activity over millennia.[55] Such inaccurate bases of policy are not only the domain of Western conservationists, however. In *Forest Guardians, Forest Destroyers*, Tim Forsyth and Andrew Walker show how simplistic and inaccurate characterizations of the Hmong of northern Thailand as forest destroyers and of the Karen as forest protectors by the Thai government and mainstream Thai media led to ineffective and counterproductive policies.[56] Thus they maintain that the presence of local communities is not as harmful as often claimed.[57] Moreover, scholars of this "new ecology" see natural environments as inherently dynamic.[58]

The evidence is at best mixed. Some studies show clear linkages between land degradation, particularly in semi-arid areas, due to the overstocking of livestock and communal grazing, and bush fires.[59] Other factors, such as climate change, also impact the environment negatively, though attributing causality to each factor is difficult. Although the conversion of forest to agricultural land clearly degrades the environment, there are intense debates about the negative effects of pastoralism and limited hunting.[60]

Buffer Zones and the Limited Economic Exploitation of Protected Areas

One way to reduce the economic deprivation that poor communities suffer from land being given over to conservation is to create buffer zones around protected areas in which local populations are allowed to engage in limited resource extraction. These buffer zones have also met with mixed success in terms of environmental outcomes, community compliance with regulations, and official law enforcement.[61]

LOCAL COMMUNITY INVOLVEMENT

In some parts of Nepal, the creation of buffer zones around national parks where local communities can collect grasses for thatching and other products, combined with transferring a portion of park revenues to local communities, has had positive effects. Rates of illegal resource extraction and human encroachment are lower in areas with buffer zones.[62] Indeed, forest wardens, community members, and conservationists all prefer the current policy to the previous "fence and fine only" approach of excluding forest communities from protected areas, removing their customary rights, and completely restricting their use of forest resources. Such a policy prevailed when the government was the sole owner of forests.[63]

Yet many problems beset Nepal's current regulatory arrangements and affect the effectiveness of buffer zones. In theory, local communities are given up to 80 percent of national park revenues. However, in the southern Terai region and the Himalayas in the north, local community representatives reported receiving far less and in some years nothing, with park revenues reportedly diverted to other priorities (and possibly for personal profit).[64] Frequently, neither the natural resources available in a buffer zone nor alternative livelihood projects for communities living in or around national parks have been sufficient to ensure subsistence needs are met, nor have they prevented continued illegal resource extraction deeper in national parks. Forest communities in particular are often dependent on illegal timber and the extraction of other non-timber forest products for basic livelihoods.

Particularly in the Terai, local community members frequently complain about forestry officials and park wardens not being responsive to their needs, and of the limited accountability of village development committee (VDC) members. VDC members are often local elites, with women and disadvantaged groups rarely represented. Since committee members can obtain significant political access by virtue of their position on the committee, and hence cultivate their own patronage networks by selectively disbursing park revenues and compensation to their clients, conflict between VDCs and the broader community is not infrequent. The fact that as of the end of 2016 no local elections have taken place in Nepal since 1997 compounds the lack of accountability at the local level.[65] Forest and buffer zone communities thus lack power to modify existing rules and function under constraint from park management bodies.[66]

Poaching and the illegal trade in wildlife, while serious problems in Nepal, have been less intense than elsewhere in South and Southeast Asia—perhaps as a result of the deployment of the military in national parks to enforce regulations. Yet enforcement away from protected areas tends to be meager, with minimal punishment for occasionally captured traffickers and their political patrons. Interestingly, elephant poaching is very limited, despite the fact that wild elephants often cause substantial damage to local communities and frequently kill villagers,[67] and despite the fact that Nepal sits next to China, with its voracious market for ivory.

Although illegal logging in Nepal has reached nowhere near the level of industrialization and intensity that it has in Indonesia, its scale is nonetheless substantial.[68] By some accounts, 30,000 hectares of forest were destroyed in 2010 alone.[69] A research survey by the conservation group Resources Himalaya estimated that four trees were cut in Terai every 20 minutes.[70] *Saal* (*Shorea robusta*) is the timber species primarily logged in the Terai, while various pine species are illegally cut in the Himalayas. The cut timber is both consumed domestically and exported—in the north to China and in the Terai to India. Endangered and highly-prized sandalwood is also smuggled via Nepal from India to China.

Indeed, illegal logging in Nepal likely employs greater numbers of people than the illegal cultivation of cannabis and opium poppy. This particular imbalance in labor between illegal economies is unusual as most of the time drug cultivation is by far the most labor-intensive of illicit economies. The fairly high labor intensiveness of logging in Nepal can be partially explained by the fact that tree-cutting methods in Nepal remain archaic, with trees felled by handsaw instead of chainsaw and removed from the forest by bullock carts. Nonetheless, sawmills are emerging in southern Terai near areas of particularly intense illegal logging, and a switch to the chainsaw and hence a rapid expansion in the scale of illegal logging can be expected. In contrast, mountaineering and tourism in the Himalayas in the north have significantly improved the economic status of the Sherpa community. Illegal logging and wildlife poaching are also reported, but at far lower levels than in the southern Terai.[71]

In buffer zones and possibly even in core parts of protected areas, it may make good sense to allow the limited hunting of non-endangered species and the limited sustainable extraction of natural resources in order to mitigate food insecurity and income loss among poor local communities. In Burkina Faso's Nazinga Game Ranch, for example, local communities are allowed to hunt no more than 5 percent of the wild animals in and around the ranch, a policy with apparent conservation success. Indeed, protected areas that increase human wellbeing through sustainable use tend to be correlated with better environmental outcomes.[72]

Thus it is crucial to make case-by-case assessments of what type of hunting is sustainable and at what levels. Careful monitoring and reassessment need to be conducted regularly to reassess whether wildlife populations and ecosystems are bearing up well, since the level of impact may change over time—whether as a result of an increased humans-to-land-and-wildlife density or the arrival of external poachers. However, even limited hunting requires monitoring by scientists and wildlife law-enforcement officials, and this is resource intensive. Often limited hunting will be the right policy, and it should be the default policy. However, such policies come at the cost of complicating law enforcement since the regulations require sorting out what hunting is legal and what is not, and which hunted animals that end up in local markets were hunted with permits and which not.

Voluntary Resettlement?

In places where hunting exceeds sustainability or where the presence of local communities results in the conversion of land to agriculture, the most contentious issue—resettlement—arises again. In March 2016, for example, the NGO Survival International, a supporter of the rights of indigenous groups, alleged that wildlife law-enforcement units in Cameroon razed to the ground the camps of Baka hunter-gatherers to get them to move out of the protected areas where they traditionally lived.[73] Such conflicts have flared up repeatedly around the Serengeti since 2009. In Zimbabwe, Grace Mugabe, wife of the country's President Robert Mugabe, and who has extensive investments in ecotourism, was accused in 2015 of orchestrating the forcible expulsion of local communities from her private ranch by burning their houses and property.[74] Burning down the houses of those who encroach on protected land is also a common practice in Zambia.[75] Disputes lasting over two decades about the displacement of communities from the Mount Elgon National Park flared up again in Uganda in 2016.[76] Conflicts over land use and resettlement have also increased in Thailand. Since 2014, the Thai government has arrested well over one thousand people on charges of land encroachment, and it has also cut down their rubber plantations in forests.[77] In India's Kanha Tiger Reserve, famous for its declining tigers and for being the presumed setting of Rudyard Kipling's *The Jungle Book*, members of the Baiga and Gond tribes have been forcefully evicted from their land in the name of tiger conservation. The eviction was carried out despite the tribes having legal title to their land and despite Indian law mandating that people can only be relocated for conservation reasons if they consent to resettlement. Such evictions continue to take place elsewhere in India, such as at the Similipal Tiger Reserve, with promises of food, livestock, and land as compensation only sometimes materializing.[78] The list goes on.

Critics of resettlement policies have highlighted not only issues of human rights and economic livelihood, but also problematic environmental outcomes. Communities have killed wildlife out of anger or in protest, or to remove the reason for creating protected areas, such as in Kenya's Amboseli or Uganda's Mburo National Parks. Cases of farmers or land owners preemptively cutting down forests or extirpating endangered species on their property to avoid restrictions on their land use have also occurred in the West, from the United States to Norway. Moreover, if resettlement is done in such a way that resettled communities do not have assured legal livelihoods, resettlement can compound the unsustainable use of natural resources outside protected areas and cause greater environmental degradation.[79]

Yet resettlement has occasionally been done in a way that does not exacerbate these problems. The relocation policy in India's Bhadra wildlife sanctuary stands out as a beacon of hope.[80] While living in the reserve, the local community frequently suffered damage to their crops and homes, and even faced risks to their lives, from encounters with wildlife. The compensation provided after relocation was considered fair: with input from local families, households were given irrigated fertile land. The area to which they were relocated had a water supply, schools, electricity, and markets, and many residents appreciated the vastly improved opportunities for their children. The distribution of housing, land titles, and access to new resources was considered equitable. And since the relocated people were primarily farmers, they did not have to alter their lifestyle. Interaction and consultation between the local people and the Indian state and NGOs supporting the relocation were frequent and sustained after the move.

But the distressing stories continue to drown out the good ones. Established in 2001 and connecting Kruger National Park in South Africa with protected areas in Mozambique, the Great Limpopo Transfrontier Park was to be a showcase of environmental conservation and human rights. Community relocation was to be voluntary, and the relocated groups were to benefit from ecotourism and other alternative livelihoods, and from the delivery of schools, health services, and public transportation. Both environmental benefits and economic benefits for local people did materialize to some extent, though not as much as hoped. Some local people benefitted economically from the free movement between Mozambique and South Africa that the park brought to the communities, and from investments in dryland agriculture.[81] Many employees of the park and private concessions surrounding it were able to access new job opportunities as wildlife guards and trackers, and as workers in lodges.

But some communities could not adequately access the alternative livelihoods proffered. The park's environmental successes also increased people–wildlife conflict, with a rise in lion attacks on livestock and crop destruction by elephants. Only some of the damage was compensated for by park authorities.[82] Moreover, many of the Mozambican villages relocated outside the park became hubs of intense rhino poaching in Kruger, with some of the previously described "local rhino kingpins" originating from them.[83] However, a crucial question concerns how many members of local communities would have ended up joining poaching gangs even in the absence of the park's establishment and the resettlement policies related to it.

In Ethiopia's Simien Mountains National Park, another prominent relocation effort took place several years ago. The local warden at the time persuaded the United Nations Educational, Scientific, and Cultural Organization (UNESCO), which normally does not provide money for relocation efforts, to pay to relocate some 150 families outside of a key wildlife corridor, though not altogether out of the national park. The warden demonstrated that each family kept several farms for itself in the national park. The government of Ethiopia also built a new school and clinic in the area to which people were relocated. The local community was persuaded to cooperate by being told that otherwise its members employed in ecotourism as guards, guides and mule operators, and in the lodges, would lose their jobs, since tourists would not come to see a ruined park.[84]

The ecotourism project generated only limited resources, and only because it was "forced" onto the tourists, who would otherwise not have hired guides, finding them of little use. Though they often had to take two years of tourism lessons after high school, the guides were not motivated to spend time in the park and gain knowledge about it, and by and large they were unable to identify species beyond four big mammals and five plants. To persuade tourists to hire them, they claimed that it was mandatory to have a guide, even though legally it was not the case, and only the park guards (known as scouts) were mandatory. Tourists who hired guides (the vast majority are young men) paid them about $10 per day, of which the guide was supposed to pay $2.50 in taxes. In the high season of September to January, each guide worked perhaps fifteen days a month, while during the rest of the year they tended to work only seven days a month. Even so, the guides claimed that working in the tourist trade in the area was a good job, second only to working for the government. In fact, the land and soil in the park and surrounding areas were so degraded that farming had become unprofitable, with local farmers struggling to meet subsistence needs. Many were keen to move out, and young men had been leaving the area for South Africa if they could, or even for Sudan.[85]

I found that the park's official scouts (about seventy in all), whom tourists were required to hire, were equally bored and unmotivated, not bothering to enforce rules against illegal activities in the park by local people, who were often their relatives.[86] Trotting around with guns, many were former soldiers who earned $4 a day when they managed to snatch a foreign tourist to "protect" in the park. Although they could not articulate at first what they were protecting the tourists from, they eventually declared that they were protecting the park and tourists from a possible invasion by Eritreans. Still, however questionable the ecotourism employment scheme in the Simien Mountains National Park is—and however extensive the habitat destruction by the local community which resides in the park and now numbers in the thousands—the park can claim an important conservation achievement: the population of the endemic Ethiopian ibex grew from 150 in 1991 to about 900 in 2010, with about 100 endemic Ethiopian wolves also present. Compared to other protected natural areas in Ethiopia, the park is the darling of international donors. At the time of my research in 2013, a new Simien Mountains National Park Integrated Development Project sponsored by the Austrian government was planned, which was to include new headquarters for the warden, new housing for scouts inside the park, a training program to teach farmers how to grow highland food crops, and a program of planting eucalyptus to help stop the felling of the last remaining native trees, which are used for charcoal.

Similarly, in Nechisar National Park, Ethiopia's government has tried to persuade the resident pastoral communities to settle down and engage in high-intensity agriculture to reduce pressure on land and animals. They were told that if they settled they could send their children to school and take advantage of medical services. While it is unrealistic to persuade these communities to abandon cattle rearing altogether since cattle ownership is seen as a symbol of prestige and wealth, a mixture of intensive farming and more limited cattle ownership could ease the pressures on the reserves caused by full-time pastoralism. At the time of my research there in 2013, the environmental results of the government effort were yet to be seen.[87] Meanwhile, for many human rights advocates and critical scholars of conservation, Nechisar has become a *cause célèbre* and high-profile case for opposing the injustice of eviction and resettlement. A few years ago, a prominent environmental NGO, African Parks, agreed to manage Nechisar but put a clause in the contract that it would only take over once community resettlement outside of the park was complete, so that it did not have to get involved in complex political issues concerning human rights, compensation, and alternative livelihoods.[88] Opponents of resettlement policies decried the arrangement as passing the buck of abuse onto local authorities and the national government instead of showing a strong commitment to ensuring resettlement programs were done properly.

The issue of voluntary resettlement also looms large in the drug field. Historically, drug-crop farmers have been displaced involuntarily as a result of eradication of their crops or violence perpetuated against them by drug-trafficking and militant groups. However, many areas where coca or opium poppy are cultivated, such as those deep in the jungles of Colombia and water-poor areas of Afghanistan, simply do not allow for the development of legal economic activities. The provision of legal alternative livelihoods is unviable. However, getting communities to relocate without perpetrating further violence against them is challenging—not only because of their attachment to the land, but also because many lack the skills and political connections needed to compete for jobs with residents in the areas to which they are expected to move.

Transferring Resource Ownership to Local Communities

Community-based natural resource management approaches often employ many of the same methods to generate revenue and provide legal livelihoods to local communities—such as ecotourism—but they always transfer rights to local communities and devolve power. The extent of the transfer of rights and devolution of power is a matter of philosophical and political predisposition and particular designs. Rarely have local communities been given full authority over lands and local natural resources. But the basic idea is that if local communities own local wildlife and resources they will manage them well and sustainably.

Evaluations of the environmental outcomes of CBNRM are surprisingly sporadic and controversial. Many assessments focus mainly on two objectives of CBNRM—social and political empowerment and poverty alleviation—and less on biodiversity enhancement. Even so—whether positive or negative—usually they are hotly disputed.

The CBNRM environmental success stories most often pointed to are Zimbabwe's CAMPFIRE program, the devolution of control over natural resources to local communities in Namibia, several programs in South Africa, and the case of a local community's ownership of vicuña herds in Peru.

LOCAL COMMUNITY INVOLVEMENT

The CAMPFIRE program was established to protect an area of land used for wildlife production by communities that is roughly equivalent in extent to Zimbabwe's state protected areas. Although the overwhelming trend in Zimbabwe is toward the conversion of wild land to farmland, the twelve CAMPFIRE sites covering some 40,000 square kilometers have maintained substantial areas of land for wildlife. In some areas under CAMPFIRE schemes, wildlife populations increased by 50 percent, with elephants doubling to 8,000 by 2003.[89]

Economically, CAMPFIRE has not done badly either. In some years, the economic returns of CAMPFIRE amounted to $390 per household, surpassing potential income from cotton by several times.[90] However, the economic utilization of wildlife, including the hunting of elephants and trade in their body parts, was very controversial with some conservationists, as 90 percent of CAMPFIRE revenues have come from trophy hunting, not ecotourism.[91] Although CAMPFIRE generated some $20 million in revenue for local communities and district governments between 1989 and 2001,[92] this was still less than that received by way of donor support, which amounted to at least $35 million between 1989 and 2003.[93] And although the original proposal of CAMPFIRE was to fully devolve rights over land and resources to communal conservancies, such devolution did not materialize.

Moreover, the environmental outcomes have varied, depending on wildlife population densities and local communities' interests. Some communities invested some of their revenue in agriculture and in livestock that compete with wildlife.[94] Some members of the CAMPFIRE effort wanted wildlife to retreat and more people to come in, as well as more shops and services.[95] In other words, some communities simply wanted more wildlife conservation, while other communities wanted greater access to elements of modernity and a reduction in wildlife populations.

The Namibian case is a more straightforward success story. There, the establishment of communal conservancies based on CBNRM models led to the recovery of black rhinos, elephants, Hartmann's zebra, and other wildlife species in the country's north. Without the communities' commitment to wildlife conservation, the black rhino, for example, would have been extirpated from communal lands.[96]

CBRNM efforts in Namibia now include fifty communal land conservancies, covering over 14 percent of the country's land, and earn

$2.5 million a year from ecotourism and trophy hunting.[97] The distribution of game meat from hunting operations is a key element. Resources are shared equitably within the community. The rights of communities to conservancy land and wildlife have been conditional and can be revoked, but they are not limited by a specific amount of time, such as a ten-year program, for example. No state tax has been placed on the conservancies, 100 percent of the revenues being retained by local communities.

Conditions in Namibia may be uniquely auspicious due to the low people-to-wildlife density and high aridity favoring wildlife over agriculture. Biodiversity is high, and the ecosystem features iconic mammals that many tourists are willing to pay substantial amounts of money to see and hunters to hunt.[98] Corruption in Namibia is also comparatively low. Finally, in various parts of Namibia where CBNRM programs have been implemented, such as the Kunene region, local communities place high value on wildlife preservation and its retention for future generations. Even so, competition for land use from agriculture and livestock has not been absent: not all communities embrace CBRNM uniformly, and such factionalism has threatened some projects.[99]

While poaching by organized wildlife-trafficking gangs has not been anywhere as intense in Namibia as in South Africa and East Africa, it has emerged there, threatening some of the successes achieved by CBNRM. In 2014, at least twenty-four rhinos were poached, while eighty were poached in 2015.[100]

Kenya's northern Laikipia district, including the Lewa Conservancy, is the exemplar of CBNRM in that country. Zebra numbers have increased fivefold between the 1960s and 1990s, the African wild dog population has likewise increased, and lions and hyenas remain widespread. With local communities (as well as private ranchers) setting land aside for wildlife, Laikipia has developed a strong wildlife tourism industry.[101]

The Makuleke community in South Africa is often pointed to as an exemplar of how CBNRM approaches can right past wrongs, economically and politically empowering communities while preserving the natural environment and allowing for sustainable development. Having suffered forced displacement from Kruger National Park decades ago, the community was able to win its land back in a landmark deal in

1998. But instead of converting the land to agriculture, it worked out an arrangement with South African National Parks, comanaging the land with them and receiving a substantial portion of the ecotourism income from a part of Kruger National Park that receives large numbers of tourists. The result has been lauded as "one of the most advanced programs of community involvement in conservation and wildlife anywhere in the world."[102] It should be noted, however, that in the late 1990s, the post-apartheid government badly needed a land reform success story. Subsequent restitution claims by displaced communities in South Africa have been more complicated and often remain unresolved. Political and economic empowerment did not take place equitably within the Makuleke community either.[103] Nor has the Makuleke area emerged as more resistant to outside poaching pressures than other parts of Kruger National Park.

Moreover, for all of the (at least partial) successes, there are also failures. In some areas, additional income from CBNRM projects resulted not only in greater pressures on land conversion for agriculture or greater livestock herds, but also in greater poaching, with revenues from CBNRM used to buy better weapons to make hunting more efficient.[104] In Peruvian Amazonia, CBNRM approaches centering on ecotourism led to highly divergent outcomes. While some wanted more conservation and ecotourism, other communities wanted to use ecotourism income to buy chainsaws, which they could not afford previously, thus intensifying deforestation.[105]

Proponents of CBNRM often posit that the uneven outcomes reflect not the failure of the CBNRM concept, but rather failures of implementation. Stressing that CBNRM approaches have been executed only in a small number of isolated cases, limited in their scope, and inadequately integrated into overall conservation efforts, they highlight the following obstacles.

One, those with power over setting regulations and ensuring the devolution of rights and resources do not give local communities enough rights and powers, retaining power themselves and failing to address weak property rights. Hence decisions are continually made by outsiders, while the local community's voice is limited, and its stake in and ability to manage local resources properly remain constrained. In addition, local communities do not get to keep a sufficient portion of revenues, due to high taxes or arrangements such as centralized trusts.

Two, elites within a community disproportionately capture revenues at the expense of the rest of the community, sometimes creating social tensions. On the Narok side of the Maasai Mara, the richest quartile of households captures between 60 and 70 percent of wildlife-based revenues while the bottom quartile gets only 15 percent.[106] Plagued by corruption and a lack of accountability, pastoralist group ranches—the name for a property arrangement established in Kenya after independence and assigning land ownership not to individuals, but to communities, such as a local Maasai community—have also repeatedly failed as communal governance institutions, leading local communities to seek individual land rights or at least smaller collective arrangements.[107] In Tanzania, too, corruption has pervaded not just local government institutions but also local informal village governing structures. Decentralization alone thus is not a perfect solution: decentralization may merely decentralize—and complicate—corruption. Additionally, local community structures may not only be unaccountable and inequitable, they may simply be weak.

Proponents of CBNRM argue that under such circumstances, the local community should receive outside assistance in developing accountability structures. Although vital, it is a long-term and enormously politically contentious process. Local elites—traditional, governmental, and new—will resist. Empowering the previously marginalized thus may require the long-term involvement of outsiders. Local communities also sometimes lack the technical and managerial skills to design and operate wildlife-based enterprises. Partnerships with outside consultants or big eco-businesses may be required; sometimes these work well, at other times they leave the community with disappointing revenues.

Three, how revenues are distributed also makes a big difference to the level of satisfaction local communities derive from CBRNM and the development and poverty alleviation that ensue. In southern Africa, revenues are often held in trust funds instead of disbursed as direct cash payments.[108] While this may allow for the development of public goods such as schools, clinics, sanitation, and transportation, it limits households' decision-making capacity. Consequently in some areas, such as Namibia, communities prefer direct cash payments.

Four, the greater the scope of economic revenues that a local community can obtain from wildlife, the greater the resilience of such revenue streams and the larger the benefits. Thus, bans on trophy hunting can undermine local community support for conservation. For example, when Botswana banned all trophy hunting in 2013 because of corruption in the trophy-hunting sector, it significantly curtailed the income of local communities from wildlife management. In Kenya, bans on trophy hunting have worsened land conversion to agriculture,[109] and failed to prevent poaching and wildlife trafficking. An international ban on the ivory trade—desirable as it may be for law enforcement and demand reduction—will thus negatively impact some CBNRM programs, such as the iconic CAMPFIRE.

Just as in the case of ecotourism and alternative livelihood projects, pastoralism and agriculture often bring far greater economic revenues than livelihoods derived from wildlife conservation, particularly if large charismatic animals are lacking or are not easily visible, such as in forests.[110] In areas with the greatest numbers of tourists, such as the Maasai Mara, that ratio can be reversed, with wildlife-based tourism bringing in more money than livestock. Similarly, if the land and climatic conditions are not conducive to pastoralism or agriculture, wildlife can bring in more money, as in Namibia.

CBNRM proponents thus also need to grant that some local communities will not have a strong interest in conservation. Their preference may be to make money quickly through unsustainable resource use, possibly involving wildlife trafficking.

In sum, local community ownership as a solution to poaching and environmental degradation is mostly only as good as the quality and strength of community leadership structures and the system of accountability, a community's cohesion and internalization of norms and recognition of the rule of law, the system and enforcement of property rights, the balance of consumption and non-consumption (spiritual) attachments and attitudes toward nature, the balance of long-term versus short-term economic horizons within the community, and the power of the local community vis-à-vis hostile or competitive outside interests. The basis of land and forest distribution to local stakeholders, and the balance between pure political patronage and the soundness of environmental regulations and their diligent enforcement, are also critical.

THE EXTINCTION MARKET

Conditional Cash Transfers: Paying People Not to Poach

In circumstances where ecotourism and alternative livelihoods are not viable, CBNRM efforts are not delivering robust outcomes, and where poor local communities are not able or willing to move outside protected areas, it is time to consider conditional money transfers to prevent environmental degradation. Marginalized people could be compensated for preserving instead of unsustainably utilizing nature. Direct economic assistance is designed to reduce the economic hardship of compliance with environmental regulations and the costs of law enforcement. Such direct cash payments have been proposed before,[111] and adopted in various forms by diverse countries, including Sweden and Cambodia. The outcomes have varied, and there are many challenges to effective design.

In fact, the money I paid to park guards and tourist guides in Ethiopia was a form of direct cash payment. The scouts and guide hardly provided any function at all. Their guiding and protection services were a mask for paying them not to perpetrate environmental violations, such as poaching and converting land to agriculture. Indeed, within Ethiopia's highly centralized state system, such state-imposed or ersatz state-imposed tourist guides were ever present, and all possible tourist sites, such as a town's fish market, could be visited only if tourists hired them. The guides were linked to the main political party (those who were not members would struggle to get a tourist license or be allowed to operate by party-linked guides). The scheme straddled a thin line between alternative livelihoods, direct cash payment, and state-sponsored extortion. But it provided jobs, and Ethiopia still had far less urban crime than Kenya or Uganda. Whether forcing guides on tourists in national parks reduced poaching is an open question, especially since there was little monitoring.[112]

Models and Outcomes

The Ethiopian approach is hardly a satisfactory model. However, there are other useful precedents and conceptualizations of economic incentives to discourage undesirable environmental exploitation. Compensation for injury and damage from wildlife is already a standard practice,

LOCAL COMMUNITY INVOLVEMENT

even if often inadequately provided. It obviously differs from paying people not to poach, but the two are related. In Africa and Asia, rural communities often kill animals to prevent damage to their herds and fields. Such damage can be exaggerated, but for poor communities, it can indeed represent substantial losses to their income and exacerbate their poverty.

In Kenya, for example, for several decades the government only provided compensation for people killed by wildlife. In December 2013, a new Act on Wildlife Conservation was passed to motivate greater community buy-in to conservation. It seeks to do so by increasing consultation with local communities and encouraging local ownership of wildlife management. Under the new environmental law, people can claim compensation for damage and destruction to property. Such compensation schemes are important since some 60 percent of wildlife populations often live outside protected areas, in places where they come into conflict with people and their livelihoods. Given the wide geographic distribution of animals outside of protected areas, conversion of all communal land to agricultural use would be a disaster for wildlife—in Kenya and elsewhere. It is crucial that such conversion is avoided, compensation to land owners and local communities being one mechanism to accomplish this, as is the practice in the United States.

Of course, how effective a compensation scheme is will depend also on how it is implemented. Important determinants include whether people feel satisfied with the speed and efficiency of the compensation process and the fairness of the amounts of compensation, as well as whether there is cheating, either by claimants or by inspectors. Given pervasive levels of corruption in Kenya, there are reasons to expect many problems. Thus in the 1990s, the Kenya Wildlife Society established a similar compensation scheme for damaged crops, but it was soon discontinued for being too slow, cumbersome, difficult to administer, lacking in funds, and pervaded with fraud and corruption.[113]

In Colombia, monetary compensation for legal crops destroyed by aerial spraying of illegal coca crops nominally existed during the first decade of this century when spraying was intensively practiced. Yet many loopholes existed in the process, and farmers could rarely overcome difficulties in proving title to their land or documenting the damage. Moreover, since some of the legal crop losses resulted from *cocale-*

ros intercropping coca with legal crops to hide the coca and deter spraying, they were disqualified from compensation. The result was that only a very small percentage of farmers were compensated.[114]

Small conditional grants of money have been used as a policy tool for several decades. Seeking to enable development and sometimes change behavior, they can center on individuals or communities. Some such programs, such as Brazil's Bolsa Familia, are considered a great success and have been widely adopted around the world. The results of others have been more problematic and far more controversial. Of course, verifying that people do something—such as show up in schools or at hospitals—is far easier than verifying that people are not doing something hidden, such as poaching inside forests. Problems of moral hazard and ethical objections loom large.

Deforestation and Compensated Poppy-Eradication Experiments

At the global level, prominent examples of cash transfers based on conserving nature include cash transfers to Brazil and Indonesia so that they do not cut down their rainforests.[115] Such "payments for ecosystem services" (PES) are meant to generate monetary incentives for nature preservation by accounting for hidden services delivered by well-functioning natural ecosystems. PES is premised on the fact that ecosystem services are valuable but unnoted and unpriced public goods providing undervalued positive externalities.[116]

PES outcomes too have been mixed. Reduced deforestation has not systematically materialized in developing countries, and much of the evidence is weak and lacks randomized cases and controls.[117] Yet the general concept has now been accepted and was embraced at the Paris climate conference in December 2015.[118] Throughout the world, communities are being paid not to deforest, not to drain watersheds, and not to convert rare habitats to industrial or agricultural uses—with widely different outcomes. A crucial difficulty with carbon off-set markets to reduce deforestation under schemes for reducing emissions from deforestation and forest degradation (REDD+) has been that the level of compensation is too low compared to alternative forms of production, such as timber extraction.[119]

LOCAL COMMUNITY INVOLVEMENT

In the case of drugs, schemes to pay farmers not to cultivate illegal crops have been largely unsuccessful. Of course, there is an important difference between PES schemes, such as REDD+, and paying people not to cultivate illegal crops: PES schemes provide large positive benefits that preservation of the ecosystem should bring (climate-change prevention, hydrological functions and water purification, soil erosion prevention, and so on), which are not priced by the market itself but which can be negotiated between a willing buyer and seller. Companies thus pay a tax for undermining these benefits. By some estimates, such ecosystem services were valued at $33 trillion per year more than a decade ago,[120] a number that has likely grown, and not simply in absolute dollar terms.

With schemes to pay farmers not to cultivate illegal crops, compensation is more or less only paid to prevent negative behavior, without necessarily preserving some hidden positive functions that benefit all. Moreover, drug-control policies that use compensation schemes do not require the same degree of collective management and collective decision-making as the management of water and soil resources.

One of the most ambitious drug-control schemes was implemented in Afghanistan in 2002. The United Kingdom promised to pay $350 in compensation for each *jerib* (half an acre) of opium poppy eradicated. The British government dedicated $37 million to the scheme, and another $35 million came from international aid.[121]

Major problems soon emerged. The money was not distributed to farmers directly, but given to local authorities. More often than not, these authorities simply pocketed the money. Local commanders also demanded bribes from farmers who wanted to evade eradication. As a result, the 2002/3 eradication campaign generally targeted only the poorest and most vulnerable farmers, while the plantations of the wealthy and influential were not destroyed.[122]

Other forms of fraud emerged. Local commanders claimed to have much larger areas of land under cultivation with opium poppy than was actually the case. In a context of high insecurity, there was no independent verification procedure. And local commanders benefited from the scheme by pocketing vast sums intended for compensation, by collecting bribes to forego eradication, and by strengthening their political capital with the landlords and farmers whose fields they spared from eradication.

Not surprisingly, the money quickly ran out. As a result, the amount of compensation per *jerib* shrank steadily from $350 to $40.[123] At that level, the promised compensation was insufficient for farmers to make ends meet without continuing to grow opium poppy. Moreover, the scheme did nothing to provide farmers with viable economic alternatives or to replace the economic structures, such as the opium-based microcredit system, that made poppy cultivation so attractive.

Moral hazard also emerged as a major problem of the scheme. Not only did local authorities over-declare the area cultivated with opium, but farmers apparently also started growing more than they would otherwise have done to obtain larger amounts of compensation. Paying compensation for eradication thus increased the extent of cultivation. The compensation scheme was widely considered a fiasco and suspended.

What is the evidence of the performance of direct payments for environmental conservation efforts? The concept is being instituted in several places. China, for example, pays tens of billions of dollars for "ecological compensation," such as to farmers to convert marginal cropland to forest or grassland under the so-called "grain for green" scheme.[124] In Sweden, where the scheme is considered effective, villages are given payments of around $30,000 for each birth of a wolverine or lynx in reindeer-herding areas as a mechanism to dissuade herders from killing the potential predators of their livestock.[125] In the northern plains of Cambodia, direct payments have been used to protect birds' nests since many local species were threatened by egg and chick collectors. So that they do not harvest the eggs and young birds—and so that they protect the nests—between 71 and 78 percent of local people have received a total of $30,000 in an area of 2,000 square kilometers where some 2,700 nests are to be found. Compared to control sites, the payments have significantly improved nest protection and bird fledging, and resulted in the population increase of at least three, though not all, species. For beneficiaries of the program, the direct payments represented a significant portion of their income. But not all people in the area benefited. Those not receiving payments were jealous and deliberately disturbed nests simply to vent their frustration or get on the payroll.[126]

LOCAL COMMUNITY INVOLVEMENT

Design Elements for Success

The objectives and criteria of compensation schemes would be to allow local communities to support environmental protection and law enforcement while preventing their suffering severe economic hardships. Establishing a fair level of compensation may be difficult. In principle, people could find any realistic level of compensation as inadequate and could potentially always ask for a higher amount. Nonetheless, if they faced the prospect of potential donors walking away, an agreement might eventually be reached. A fundamental problem for compensation schemes, like alternative livelihood efforts, might be insufficient funds to maintain them.

Establishing an eligibility cutoff would be politically and ethically difficult, but such decisions are no different from establishing eligibility criteria for social benefits, such as unemployment assistance, or disarmament, demobilization, and reintegration programs for former fighters.

How long to maintain such a compensation scheme would be another crucial question. Once eligibility is determined, will it ever be reviewed? Might there be circumstances when more people were permitted to claim compensation or some were disqualified from receiving it because their poverty level had declined? Such details would have to be designed on the ground and experimented with and adjusted.

It would be important to couple compensation schemes with regular community engagement and program assessments, environmental education, and other efforts to promote alternative livelihoods. Such technical assistance and community engagement should not be different from those employed within CBNRM programs, and, in fact, all alternative livelihood schemes. Providing financial incentives to combat poaching would also have to be coupled with diligent and extensive law enforcement. Those who persist in poaching would face suspension of compensation payments and perhaps other penalties.

Compensation schemes to prevent poaching are fraught with problems and challenges. Careful monitoring, assessment, and adjustments, and even suspensions, will be required. Some attempted schemes may have to be discontinued. However, if the social costs of conservation are high, if local community ownership is not producing environmentally sustainable outcomes, and if alternative livelihoods are not available, the cost of experimenting with such policies is unlikely to be prohibitive.

A decision not to compensate at all under such circumstances would imply two outcomes. One is to give up on conservation altogether and allow the local community do what it wants with nature, even if it means that the local community decides to hunt out, log out, or graze out local natural resources, or join forces with outside wildlife and timber traffickers. The second outcome is simply to rely on the brute force of law enforcement, even though that may mean very harsh and unjust economic privation. Both outcomes are undesirable. Given the paucity of effective conservation models working consistently in varied settings and withstanding external shocks, such as industrial-scale poaching driven by global demand, as well as internal shocks, such as changes in local community structures or political systems, foregoing local experimentation and out-of-the box strategies will not advance either conservation or social justice.

9

ANTI-MONEY LAUNDERING EFFORTS

The environmental community is also increasingly turning to anti-money-laundering (AML) measures to deprive poaching of funds and discourage wildlife trafficking. Similarly, drug-policy activists and officials as well as counterterrorism experts have often looked to AML to change drug-trade dynamics as well as to deprive terrorist organizations of income. For example, as drug-related violence in Mexico escalated, Mexico's presidents Felipe Calderón and Enrique Peña Nieto repeatedly suggested that if the United States did more to stop money flows from the United States to Mexico, including bulk cash flows, Mexican drug-trafficking groups would be significantly weakened. Mexico also toughened its own AML measures, limiting transactions conducted in cash and ATM withdrawals, and increasing bank due-diligence and reporting requirements.[1]

Adopting AML measures can have multiple reasons. Calls to increase AML efforts have often been made out of desperation when other measures such as law enforcement or alternative livelihoods have not produced satisfactory outcomes or generated too large negative side effects. They have also been adopted to pass the enforcement buck to another actor in the international system or to emphasize joint global responsibility. Sometimes they are made in the hope of bankrupting the bad actors and deterring illicit activities or terrorism. In other cases there is a desire to recover money stolen from poor countries by corrupt officials. Or there can be an ethical compulsion to deprive law breakers of enjoying the proceeds of their crime and corruption. Moreover, AML measures and financial forensics are yet another intelligence tool for developing a comprehensive picture of smuggling networks and, potentially, of consumers and suppliers.

Evaluating AML effectiveness in suppressing wildlife trafficking, particularly the deterrence effects of AML, is difficult since very few AML tools and operations to combat wildlife trafficking have been launched. Moreover, those prosecutions that have occurred have taken place predominantly in cases where the original wildlife offense was committed in the same country that launched the AML investigation and prosecution—in other words, few AML prosecutions to combat wildlife trafficking have been initiated in situations where the environmental crime takes place across international borders and jurisdictions.[2] In many countries, wildlife trafficking is not a predicate crime for AML efforts. A predicate crime is an action that provides the underlying resources for another criminal act. In the United States, a predicate crime most often pertains to actions providing funds for terrorism financing or money laundering. Many prominent environmental NGOs, such as United for Wildlife and the Wildlife Conservation Society, are lobbying to make wildlife trafficking a predicate crime for AML around the world. The US National Strategy for Combating Wildlife Trafficking, for example, called on the US Congress to consider legislation to designate wildlife trafficking a predicate crime and ensure that seized funding be directed back to conservation.[3] In 2015, the House of Representatives passed the legislation. In addition to changing domestic legislation, environmental NGOs are also seeking to persuade the banking sector to adopt voluntary compliance and best-practice standards and apply due-diligence approaches to the wildlife trade. Still, there is a paucity of direct evidence from the wildlife trade as to the effectiveness of AML efforts in suppressing environmental crime. But important insights as to the scope and types of effects that AML efforts are likely to produce can be gleaned from the extensive experience of AML measures deployed to combat drug trafficking and terrorist financing.

ANTI-MONEY LAUNDERING EFFORTS

Toughening the Regulations

Historically, the United States has emphasized AML mechanisms in its counter-narcotics, anticrime, and counterterrorism approaches, and has insisted that countries around the world embrace similar standards. The United States, for example, helped Colombia adopt an extensive AML regime in the 1990s that is widely regarded as a model for comprehensive AML design.

A major expansion of AML regulations and their enforcement took place after the 9/11 attacks when the United States created an expanded international regime to criminalize terrorist financing, impose global AML standards and aggressively prosecute those who provide material support to terrorist groups—a very broad category of offenses. If an entity was designated as a terrorist group, it would be cut off from the financial system and have its assets frozen. The United States tightened its domestic rules, created new tools to sanction individuals and countries, such as the International Economic Emergency Powers Act and the Financial Crimes Enforcement Network (FinCEN), empowered existing international AML bodies such as the Financial Action Task Force (FATF) and the Egmont Group (an umbrella organization sharing best practices and technical advice on how to counter terrorism financing), and pushed through new regulations, agencies, and instruments at the international level.[4] Many of these moves had been previously resisted by banks and other countries. The United States itself had not diligently gone after the financing of the Irish Republican Army because US banks did not want to absorb the costs of enforcement.[5]

Following 9/11, however, the number of countries ratifying the UN Convention on Terrorism Financing promoted by the United States rapidly grew to more than 150, with the FATF naming and shaming non-compliant countries and territories. Holdouts would be treated as pariahs and sanctioned, often subsequently yielding to the demands for international financial regulations. Thus, the United States insisted that Caribbean countries close down their off-shore financial havens and adopt the same due-diligence requirements that financial businesses located on US territory have to comply with, or else risk being prosecuted for violating US antiterrorism laws or being blacklisted by the United Nations and other international AML regimes. (Many island financial centers have objected that the supposed counterterrorism objective of AML measures is merely a ruse to level the playing field for US businesses, with dubious effects on financial flows to terrorist groups.)[6] Globally, strict due-diligence laws requiring financial institutions to know their customers and report suspicious activities to law-enforcement agencies were put in place. The USA PATRIOT Act of 2001 also greatly expanded the definition of suspicious activities. The goals were to seize terrorist money, slow down and complicate the movement of fresh funds to terrorist groups to make operations more difficult, and deter possible financiers of terrorist groups.

Less than Meets the Eye: Uneven Policy Outcomes

However, despite the immense beefing up of law enforcement and regulatory activities to strengthen AML efforts and counterterrorism financing, the effectiveness of AML measures remains unclear, and is often highly contingent on specific susceptibilities of the target. There is great variation in the effectiveness of AML efforts to constrain dictators, economically squeeze rogue regimes (such as North Korea and Iran), deprive terrorist groups of financing, and weaken criminal groups. For example, efforts to find and seize dictators' money or compel rogue regimes to change their policies through financial sanctions and AML efforts seem particularly ineffective.[7] On the other hand, US officials believe that their efforts to deprive al-Qaeda of money, and in particular halt the very visible financial flows it received from major companies in the Middle East and possibly from other semi-legal ventures, such as the African diamond trade, have been effective.[8]

But such success appears to be more of an exception than a standard outcome. The international community, for example, never succeeded in stopping the flow of money from international financiers who sponsored piracy off Somalia and funded various pirate gangs, despite repeated international calls for such measures. Going after the pirate money should have been one of the easiest tasks for AML efforts to succeed in for three reasons: The scale of the illicit financing was fairly limited compared to drug money laundering or terrorism financing. The number of financiers was relatively small and they were concentrated geographically. Instead, what ultimately suppressed piracy off the Horn of Africa was intensive naval patrolling combined with a significant increase in the robustness of targeted ships' defense measures, such as the so-called "citadels" (safe-room spaces).[9]

However, it is important to note a significant difference between laundering crime money and raising terrorist money. Illicit economies need to turn hot money, often mostly cash, into clean, ideally untraceable money. Paying someone part of that pile of cash to divide and deposit it in a way that avoids triggering AML reporting tripwires requires overcoming very different inhibitions than raising money to murder people through terrorist acts. In many poor countries that are also sources of illegal wildlife products, recruiting people to divide money up and deposit it in numerous smaller amounts is not terribly difficult.

Overall, efficacy rates of AML efforts are rarely assumed to surpass 5 to 10 percent of laundered flows, often catching as little as 2 percent of illicit flows.[10] Money-launderers have a large menu of laundering options at their disposal, such as bulk cash smuggling (a primitive but widely used method, such as for moving large sums of drug proceeds back from the United States to drug cartels in Mexico), "smurfing" (where money is divided into numerous small bundles and deposited in many transactions to avoid detection), currency exchange bureaus, trade-based money-laundering (such as under- and over-invoicing by front companies), purchases of real estate, securities, trusts, casinos, online gambling and gaming, and wire transfers to name a few. New mechanisms include virtual currencies such as bitcoin. Even if the effectiveness of AML efforts increased to 20 or 30 percent, smugglers could still easily pass the costs down to the customers, increasing retail prices, particularly for highly desirable commodities such as drugs or rhino horn, where the profit margins are large. Yet improved effectiveness of AML measures can increase the cost of smuggling, weeding out the least competent money launderers and smugglers and reducing the pool of all available launderers, outcomes that could be valuable.

To the extent that smuggling networks can operate in one currency (usually US dollars) along the entire smuggling chain, financial operations and money laundering are made significantly simpler than if they have to pay money to poachers in one currency, smugglers in another, and then deal with the integration of retail profits in yet another. In the drug trade, many drug-trafficking organizations avoid this headache as they often operate in US dollars. For example, much of the cocaine trade is conducted in this currency, even for transactions outside of the United States. Given the paucity of systematic data on the operations of wildlife-trafficking networks, and particularly on money laundering related to wildlife trafficking, we cannot yet say whether wildlife trade is predominantly conducted in US dollars, like the cocaine trade, or in Chinese renminbi, or whether it involves frequent currency exchanges that complicate money laundering. Certainly in many source countries in Africa and Asia, such as Kenya, Burma, and Indonesia, US dollars are widely accepted in the informal economy.

Looking for dirty money among a myriad of bank and other financial activities is like looking for a needle in a haystack. What is often required for success is coupling financial intelligence with other streams and sources of intelligence. Tip-offs are often necessary for banks and officials to locate illicit money. Financial interdiction, just like physical interdiction, is very difficult, and its effectiveness should not be overestimated.

Moreover, effective AML measures require intensive international cooperation that is frequently lacking. In the case of Islamic groups, the problem is further complicated by the availability of informal fund transfer systems, such as *hawala*, that easily escape monitoring.[11] *Hawaladar*s rely on networks of kinship and trust to move money, including from migrant communities back to developing counties, such as Somalia, but also within developing countries, such as Afghanistan. Money often does not change hands physically, nor is there a traceable record. The *hawaladar*s settle their accounts through a large set of transactions unconnected to a single customer.[12] The United States eventually realized that it was both impossible and counterproductive to shut down *hawala* networks since it would cut off remittance flows to some of the world's poorest, such as in Somalia, and prevent even US companies, aid agencies, and military and government actors from operating in places like Afghanistan. Moreover, illegal money flows would simply move to another mechanism. So instead, the United States sought to regulate the *hawaladar*s, demanding that they keep transparent and complete records and adopt know-thy-customer practices, and cooperate with law-enforcement agents if suspicious money entered their hands.[13] But the extent to which the United States can detect non-compliance by *hawaladar*s and punish those who disobey the rules—particularly in countries such as Afghanistan, Pakistan, Yemen, and Somalia, but also even in Kenya—cannot be assumed to be large.

Informal cash transfer mechanisms also exist in Latin America, such as the notorious "black peso" market. Dating back to the 1970s and never successfully shut down, the market allows Colombian drug traffickers to repatriate drug profits from the United States to Colombia and to do so without any actual cross-border financial transaction. The trafficker merely seeks out a Colombian entrepreneur operating in the informal economy who needs dollars. The entrepreneur then wires Colombian pesos back to the trafficker's account, all within Colombia.[14] Such black peso markets also operate in other parts of the world, including Florida in the United States.

Thus, even in the case of Colombia, despite the robustness of its AML regulatory regime, AML measures seem to have contributed little to reducing the attractiveness of the drug trade and other illegal ventures for drug-trafficking organizations, or to weaken groups' operational capacity. Direct interdiction measures—such as arrests of key traffickers and DTO middle layers—seem to have had a far more pronounced effect in weakening drug-trafficking groups and reducing their capacity to corrupt and coerce. Although Colombia's banking system appears to have been purged of the intense penetration by drug money that characterized it in the 1980s, informal banking and money-laundering systems have emerged and been used by criminal groups.[15] Penetration of illicit proceeds belonging to paramilitary groups and post-paramilitary *bandas criminales* (criminal bands) into Colombia's political system and overall economy also appears to continue unabated.

In many countries that are sources of illegal wildlife products as well as in countries where there is demand, such as in East Asia, banking and financial sectors lack integrity, and are characterized by extensive corruption and a prevalence of money laundering. As of 2005, no country in southern Africa, for example, could report more than ten successful AML prosecutions based on any predicate crime.[16] Financial diligence has improved only slightly since then. Eastern Africa too does not abound with extensive and robust AML best practices nor with prosecutions. At the end of 2013, Global Financial Integrity, a Washington-based financial watchdog, assessed the amount of illicit money entering

Kenya from crime, faulty invoicing, shady businesses, and corruption, concluding that it had increased more than five-fold between 2004 and 2013, and equaled approximately 8 percent of Kenya's economy.[17] Most Kenyan banks had filed reports on suspicious activities only sporadically, and often ignored rules about reporting currency transfers over $10,000. However, after adopting at least procedural compliance (without a significant increase in effective AML prosecution), Kenya was removed from the list of countries at a high risk of money laundering in 2015.[18]

The situation is hardly better in many key countries where there is a demand for wildlife, such as China and Vietnam, and where AML systems also tend to be weak. The weakness of AML protocols and enforcement also constitutes a difference in comparison with drug trafficking. Regarding drug trafficking, countries which are a source of demand for illegal drugs, such as the United States and countries of Western Europe, have extensive bank regulations and oversight. In addition, they have a history of far more determined efforts at stopping, and the capacity and will to pursue, money laundering. Even so as the statements by Mexican presidents with which I opened this chapter show, countries which are involved in the production and transshipment of drugs in Latin America like to blame the United States for its meekness regarding AML efforts and its inability to stop the drug trade as a result. However, it is not just will but also capacities that are lacking in illegal wildlife markets. Thus the rates of AML effectiveness (however low they are) achieved in countries where there is demand for illegal drugs are unlikely to be achieved any time soon in many countries where there is a demand for wildlife such as China or Vietnam.

There is a real ethical, political, and societal value in having a clean banking and financial system. Banks and financial sectors are a crucial part of the global economy, and their instability or lack of trustworthiness could precipitate catastrophic economic outcomes, impoverishing millions of people around the world. Thus keeping dirty money out of the formal financial sector and legal economy as much as possible is important. Moreover, it can help discourage political corruption and contribute to a societal acceptance of norms, rules, and laws. Oftentimes, if people believe that the rich and powerful, such as bankers, can avoid the cost of compliance and violate rules at will, and that only the weak and poor are punished, norms and laws do not become internalized and violations increase. Yet even the broad internalization of the importance of adhering to laws against wildlife trafficking will not prevent determined smuggling groups from committing crime.

AML measures, however, have other benefits beyond the promise of reducing financial flows to belligerent actors, criminal groups, and corrupt officials. Intelligence developed by following the money can illuminate the extent of the target's network and supplement other sources of intelligence, facilitating physical interdiction operations. Among the few cases when AML tools have been used to investigate wildlife crime across international borders is a 2014 case involving timber smuggling in Southeast Asia. The timber in question was Siamese rosewood, an endangered species, which was transported from Laos to Thailand in trucks modified to hide the contraband. Investigations also revealed that ATM and other financial transfers were made between Thailand and Laos, and AML methods and evidence helped in identifying the smuggling group and generating evidence.[19] Another wildlife trader in Thailand, Jay Daoreung Chaimat, who operated a zoo and is connected to the Bach brothers, was ordered by a Thai court in 2014 to hand over $35.6 million as proceeds of crime, following raids by Thai police at her zoo and AML investigations.[20] However, in 2016, the court order was revoked. In March 2016, the first case of wildlife-related AML extradition took place in the United States. A retired officer of the Royal Canadian Mounted Police who had previously been convicted in Canada for smuggling 250 narwhal (whale) ivory tusks worth more than $2 million into the United States was extradited to the US to face money-laundering charges related to the wildlife trafficking scheme. In Canada, he was fined about US$300,000 and given an eight-month conditional prison sentence.[21] In the United States, the money-laundering offense could bring a fine of up to $500,000 and a prison sentence of twenty years. At the time of this writing, the trial was still on-going in the United States.

Adopting legislation that makes wildlife trafficking a predicate crime for money laundering allows wildlife law-enforcement officials to access additional forms of intelligence, including financial forensics, to develop further streams of evidence, and to toughen penalties.

Moreover, like the US Racketeer Influenced and Corrupt Organizations Act (RICO), which establishes culpability on the basis of merely belonging to a prohibited organization, AML legislation can facilitate prosecutions on the basis of AML law itself when evidence is not available to prosecute wildlife criminals for predicate crimes. Another powerful argument for adopting AML measures to counter wildlife trafficking is that the seized financial assets of smuggling organizations and corrupt government officials can be funneled into conservation—whether to further law-enforcement measures or to involve local communities in conservation.

Yet in many countries, wildlife trafficking is currently not considered a predicate crime for money laundering. Moreover, in many European countries, investigations will only cover both money laundering and the predicate crime if the predicate crime was committed inside the national territory of the country in question, or else they require that the predicate crime be considered a criminal offense both in the country of origin and in the country of prosecution.[22] Even in countries that consider any crime a predicate crime for launching AML operations, financial intelligence units rarely focus on wildlife crime, devoting scarce resources to predicate crimes that national governments consider of higher importance.[23] Thus, there is currently a very limited evidence base on the links between wildlife crime and money laundering. That does not mean that money-laundering methods are not being deployed to hide profits from wildlife trafficking—obviously they are—only that a greater understanding needs to be developed about the preferred and varied money-laundering tools used by specific smuggling networks.

Problematic Side Effects

It should be noted that stringent AML measures, such as those against terrorist financing adopted without careful consideration of the full scope of their impact can and do have undesirable side effects. Because auditing the vast number of financial transactions that take place daily is extraordinarily resource intensive, governments have passed the costs onto the banking sector, which has a varied record of interpreting and implementing the new regulations. Thus, even as late as 2013, the United

Kingdom's HSBC, one of the largest banks in the world, was accused of failing to monitor more than $670 billion in wire transfers and more than $9.4 billion in purchases of US currency from HSBC Mexico, which permitted money laundering by Mexican drug-trafficking groups.[24] Mexico's Sinaloa cartel and Colombia's Norte del Valle cartel were able to take advantage of insufficient due-diligence controls to move more than $881 million through HSBC's US unit from 2006 to 2010. US prosecutors also alleged that HSBC reduced its AML programs to "cut costs and increase profits." HSBC and prosecutors finally made a settlement deal, with HSBC agreeing to pay a $1.25 billion forfeiture and $665 million in civil penalties.[25]

The HSBC case is more an exception than standard practice. Prosecutors, auditing bodies, and international oversight boards have been frequently accused of double standards, going only after small banks in strategically unimportant countries, such as in recent FinCEN actions against Cyprus and Andorra, and often with heavy-handed tools.[26] The scope of punishment can be all the more questionable since FinCEN 311 designations made by the US Treasury—that a foreign financial institution, foreign jurisdiction, a type of international financial transactions, or specific accounts are of primary money laundering concern lack transparency, with no requirement to make detailed evidence public or available to a court.

Often, however, banks have reacted in the opposite manner to HSBC—overreporting suspicious activities to law-enforcement agencies to avoid precisely this kind of liability. Yet filing too many suspicious activities reports introduces too much noise into the system, passing the burden back to law-enforcement agencies and making the amount of information for them to sort through even bigger.[27] But it is not just the banking sector but also states that have a highly varied capacity and will to monitor financial transactions and implement AML regulations. The problems of overreporting and the resulting meager amount of AML prosecution and effective action are also very much in evidence in China, one of the most important markets for illegal (and legal) wildlife products, where effective AML policies could at minimum facilitate developing intelligence of the smuggling network. In order to avoid sanctions, bank employees in China overreport suspicious activities to avoid liability, overwhelming China's central bank's capacities. According to the FATF, between 2010 and 2012, the central bank received eight million suspicious activities reports! Yet money-laundering investigations and prosecutions are infrequently mounted in China because of a lack of investigative and legal resources,[28] as well as limited state interest.

Apart from the sheer scale of the monitoring tasks and the alleged effect of putting businesses at a competitive disadvantage, extensive AML measures can in fact reinforce the informal economy in a country by pushing everyday legal transactions from the formal banking system into an informal one, along with illegal transactions. In countries where the informal economy equals or surpasses the size of the legal economy, and where the fiscal capacity of the state is very limited, such a reduction in the formal economy can be more detrimental overall to the state than the presence of illegal money. AML measures thus need to be designed in ways to reinforce and enlarge the formal economy. The question nonetheless is whether undertaking AML efforts to combat wildlife trafficking also significantly increases the marginal costs of doing AML for other reasons, such as to combat terrorism or chase drug money, or whether it significantly increases the negative side effects of existing AML efforts. To the extent that it does not worsen the negative side effects very much, expanding the frequency of wildlife-crime-based AML efforts makes good sense even if one needs to be realistic about their limited effectiveness and impact as a deterrent.

Importantly, however, AML efforts can and do hurt the innocent and marginalized. Efforts to counter the financing of terrorism through AML measures have, for example, negatively affected non-governmental groups and charitable organizations trying to provide assistance to war-torn countries, such as Somalia, and war-battered and desperately poor populations. When the famine in Somalia that claimed at least 260,000 people broke out in 2011,[29] many prominent humanitarian NGOs were deterred from working in the country since they feared that an inadvertent and unavoidable leakage of their assistance to the terrorist group al-Shabaab would be construed by the US government as violating the law against providing material support to a terrorist group. Indeed, the Obama administration, having strongly taken this position just before the outbreak of the famine, had to scramble to reassure NGOs that they would not be liable. Meanwhile, assistance was delayed by months while many Somalis who might have been saved died. Undesirable blanket effects of this kind could under some circumstances even worsen poaching and wildlife trafficking if desperate and starving local populations, unable to access foreign humanitarian aid, were to choose to intensify hunting and natural-resource extraction, such as deforestation, as a strategy to ensure their subsistence and survival.

Fundamentally, pushing money from the legal banking sector into informal channels also makes intelligence gathering more difficult: banks maintain traceable digital records of all of their transactions, and thus law-enforcement and intelligence agencies can conduct forensic accounting exercises and learn about the shape and size of a criminal network and its nodes. If money laundering takes place through online gaming, such forensic inquiries become much more difficult, if not altogether impossible. That may then lead to fewer (and less important) wildlife traffickers being arrested.

Designers of environmental regulations and policies to combat poaching thus need to carefully consider what objectives they seek from AML measures and what the side effects will be. The environmental community, meanwhile, needs to be realistic about what AML measures can and cannot accomplish, realizing that those measures can only supplement at the margins, not replace, the primary tools for protecting wildlife.

Inducing bank cooperation in checking for and interdicting proceeds from poaching can carry the important normative value of having all hands on deck, so to speak. This kind of financial intelligence can facilitate success in seizing the assets of prominent poachers and corrupt government officials. Some potential customers may be deterred from poaching as a result, while many are likely to be deterred from using banks for money laundering, turning instead to trade-based laundering, online gaming, and other money-laundering mechanisms easily available outside of the banking system. Overall, deterrence effects are unlikely to be extensive as the effectiveness of deterrence is crucially dependent not merely on the severity of punishment, but also, crucially, on the likelihood of detection. It is useful, nonetheless, that seized assets can be channeled into conservation, increasing resources for other tools to combat poaching and wildlife trafficking.

Perhaps most important, tracing money flows can help law-enforcement officials develop a comprehensive picture of a smuggling network. This is a very useful tool that should be available to wildlife law-enforcement agencies. There is thus an important value in making poaching and wildlife trafficking a predicate offense, providing the basis for deploying financial intelligence tools and asset seizure. But AML measures are not likely to remove enough funding to make wildlife trafficking unprofitable, or dissuade enough wildlife traffickers, just as they have not bankrupted all terrorists nor shut down illegal drug trafficking and other illicit economies.

10

DEMAND REDUCTION

The previous chapters show that neither bans combined with intensified law enforcement, nor licensing, nor alternative livelihoods are sufficient to stem the illegal trade in drugs. That is also the case for wildlife trafficking. Community-based natural resource management (CBNRM) can improve species conservation and habitat preservation in some areas, but success is highly contingent on the cohesion, accountability structures, and interests of the local communities involved as well as issues of political economy at the national level. Additionally, CBRNM efforts are not necessarily capable of resisting or preventing large external pressures, such as organized wildlife trafficking at the international level. While bans, licensing of supplies and suppliers, and community-based conservation all contribute to conservation objectives with varying effectiveness, tackling the demand for wildlife is absolutely critical in order to reduce poaching and wildlife trafficking.[1]

Not only does reducing demand address the underlying cause of illegal trade in wildlife, it also facilitates law enforcement and alternative livelihood efforts by reducing the incidence of illegal behavior and the economic, political, and cultural significance of wildlife use and consumption. The fewer violations of a regulation there are, the easier for law-enforcement agents to identify remaining violations and effectively prosecute them. The less demand for wildlife there is, the more customers and local communities near supply and transshipment centers may be willing to cooperate with law enforcement.

But reducing demand can also interfere with and undermine licensing schemes and even community-based conservation approaches insofar as they are premised on the owner's or community's decisions to hunt and sell locally available species. A successful strategy to reduce demand for ivory may decrease organized poaching and trafficking, but it could also undermine the livelihoods of local communities or businesses based on the ivory trade.

Ideally, policies to reduce demand will only reduce demand for illegally sourced wildlife products while maintaining or even increasing the demand for legally and sustainably sourced wildlife products. Yet, as elaborated below, precision in influencing demand is extraordinarily hard to engineer.

Successful Examples of How to Reduce Demand for Wildlife Products

In the United States and Western Europe in the early 1980s, an energetic campaign against seal hunting run by the International Fund for Animal Welfare (IFAW), directed mainly against the killing of harp and hooded seal pups on humanitarian grounds, succeeded in creating a situation in which wearing seal fur was seen as morally unacceptable, and resulted in a serious decline in the market for it.[2] The effect of the campaign seems to have held over the years. In 2010, for example, despite a permit to harvest 330,000 seals, hunters killed only a fraction of that since only one of the established purchasers was buying the pelts.[3] In Litang in China's Sichuan province, a key wildcat fur market, awareness campaigns launched in 2005 culminated in people gathering from across the region to burn their furs. Subsequently, demand for new furs declined, prices also declined, and the market in Litang decreased. However, since the awareness campaign was limited to a specific locale, the market simply shifted to Kanding and Batang in Sichuan.[4]

Even in entrenched markets where consumers have a deeply ingrained taste for wildlife, such as in Asia, there has been some change in attitudes. The intense demand for shark fin in Hong Kong and mainland China, where the fins are made into soup, has made them global hubs of the shark-fin trade. Other Asian countries, such as Singapore and Taiwan, also greatly indulge in shark-fin soup: serving it, often at the cost of $200 per bowl, is considered an important sign of prestige and status, and it has long been associated with weddings and official banquets. A decade ago, some 73 million sharks of all types were estimated to have been killed annually—more than three times the official catch reported to the United Nations. By 2016, the number of sharks killed yearly increased to 100 million.[5] Three-quarters of the market volume is in China.[6] Nonetheless, particularly in Hong Kong, but also increasingly in China, opposition to the consumption of shark fin has emerged and appears to be expanding because environmental campaigns have highlighted the devastating consequences of demand for shark fin on shark populations around the world and on marine ecosystems, as well as the brutality of the practice.[7] The price of dried shark fin in Guangzhou, an important hub for shark fin and other wildlife products, fell from 3,000 yuan ($470) per kilogram in 2010 to 1,000 yuan ($164) in 2015, presumably due to reduced demand.[8] With their high price levels, wildlife products can be even more profitable than drugs. Between 2010 and 2015, official imports of shark fin in Hong Kong dropped by 42 percent.[9] A 2015 Bloom Association survey found 92 percent of respondents said they found it was acceptable that shark-fin soup not be served at public receptions, such as wedding banquets, and 70 percent claimed to have reduced or altogether stopped their consumption of the soup.[10]

The decision of China's President Xi Jinping to combat corruption, extravagance, and opulence among party officials—a political move to consolidate his power after assuming the presidency by eliminating possible rivals within the Communist Party, as well as to increase the party's popular legitimacy—also helped to significantly drive down the demand for shark fin. As part of the anti-opulence campaign, the new leadership prohibited serving shark-fin soup at official functions. In fact, to indicate simple meals, the campaign was labeled "four dishes and one soup." While violations have obviously occurred,[11] in the context of China's overall law-abiding society and Xi's major crackdown on all kinds of corruption and illegal markets, including drugs, prostitution, and unregistered models and actresses, the anti-shark-fin-soup message of the Chinese president strongly resonated among affluent consumers.

For a while, environmental NGOs rejoiced that consumption of shark-fin soup was declining and believed that it was a key indication that consumer attitudes in Hong Kong and China were finally changing as a result of public awareness campaigns directed at the pernicious consumption of wildlife products and at encouraging greater environmental consciousness. But subsequent research in January 2016 threw cold water on this upbeat note when it was revealed that at the during the Chinese Lunar New Year, 98 percent of Hong Kong restaurants still served shark-fin soup and that consumers were buying it.[12] Even though surveyed consumers would state that they found eating shark-fin soup unacceptable, in practice they would yield to peer pressure, such as from family members, bosses, and business colleagues. In short, the outcomes show that an increase in environmental awareness, important as it is, does not necessarily directly translate into actual decreased demand, and other obstacles beyond a lack of consumer awareness and empathy for massacred species need to be overcome.

Indeed, tackling demand for wildlife, especially in Southeast and East Asia and among East Asian communities elsewhere in the world, is unfortunately not easy. Overall, wildlife consumption is deeply entrenched and environmental destruction generates little empathy among consumers. Even though in China environmental consciousness and appreciation of pets and concepts of animal welfare are rising—a crucial development since China is the world's largest consumer of wildlife, including poached animals—the baselines are low. Moreover, in other parts of East Asia, such as Vietnam, Thailand, and Indonesia, the trends are less auspicious.[13] Historically, in much of the region, nature and wildlife have been seen predominantly in terms of consumption, with nature as something to be controlled and consumed, not to be preserved in its wild (often seen as dangerous) state. Wildlife use and consumption are so culturally ingrained in many parts of Asia, and linked to social status, that finding the right triggers to reduce demand is difficult. In some cases, such as among poor, uneducated, and isolated hunters or processors in Myanmar or China, people are simply not aware of the impact of their actions on wildlife.[14] But simply spreading awareness about the illegality of trade in and consumption of wildlife has often proven insufficient, and improved awareness has not resulted in a substantial decline in illegal wildlife consumption.[15] Often owning an illegal animal can result in higher social status. In Indonesia, for example, ownership of a rare and protected bird is a "popular way of showing that one is sufficiently important and powerful to be immune from prosecution."[16] Other times, consumers are simply indifferent to acting illegally and to their actions having a detrimental impact on wildlife and biodiversity. Indeed, the main consumer groups in China are young males with good incomes and high levels of education, people who could hardly claim not to be aware to some extent of the impact of their behavior as consumers.

Moreover, studies of consumer preferences even among groups aware of the environmental impacts of buying wildlife products show that people often talk the talk, but they rarely walk the walk. Nor is their talk all that credible. It is safe to assume that not all people disclose illegal or problematic behavior even in anonymous surveys, and that such surveys underestimate the size of the illegal market. For example, a 2015 survey of Hong Kong residents commissioned by the environmental NGOs WildAid, the African Wildlife Foundation, and Save the Elephants produced hopeful numbers: 75 percent of Hong Kong residents supported a ban on ivory sales in their territory, and 91 percent claimed they did not own any ivory products.[17] A National Geographic Society and GlobeScan survey similarly revealed that 79 percent of Chinese people surveyed also nominally supported a total ban on ivory trading. However, it also showed that 35 percent of Chinese respondents at the same time wanted to buy ivory and could afford it, while an additional 20 percent wanted to buy ivory and could not afford it. (In the US, those two respective categories were 13 and 22 percent.)[18] Indeed, other surveys of Chinese customers also show more ambiguous outcomes, with rising environmental awareness coinciding with persistent, if weakening, demand for wildlife products. For example, a 2012 survey by Li Zhang and FengYi showed that 52.7 percent of respondents agreed that wildlife should not be consumed, a significant increase from their 2004 survey which showed 42.7 percent with that attitude. In addition, the number of respondents who believed that wildlife could be used decreased from 42.8 percent in 2004 to 34.8 percent in 2012. However, the number of people who actually stopped consuming wildlife declined much less, only to 29.6 percent in 2012 from 31.3 percent in 2004.[19] As the authors aptly say, conservation awareness in China has a long way to go.

Nonetheless, these numbers also beg the question of how much compliance is sufficient and how much demand needs to be reduced in order to achieve the robust sustainability of vulnerable species. Neither demand-reduction campaigns nor law enforcement need to achieve 100 percent compliance and prevent every single act of poaching and wildlife trafficking. Indeed, whenever resources to achieve compliance are limited (which is almost inevitable, and always the case in reducing wildlife crimes), identifying how much compliance is good enough, and what the most "mission-critical" elements are, focuses resources and enhances effectiveness.[20] One encouraging element is that wildlife consumption does not have to tackle addiction, so presumably both the speed and extent of demand reduction for wildlife products could be faster than in the case of reducing demand for drugs or cigarettes.

Messaging Matters

Experience from the wildlife trade, from other environmental campaigns, and from campaigns focused on drugs and cigarettes shows that it is often not highly effective to alter consumer preferences through messaging campaigns that rely solely on people's altruism—whether toward other humans or other creatures. This is particularly so when the cost of persistent undesirable behavior is diffuse and delayed, and affecting global conditions rather than the user. For example, one post-9/11 drug demand-reduction campaign which centered on telling drug users not to consume because they were funding terrorism did not resonate with users and was dropped. The United Kingdom adopted a similar altruism-based campaign to reduce cocaine use, telling users that their use of the drug kills other people and leads to deforestation. Started in December 2015, its results will not be in for some time, but there are reasons to be skeptical about the campaign's effectiveness beyond influencing the behavior of a small segment of highly socially conscious users (who are not addicts).[21] The government of Colombia sought almost a decade ago to launch an almost identical campaign—to encourage a reduction in cocaine use in the United States and Europe by emphasizing the link between coca cultivation and deforestation in Colombia, but at the time, the campaign did not get traction in Europe or in the United States.[22] Altruism-based campaigns and indeed all mass-media campaigns become less effective and sustainable when they have to compete with the "pervasive marketing

DEMAND REDUCTION

of competing products or opposing messages."[23] All the more reason to push wildlife-trafficking markets underground and shut down websites promoting illegally sourced products, thus reducing the visibility of and easy access to retail markets and the ability of their operators to advertise their goods.

The most effective campaigns have often been ones in which people perceive a particular behavior as a direct threat to their personal interests, especially within a relatively short time, such as severe negative health effects or a reduction in sexual vigor. In Montana, one very effective anti-drug-use campaign centered on the bad effects of methamphetamine on dating and sex: billboards showed ill-looking teenagers without teeth and argued that if they consumed the drug they would become hideous, and would not be able to get boyfriends or girlfriends. In the 1970s, a US campaign to treat Mexican marijuana and poppy fields with paraquat temporarily cut demand in the United States for marijuana since the paraquat used to kill the cannabis plant was widely believed to be carcinogenic. However, new suppliers of marijuana moved in, stamping their product with the message Not Made in Mexico (thus implying that their pot was unaffected by paraquat) and recaptured the market, at least temporarily. When paraquat spraying was suspended, Mexican suppliers again robustly entered the US market.

Anti-smoking advertisements that emphasize the long-term negative health effects of smoking appear to have less effect on teenagers than on adults because teenagers tend to live in the present, consider themselves invulnerable, and/or believe that they can quit before they ruin their health.[24]

Similarly, some of the environmental campaigns that resulted in great changes in regulations—some inducing change in industry behavior at a global level concerning issues such as acid rain, mercury pollution, DDT, and ozone depletion—centered on the immediate and vastly negative health effects of specific pollutants to people. Indeed, an often underemphasized but crucial part of the success in reducing shark-fin soup consumption was the discovery and the spread of information about its high mercury and cadmium content, which represents dangers to human health, such as cancer and birth defects. Awareness of the dangers of mercury and cadmium levels also helped build support for the seeming decline in China's demand for manta ray gills and the possible prohibition of that market.[25]

The difficulty of structuring environmental messages in Asia this way is that people often believe that consuming wildlife will actually enhance their health and sexual potency, rather than causing them to worsen. Predominantly, such claims are bogus: the keratin in rhino horn no more cures cancer than consuming a tiger's penis enhances sexual performance, but these myths are widely believed. Merely exposing their spuriousness with scientific evidence has not been sufficient to dissuade consumer behavior. Campaigns which emphasize actually detrimental effects to people's health and sexual capacity can be far more effective. But it is not enough to emphasize a general threat to one's health: the targets of a campaign must believe that they are personally at risk and can take steps to mitigate that risk.[26]

Messaging can portray the consumption of wildlife products or possession of ivory trinkets as a sexual turnoff. An anti-rhino-horn-consumption campaign video, for example, could portray an attractive young woman rejecting the advances of an affluent man who had consumed a rhino-horn elixir or drunk from a rhino horn libation cup, and show her walking away with a man who did not consume such potions or show off prestige items derived from the wildlife trade.

A second leverage mechanism that has proven effective in the case of cigarettes and drug use, for example, is to stress that smoking or drug use is not "cool"—hence reducing the social status of the behavior and working to counter peer pressure. To the extent that consumers such as teenagers care about autonomy and resisting authority, campaigns can be structured along the lines of: "Don't be a patsy for the cigarette industry, don't puff." Once again, this approach has its limitations in the case of Asian wildlife consumption, where the practice is deeply ingrained and socially very significant. Pressure from parents to consume shark-fin soup or from business colleagues to give ivory as a gift have proven difficult to resist for many of those potentially swayed by altruistic sentiments toward wildlife. Especially if the messengers are foreign, such as Western NGOs, the message runs the risk of being dismissed as a form of imperialism, which has happened, for example, with efforts to reduce foot binding in China and female genital mutilation in Africa. In places like China (or Russia, where wearing fur is a status symbol of the newly rich), consumers may not care about what the West considers normatively appropriate. Messengers drawn from

DEMAND REDUCTION

people's own cultural community, such as local NGOs or, better yet, local celebrities and government officials, may be far more effective. Indeed, local celebrities have been rallied by environmental NGOs to spread anti-wildlife-use messages. Such influential opinion leaders may include international sports and movie celebrities, religious authorities, tribal elders, and other respected personalities.

Maintaining Messaging Over Time and Tailoring it to User Subgroups

Effective messaging requires a highly nuanced understanding of demand around the world. What needs answering in each specific location includes who the existing customers are, who new potential entrants can be, why and how the demand for illegal wildlife products exists, and why and how any consumers stop using. Understanding the reasons behind consumer choices is also very important as it can illuminate the elasticity and cross-elasticity of demand.

The most pernicious markets are those where demand is driven by conspicuous consumption based on signaling the ability to afford something that is expensive. In that case, scarcity drives up value, not just price, and the market does not stabilize but leads to unsustainability and depletion. Tragically, that has been the case with ivory, tiger products, and rhino horn over the past decade, and one of the reasons why it currently makes sense to shut down trade and consumption in these products and outlaw remaining domestic markets in them, such as in China and Japan, as called for at the 2016 Conference of the Parties to CITES.

Anti-use messaging must be maintained over time and adapted to new subgroups of consumers, otherwise previous accomplishments may wither. Despite the effectiveness of campaigns that reduced demand for wild-caught parrots and supplied the markets with captive-bred parrots, demand for poached parrots from Latin America is once again rising (or at least persisting) in the United States. Following the outbreak of SARS in 2003 and bird flu in 2004, some of China's previously legal wildlife trade was temporarily banned as medical experts mounted campaigns to point out that wild animals, such as primates, rodents, and ungulates share more than 100 diseases with humans, and the public became more aware of the negative impact wildlife consumption could have on their health. However, as the epidemics and surrounding media coverage tapered off, consumers often returned to their previous consumption patterns.[27] Indeed, robust evidence suggests that the immediate salience of a problem is often crucial for increased compliance, and as salience wanes so does compliance.[28]

The cigarette market provides another important example. After decades of great success in the United States in the campaign to discourage smoking, with smoking rates declining by half between 1964 and 2014,[29] cigarette use among teenagers has started growing again, despite high taxes on cigarettes and high prices. One cause of this new negative trend within the overall successful antismoking campaign was overestimation of the extent to which teenagers had internalized awareness of the health effects of smoking. Research revealed that an increasing number of teenagers were resorting to cigarettes while being well aware of the long-term health effects because they believed that smoking would keep them from getting fat, hence increasing their physical attractiveness.[30]

For wildlife-protection messaging to be effective, it too needs to be specifically tailored at particular times to particular consumer groups, among whom there is great variation and hence varied receptivity to particular demand-reduction signals.[31] Some groups may be highly susceptible to altruism-based messages, others not. Moreover, demand for particular wildlife products varies over time and over generations. Ethnic groups in a particular country may experience a rise in their disposable income and become new consumers of a wildlife product the demand for which was believed to have been effectively suppressed. Chinese citizens tend to be highly rule-abiding; other cultures, such as in Latin America, may celebrate law evasion.

Where a lack of protein drives bushmeat consumption, which still occurs in some parts of Asia and large parts of Central and West Africa, demand-reduction strategies centering on normative arguments, rather than on delivering a legal source of protein, will be ineffective. An appropriate demand-reduction strategy would be to provide protein from legal sources (hopefully without converting forests to agricultural land) and emphasize the fact that consuming primates spreads diseases such as Ebola. Unfortunately, efforts to replace bushmeat among forest populations in Africa with chickens or goats have not been highly effective, being "too great a change in the dietary habits of the settlers."[32]

DEMAND REDUCTION

Demand-reduction strategies in such cases need to examine why people are not willing or able to switch to alternative sources of protein. Other efforts in Central Africa have attempted to encourage the rearing of some wildlife species commonly hunted for bushmeat that are not endangered. Again the results have often been disappointing, with the cost of production too high to be viable for poor households. The projects could be commercially viable, however, if geared toward supplying the urban rich in Central Africa, who prefer to consume bushmeat out of choice.[33] An alternative strategy for dealing with subsistence dependence on bushmeat protein is to disperse non-endangered bushmeat prey into buffer zones near protected areas where local people are allowed to hunt within the limits set by sustainable quotas. Such an approach is being tested at the Parc National Nouabalé-Ndoki in the Republic of Congo.

What works in one place may be counterproductive elsewhere. Demand-reduction programs for reducing snake consumption in Vietnam will be misdirected if focused on the lack of available protein. Reducing the appeal of snake-blood elixirs and countering messages about the presumed effects of these elixirs on sexual potency and longevity, and challenging ideas about the link between the consumption of these elixirs and social status, will be more relevant there. In short, the design of effective demand-reduction strategies requires investigating and understanding the dynamics of each consumption market. The Chinese market in ivory, buoyed by a large number of ivory investors and speculators, is likely to respond very differently to the market in furs from big cats among Buddhist monks.

Preventing Use

Experience with drug use also provides important, and somewhat sobering, insights into demand-prevention policies designed to preserve endangered wildlife populations. The overwhelming evidence suggests that prevention policies aimed at drug use are not very effective, though they are more cost effective than other interventions. High-school anti-drug programs such as DARE, targeting teenagers and emphasizing the long-term costs to individuals and societies, often presented by police officers, show little indication of effectiveness in the United States.[34] The most effective among drug-use prevention strategies seem to be those which focus on intervention with young children (as young as five or six years) and are not specific to drugs but seek to build self-confidence and peer-pressure resistance. Such effective interventions combine messaging against obesity, bullying, and other undesirable behavior.[35]

Some of these lessons can be incorporated into policies to prevent wildlife consumption. While increasing, environmental awareness in China (as well as Vietnam) is still at a very low level. Much environmental activism in China has centered on unrestrained water, soil, and air pollution and their negative health impacts, giving less emphasis to species conservation or broader natural habitat preservation. Significantly, some research shows that the consumption of wildlife, while widespread, is also concentrated among particularly "heavy users" initiated into consumption in their teens or early adulthood, and that prevalence of wildlife product use among the overall population is declining, with use of wildlife products increasingly concentrated among those few heavy users.[36] These findings suggest that early-childhood intervention and awareness programs can be an important element of discouraging demand.

Yet the findings need to be rigorously tested. Evaluations are difficult since the effects are unlikely to be manifest for two or three decades, when people who received environmental education while they were children have the means and opportunity to engage in the consumption of wildlife products. One measure of effectiveness of these early-childhood education programs is whether the children who receive environmental education seek to discourage their relatives and neighbors from the use of problematic wildlife products. This would require the children to have actual knowledge of the use of such products in their proximity (which may often not be the case) and be willing to disclose that their relatives or friends engage in behavior deemed by some to be undesirable.

Swift, Certain, and Mild Punishments and Ways to Reduce Noncompliance

A punitive regime aimed at discouraging drug use by employing swift, certain, and mild but increasing levels of punishment (such as one night in jail, then two nights in jail, and so on) has been developed by criminologists and implemented in Hawaii, where it is known as Project Hope.[37] Each time a problematic user of alcohol tests over a certain limit, or each time a drug-user tests positive, means an increase in the level of punishment. Identified problematic users are mandated to test on a regular basis or face similar punishments or possibly lengthier prison terms. Compared to other demand-reduction programs, programs such as this one often exhibit far greater effectiveness rates as well as producing other highly desirable benefits, such as reducing drug or alcohol-use-related mortality and domestic violence.[38]

Drug-use reduction programs featuring graduated but swiftly applied punishment provide possible models for enforcing prohibitions on wildlife consumption. For example, once China and Hong Kong fully outlaw all ivory sales, small-level buyers of ivory trinkets who can be identified and listed on a database (whether private or public) could be subject to escalating fines or sentenced to a day or two of public work. The shaming effect of such punishments could be particularly strong, discouraging use. Of course, it would be unwise to flood prisons with low-level offenders. But, if such programs managed to identify and apprehend heavy consumers of prohibited kinds of wildlife, who, if they continued in their illegal activities, would face increasingly severe penalties including prison terms, there could be a substantial effect on the wildlife market.

Admittedly, in the case of drugs, approaches employing swift and initially mild levels of punishment tend to have an important advantage. Many users are identified because they engage in visible problematic behavior, such as drunk driving and disturbing the peace, for which they are apprehended. To the extent that wildlife product users merely engage in illicit purchases, policing the market in wildlife products would have to be intensive, particularly if there were many small dealers or if purchases took place behind closed doors or over the internet. Thus the number of users who are identified, punished for, and possibly dissuaded from consuming wildlife products as a result of swift, certain, and mild punishment approaches might well be far lower than that which pertains to drug or alcohol use—thus weakening the effectiveness of such demand-reduction strategies.

The reasons that programs like Project Hope and a program known as the 24/7 Sobriety Project (an equivalent program for drunk driving and alcohol abuse) work is that alcohol and drugs linger in the blood stream and the test to determine whether they have been consumed is cheap and reliable. Obviously, there is no equivalent way to test whether someone purchased an ivory *hanku*; traces of shark-fin soup might stay in the saliva but could be removed by using a mouth wash. Moreover, in the case of the consumption of illegal wildlife products, intrusive test measures may be socially unacceptable.

Detection requires that the violator be observed purchasing the prohibited product—such as purchasing meat derived from an endangered species in a restaurant or an ivory product from a seller. That requires the resource-intensive policing of known markets and a constant search for new areas or online domains to which markets have relocated as a result of enforcement.

Yet the smaller capacity of law-enforcement agencies to monitor wildlife retail markets in comparison to their capacity to monitor drunk driving (because problematic consumption of wildlife is not visible in the same way that drunk driving or drug use on the street is), and hence their smaller capacity to apprehend users of outlawed wildlife products, do not negate the benefits of this approach as a means of managing wildlife consumption. Even when enforcement is less than perfect and certain, it increases compliance, and consistent and mild punishments are certainly far preferable from both human rights and resource-demand perspectives, as well as more politically sustainable, to imprisoning users of illegal wildlife products for lengthy periods. Moreover, mild (but escalating), swift, and certain punishments can be meted out to restaurant owners who sell exotic meat from endangered species and to dealers in Traditional Chinese Medicine (TCM) who sell potions derived from such species. Identifying violations by dealers will be easier than identifying customers who order illegal products.

Still, there are complex normative issues. Should the subsistence-driven, consumption of bushmeat, such as from gorillas and chimpanzees, among a forest-dwelling family in the Congo carry the same penalty as the consumption of that meat by a rich businessman in Vietnam? In the latter case, mild punishment to discourage use could be broadly socially acceptable and normatively just. In the former, other interventions such as delivering legally sourced protein (as discussed above) would be the appropriate form of intervention. In other words, the reason for the violation of a law should result in different kinds of intervention.

If the high cost and limited availability of painkillers and cancer medication drive customers to consume rhino horn and other TCM, widely distributing cheap non-TCM medication can dry up demand (assuming someone is willing to pay for the distribution of such drugs and legal distribution channels can be established). Of course, handing out a limited number of pills is easier than establishing accessible medical care, though the latter might be as important as the cost and availability of the pills.

Green Certification and Redirecting Users to Sustainable Resources

However, efforts to reduce the consumption of illegally sourced wildlife need to be carefully designed so as not to eliminate all demand for wildlife products if and when conservation policy is critically built on economic incentives for conservation. If a conservation policy is based on giving the local community an economic stake in conservation, a significant reduction of demand for that particular wildlife product, not just demand for illegally sourced wildlife, could undermine the conservation effort. Pressures for clearing the natural ecosystem and turning land over to farming or cattle ranching could once again proliferate.

Achieving finesse in messaging concerning demand is very difficult. A crucial question is whether overall demand has reached a plateau and consumers will merely switch between illegally or legally sourced wildlife, or whether overall demand continues to rise. In the latter case, promoting environmentally certified products may increase overall demand, and thus also demand for illegally sourced wildlife products. As Kent Weaver aptly points out, "Tailoring information to specific sections of the target population may improve compliance results, but it is also likely to increase confusion as well as the costs and complexity of implementation."[39] Other crucial questions include: Do consumers (and those who issue certification) have the capacity to accurately distinguish legal from illegal products? Do customers in fact care about green certification, wildlife sustainability, and ecosystem preservation? What are the price differentials between legally and illegally sourced products?

Directing consumers to the right choice is complicated by the fact that they often face a bewildering array of labels supposedly catering to responsible consumerism, with consumers themselves having the burden of assessing the quality of these labels and/or prioritizing among "green" and "fair trade" and other positive claims. The more such labeling becomes prevalent, the higher the odds that consumers will make random, erroneous, or indifferent choices. The credibility of certification also cannot be taken for granted, as many a dirty company will seek to create its own positive marketing label. Moreover, even well-motivated certifiers face tough obstacles in ensuring that certification is reliable.

So there is likely to be great variation in the extent to which encouraging environmentally aware demand can be made consistent with reducing demand for illegal wildlife, and the extent to which efforts at differentiation and eco-labeling result merely in more intense greenwashing practices and a reduction in consumer awareness. Diligent monitoring of specific wildlife markets along the entire chain of production and distribution will be required, as well as adjusting policy designs in order to reduce greenwashing and the appropriation of eco-labels.

Discouraging the use of rhino horn for the making of Yemeni daggers (known as *jambiya*s) is an example of an important success in redirecting demand toward greener, more sustainable practices—in this case from rhino horn to an artificial gum. (It should be noted, however, that the campaign coincided with a large economic downturn in post-unification Yemen that made rhino horn far less affordable to most customers.) Like the ownership of birds and daggers in Indonesia, the ownership of a *jambiya* has long been a crucial symbol of prestige and honor for Yemeni men, with virtually all men wearing one. *Jambiya*s have been used as gifts to cement family and business relations and friendships. The most desirable *jambiya* has long had a hilt made of rhino horn, not because this was thought to possess qualities regarding healing or sexual potency, as it is in East Asia, but because hilts made out of rhino horn are more resistant to damage and wear and tear.[40]

Yemeni demand for rhino horn from East Africa soared in the 1970s and 1980s, to the point where, by the early 1990s, some scholars identified Yemen as a primary culprit in the near-elimination of the northern white rhino and black rhino.[41] Since then, *jambiya*s using buffalo horn have been made in increasing numbers, though they continue to

DEMAND REDUCTION

be considered less desirable because, although they are much more affordable, the hilts break easily. Even the government's relaxation of import duty on buffalo horn from India (aimed at reducing the use of rhino horn) did not fully motivate craftsmen and consumers to switch to what was seen as a substandard product.

In North Yemen, the use of rhino horn for hilts persisted for three decades after the government banned rhino horn imports in 1982. This legislation was followed by a ban on all domestic trade in new rhino horn in 1992, and a *fatwa* (religious ruling) was issued, to the effect that it is against the will of Allah to kill rhinos for their horn.

After the unification of North and South Yemen in 1990, Yemen remains one of the poorest countries in the Middle East, with very low life expectancy and low literacy rates, authoritarian and ultimately unstable governance, and the country is plagued by violent secessionism, jihadism, and, since 2015, a civil war involving the intervention of external interveners.

Yet the use of *jambiya*s has remained robust in the north of the country, with rhino horn considered the most desirable material for the hilt until about 2013. Even so, smuggled imports declined steadily until 2002. In the late 1980s, at least 500 kilograms of rhino horn were smuggled into Yemen, dropping to 50 kilograms in 1995.[42] From 1995 to 2002, the quantity of rhino horn smuggled into Yemen was 30 kilograms a year, but between 2003 and 2006, demand rapidly escalated again and the amount of smuggled rhino horn doubled.[43] The amount of rhino horn nonetheless still allowed for the production of only 200 *jambiya*s with rhino-horn handles, compared to the 1970s when many thousands were produced yearly. In the early 2000s, 90 percent of Yemen's *jambiya*s were made using buffalo horn, even though its fragility meant the product was seen as substandard. Despite the substantial increase in the quantity of rhino horn entering Yemen, the price of raw horn went up by over 40 percent (from $1200 per kilogram in 2003 to $1700 per kilogram in 2007), indicating that demand was expanding far more than the smuggled supply. The price of *jambiya*s with rhino-horn hilts similarly escalated from $255 for a small dagger and $824 for a large dagger in 2003, to $450 and $1670 respectively in 2007.[44] Surprisingly, the spike in demand and rise in price took place even though per capita income continued to stagnate or deteriorate. Three years later, large and particularly prized rhino-horn daggers could sell for as much as $29,000.[45] The causes of the rise in demand among Yemenis are not well understood.

Thus, before the 2015 civil war started, efforts to eliminate rhino-horn smuggling into Yemen centered on explaining to people the damage that the use of rhino horn causes to rhino populations—most Yemenis had not seen a picture of a rhino nor had a concept of the animal, referring to it with the generic *zarraf* (meaning "giraffe" in Arabic)—as well as finding an acceptable hardy substitute. The campaign to promote general awareness about rhinos placed banners about them and their horns at entrances to zoos, and various ministries and educational facilities were engaged in the campaign.

The big breakthrough came when *jambiya* craftsmen started using an artificial product, a solidified gum, to make dagger handles. The first daggers of this kind appeared on the market in 2008, and their use by craftsmen and buyers has expanded rapidly since. The gum is as resilient to chipping and breaking as rhino horn, is much cheaper, and it can be made to look a lot like rhino horn when a grainy material is mixed into the gum.

The effect of Yemen's economic downturn should not be dismissed, even though by itself it did not halt the spike in demand during the early 2000s. One of the reasons why demand for rhino horn greatly escalated in Yemen in the 1970s and the 1980s was the influx of oil revenues to the country.[46] The political instability and economic crisis that have affected the country since 2010 have resulted in a rise in the cost of living and a decline in disposable income among Yemenis, and many have resorted to selling their *jambiya*s to buy air tickets to leave the country. The relative decline in disposable income, and thus in demand for rhino-horn *jambiya*s, has been further compounded by escalation in the price of rhino horn internationally, diverting smuggling and sales from Yemen to China and Vietnam.

Surprisingly, political instability and insecurity also contributed to a reduction in demand for and a reduction in the display of rhino-horn daggers. Often insecurity hampers all manner of public policy interventions: preventing effective law enforcement, alternative livelihood measures, CBNRM strategies, as well as demand-reduction efforts by limiting access, depleting resources, and pushing people to desperation. In

DEMAND REDUCTION

this case, as insecurity and crime increased, men came to consider it unwise to display a rare and valuable commodity for fear of being attacked and robbed. *Jambiya*s with hilts of artificial-gum, prevalent and cheap as they became, were not considered likely to incur such risks, and they allowed men to showcase their honor and tradition without the risk of robbery.

Interestingly enough, the communist government of South Yemen was far more effective in eliminating demand for rhino-horn *jambiya*s by eliminating the demand for all *jambiya*s. It banned the possession of all weapons and aggressively collected them. In 1972, the *jambiya* ban was thus accompanied by a massive campaign to rid the country of them, with even rich and influential families targeted and forced to sell their daggers to North Yemen and Saudi Arabia for low prices.[47] When the two countries reunited, the people of former South Yemen readily took to capitalism and enjoyed the consumption opportunities it offered them. Islamic religiosity also spread quickly in the south, but although they adopted both capitalism and Islamic mores from the north, Yemenis in the south continued to look down on the wearing of *jambiya*s. Thus, the wearing of *jambiya*s there did not escalate, even as it persisted in the north.[48] In fact, the effect of peer pressure was, interestingly enough, eventually manifest in the opposite way: most northerners soon gave up wearing *jambiya*s in Aden, the former capital of South Yemen, when they learned that no one else wore them there. (That does not mean they discarded the *jambiya* or refrained from displaying it when they traveled north again.)

In this case, several factors pulled in the same direction: the ban on *jambiya*s in the communist South, which was not only effectively enforced by the government but ultimately internalized by the country's population; demand substitution and reduction strategies in the North; also economic decline throughout the country after reunification; and even the systemic decrease in personal security as a result of the civil war. Together, these developments made it increasingly hard for men to afford to wear *jambiya*s with rhino-horn hilts. The robustness of the changes taking place (at least before the civil war started in 2015) is also illustrated by the fact that rich Yemenis who could still afford rhino-horn *jambiya*s chose instead to give daggers made with hilts of agate to foreign dignitaries, and also to wear them themselves instead of rhino horn ones. Mined in Yemen, agate is rare and expensive, and agate *jambiya*s cost around $1000. Thus, by 2013, Yemen was no longer considered by researchers to be a threat to rhinos.

Such substitution efforts and moves toward the use of greener products are not always a happy story, at least not in the long run. In a similar effort to curb the demand for rhino horn in the 1990s, conservationists promoted the use of the horn of the saiga antelope, mostly found on the steppes of Russia, Kazakhstan, and Uzbekistan. The saiga antelope was abundant at the time, and its horn resembled that of the rhino, thus making it seem a viable greener substitute. The demand substitution effort succeeded—but only up to a point. The demand for saiga antelope horn indeed shot up, with China, Singapore, and Japan becoming principal consumers, and Malaysia, Hong Kong, and Singapore the market's distribution hubs. Along with increased demand for the horn came vastly increased legal and illegal hunting of the saiga antelope—to the point that hunting it became unsustainable. The saiga antelope population shrank dramatically: in the 1970s they numbered about 1,250,000, and over 1,000,000 remained in the early 1990s; today, there are roughly 30,000 to 50,000—and most of these are (hornless) females, as it is only the males that sport the horns.[49] Some biologists doubt whether the species can ever recover. When a TRAFFIC study subsequently sought to identify viable substitutes for saiga antelope horn, it found that some TCM traders were recommending rhino horn.[50] Clearly, the substitution worked rather well; however, it did not produce desirable environmental outcomes.

Thus, as with all other policy tools, there is likely to be large variation in the extent to which encouraging environmentally conscious demand can be made consistent with reducing demand for illegal wildlife products, and the extent to which such efforts at differentiation and eco-labeling result in further greenwashing practices and a decline in consumer consciousness. Diligent monitoring of specific wildlife markets, as well as adjusting policy designs, will be required in order to reduce greenwashing practices and to achieve positive outcomes that avoid insufficiently anticipated and undesirable consequences.

Allowing for policy failure is key. If monitoring and self-monitoring reveal that a demand-reduction program is not working, its designers and implementers should be given a chance to redesign the program to

DEMAND REDUCTION

address its deficiencies instead of being simply deprived of funding. Otherwise, there will be a strong tendency to obfuscate assessment measures, and the species of wildlife that is under threat will continue to decline irretrievably. Many of the factors facilitating or preventing compliance discussed in this chapter can change over time. Notions of what is "cool" can change as a result of factors exogenous to demand-reduction campaigns. Enforcement intensity, whether actions are conducted by local communities, NGOs, or law-enforcement bodies, may decline as attention shifts elsewhere. The effects of peer pressure may change, such as when authorities become discredited or as a result of demographic changes. Thus a strategy that seems to work in one context might not be as useful in another.

11

CONCLUSIONS AND POLICY RECOMMENDATIONS

The precipitous collapse of species due to wildlife poaching and trafficking requires urgent policy action. Once a species is gone, it is gone, sometimes with vast negative effects on the rest of an ecosystem. Sadly, the overall prospect for interventions that can achieve substantial reductions in the illegal wildlife economy is not good. There are no silver bullets, nor even universally appropriate ameliorants. Policy outcomes are highly context specific and contingent. All strategies—bans and interdiction, legalization and licensing, community-based natural resource management, and demand reduction measures, as well as supplementary tools such as anti-money-laundering efforts—face structural and resource constraints. In addition to these limitations, the strategies also come with direct downsides. Bans on hunting and local resource extraction can reduce sources of income and lower the standard of livelihood among poor local populations. The imposed relocation of communities and other coercive measures can generate deep resentment and a rejection of conservation, as has often happened in Africa, Asia, and Latin America. Licensing legal trade, however, significantly complicates law enforcement and facilitates traffickers' strategies of evasion.

Bans or licensing can help in particular circumstances, depending on local cultural and institutional settings and the ecological requirements and circumstances of particular species. But there are limitations on how much even greatly intensified law enforcement can halt poaching and wildlife trafficking, create deterrence effects, and suppress supply in the absence of a dramatic reduction in demand. Licensing and regulating the hunting and wildlife trade equally only works under some circumstances. Sometimes it gives various participants a stake in conservation that would otherwise be absent, serving to protect not just a species but also its habitat and the broader ecosystem. At other times, a legal supply of a vulnerable species boosts overall demand, including for poached animals, complicates enforcement, and enables the laundering of illegally sourced products. Community-based natural resource management (CBNRM) can significantly motivate local communities to support conservation and resist poaching and trafficking, but just as with other tools, the outcomes have varied widely. Going after consumer demand is crucial, but demand-reduction measures are complex and take time. Anti-money-laundering (AML) efforts provide an additional tool, but they will not bankrupt the illegal wildlife trade.

Moreover, like counter-narcotics policies, conservation policies generate their own costs. Historically and even today, poor marginalized populations have been forcibly excluded from protected areas. Restricting their hunting and other resource extraction activities can significantly undermine their human security, including access to food, and their human rights. Conservation policies are thus often merely enforced and not internalized by the community—an outcome that significantly increases the resources needed for law enforcement and undermines the ethical justification and political sustainability of conservation regimes. What is illegal may well be seen as legitimate by a local community, and as their right. Sponsors of illicit economies on which local livelihoods depend can accumulate significant political capital. Meanwhile, bans and restrictions on trade and consumption can be seen as Western impositions that unfairly try to change the habits of other cultures. In short, if conservation policy ignores the plight of poor poachers and wildlife-dependent communities, its effectiveness will be at best limited, and at worst counterproductive.

In this book I have analyzed the promise, outcomes, and policy effectiveness of various anti-poaching and anti-trafficking tools, and drawn lessons from the effectiveness of those same tools in drug policy efforts. Both assessments reveal large variations in outcome across time

CONCLUSIONS AND POLICY RECOMMENDATIONS

and space, and thus great policy complexity, with effectiveness highly contingent on local settings. Drivers internal to the illegal market itself—such as fads and fashions among users, dealers' marketing strategies, balances of power among criminal groups, and the behavioral socialization of poachers and traffickers—are all influenced by policy interventions, but also operate independently from them.

Summary Assessment of the Various Policy Tools

To recap, the key findings with respect to each policy tool discussed in the earlier chapters of this book are highlighted below.

The Effects of Prohibition, Bans, and Law Enforcement

Wildlife policy enforcement efforts are often inadequate. They are severely under-resourced and given low priority. A significant increase in the diligence and resources dedicated to the enforcement of wildlife regulations is certainly warranted. However, it is unlikely to halt poaching, wildlife trafficking, and other environment crimes.

The objectives of prohibition and interdiction are to prevent or at least restrict illegal supply and discourage use by creating barriers to entry for both sellers and buyers, boosting prices, limiting commercialization, and creating a normative set of values against threats to vulnerable wildlife ecologies. Just because not every user and supplier is deterred by illegality does not mean that removing illegality will avoid increasing supply and consumption. Nonetheless, if consumption is driven by a desire to display status, power, and wealth, such as by wearing ivory bangles or coats made from endangered species, a ban that discourages such ostentatious display may well shrink demand and be highly valuable. But the demand-suppression dynamics will be very different if illegal bushmeat is consumed not as an exotic luxury indulgence but for subsistence. In the latter case, without alternative protein sources being made available, demand may not go down at all.

Implementing interdiction effectively is hard and very resource intensive. Effective interdiction designed not just to incapacitate poachers and traffickers but, importantly, to deter them requires knowing specific structures of particular wildlife smuggling networks. Yet analyses at times exaggerate the extent to which poaching is undertaken by organized criminals and militant groups. While the latter may be involved, such cases often represent only a sliver of poaching activity. In fact, there is a large degree of variation in the structures of poaching and trafficking networks, and there are many atomized small-level traders and poor poachers also in the business. Exaggerated and simplistic characterizations also divert policy attention away from corrupt practices among legal actors, including licensing entities, ecolodges, and top environmental officials.

Poaching and trafficking networks are often far less vertically integrated than many interdiction advocates imagine. Moreover, even top traffickers and entire wildlife trafficking networks are easily replaceable as long as demand stays robust. Nonetheless, knowing what the poaching and smuggling structures actually look like in a particular place, rather than imagining what they might be, is crucial for making law enforcement and other policy interventions effective in that locality.

Even with a very large dedication of resources, rarely does drug interdiction surpass a 50 percent effectiveness rate, often remaining much lower. Such interdiction levels may be insufficient to prevent the collapse of particular species. What crucially facilitates the effectiveness of interdiction, law enforcement, and bans is a reduction in the demand for a commodity—whether as a result of the ban, purposeful demand reduction strategies, or exogenous factors. Paradoxically, the more effective law enforcement becomes, the more the value of a smuggled animal or wildlife product goes up. Perceived scarcity—whether as a result of species depletion or more effective law enforcement—increases the financial benefits of smuggling.

Thus: It is necessary to implement and sufficiently resource law enforcement efforts before a species is announced as rare or endangered, or an area is declared as protected. Prohibition of trade without enforcement can stimulate an increase in poaching. Effective interdiction also creates undesirable balloon effects, displacing illegal supply (poaching) and smuggling routes to new areas, and diverting illegal trade to underground markets or less controllable routes, such as the internet, thus making interdiction even more difficult the second time around.

For law enforcement to start producing deterrent effects, it must be able to effectively interdict and prosecute a sufficient number of suc-

CONCLUSIONS AND POLICY RECOMMENDATIONS

cessful and powerful traffickers in addition to capturing the weak, vulnerable, and incompetent ones. However, even an effective high-value targeting strategy will not end poaching, just as it has not ended drug trafficking. Far more so than leaders of terrorist groups, crime "kingpins" are easily replaceable. Difficult as it is operationally, targeting the middle operational layer of a smuggling group can be far more effective in disrupting its operational capacity than going after the kingpins, particularly if much of the middle operational layer can be brought down at once. Opportunistic, piecemeal interdiction will not achieve systemic deterrence effects.

Even with increasing transfers of high-power firearms and other technical assets, basic wildlife investigation and corruption skill sets continue to be deficient among rangers and wildlife enforcement officials in many countries. Lack of rehabilitation facilities not only hampers the release of seized animals back into the wild, but also discourages interdiction. But addressing pervasive and increasing corruption among wildlife law-enforcement officials in source and transshipment countries is important, though difficult, particularly when corruption also involves high-level politicians and government officials, including those tasked with wildlife protection. Even though special interdiction units focused on seizure and disruption are often the favored interdiction tool, they can end up going rogue and their success ultimately depends on steadfast support from high political levels.

En route interdiction of smuggled wildlife can create highly undesirable and counterproductive effects by encouraging smuggling networks to poach more to compensate for their losses. In this way, the wildlife trade differs significantly from the drug trade, where seizures higher up the supply chain are more effective than seizures close to the source. In the case of wildlife, the opposite holds: *in situ* law enforcement is crucial.

Dismantling wildlife trafficking networks is normatively important, and it can create temporary, short-term interruptions in supply, and even potentially long-term ruptures, but unless demand goes down as well, it will not by itself stop wildlife trafficking.

Effective incapacitation and deterrence require not merely sufficient arrests at the appropriate level of a smuggling network, but also effective prosecution and punishment. However, while baselines matter, the deterrent effects of increasing penalties maxes out at some point. It is not desirable to flood prisons with low-level poachers or corrupt rangers who will leave prison in a few years, with their lives destroyed and most likely as hardened criminals who resume poaching to make ends meet. Punishment that deters but also facilitates rehabilitation of former poachers is what is needed. Effective deterrence is not just about severity of punishment, but critically about consistency and certainty of punishment.

The Effects of Legal Supply from Captivity or Certified Sources, and Managed Legal Hunting

Allowing a legal supply of animals, plants, and wildlife products is equally fraught with dilemmas and imperfect outcomes. The arguments for legalizing drugs are substantially different from arguments for permitting legal trade in wildlife products. Critics of the war on drugs argue that since drugs cannot be eliminated from the world, criminalizing drugs overburdens law-enforcement and justice systems, empowers criminal groups, severely undermines human rights, compromises public health, and undermines anti-militancy efforts. Legalization advocates promise that legalization will bring resources to the state in the form of taxes and license fees, and that it will undo the negative effects just mentioned. Yet many critics of the war on drugs dispute whether drug legalization will in fact undo all of the negative effects, and they do not necessarily support outright legalization, preferring decriminalization and differently designed enforcement measures instead. The extent to which legalization will increase problematic use and addiction is also disputed.

The regulatory arguments for permitting legal wildlife trade are fundamentally different. First, farming can take pressure off wild resources. Second, allowing some level of trade can give hunters, ranchers, and others close to traded wildlife resources a stake in preserving species and entire ecosystems, and managing them sustainably. Third, regulated trade can also raise money for conservation.

But just as with bans, the effectiveness of licensing, eco-labeling, and creating a legal supply of wildlife depends crucially on the capacity of law-enforcement bodies to effectively monitor and enforce the regula-

CONCLUSIONS AND POLICY RECOMMENDATIONS

tions and restrictions concerning legal trade in all the countries involved in it; additionally, it depends on consumers preferring legally and sustainably produced items, and on them having the capacity to distinguish genuine from fake labels.

As with bans, many of the presumed or hoped for positive outcomes of legal trade do not always fully materialize in reality. A legal supply does not guarantee sustainability, nor does it necessarily take pressure off the wild supply. If a legal supply is expensive (because of heavy taxes to discourage demand) and difficult to produce (because breeding animals in captivity is hard), then poaching will likely persist. Captive-breeding and licensing schemes often do not prevent the leakage of illegally caught wildlife into the supposedly legal supply chain. Frequently, neither customers nor law-enforcement officials have the capacity or motivation to determine whether a wildlife product was obtained from the wild or from a captive-breeding facility, or from a legal supplier. Permitting a legal supply greatly increases the burden for law-enforcement bodies to differentiate between illegally and legally sourced products. However, while total bans on wildlife trade do away with the need for law-enforcement officials to distinguish between the legal and the illegal, they do not necessarily reduce the overall resource requirements. Legal supply can also reduce moral opprobrium surrounding the trade in particular species, thus inadvertently boosting demand for illegally sourced wildlife and whitewashing consumer consciousness.

Moreover, the wildlife revenues from a legal supply—such as breeding facilities or trophy hunting—do not necessarily go to local stakeholders and communities. Other actors, including local or national elites, large eco-businesses, or distant breeding facilities, can capture them through corruption, problematic regulatory redesign, or natural market dynamics. Local stakeholders may not benefit from conservation as a result.

Whether to design demand reduction campaigns as blanket opprobria to discourage all use of a particular wildlife product, or to build in enough nuance to persuade consumers to buy only a certified version of that product, depends on how malleable demand is to start with.

Eliminating the value of the illegal market via a massive expansion of legal supply—either regular or as a result of a one-off sale—or the supply of artificial replacement products can only work well if several conditions coincide: if the price of the legal product is considerably lower than that of the prohibited product and can be maintained so indefinitely; if consumers either prefer the legal product or cannot distinguish it from the prohibited one; and if law-enforcement officials can distinguish between the two and seize the illegal one. In other words, if all the problems of laundering and indifferent customers went away, such schemes could work. That does not mean that the substitution of less harmful products cannot work. In fact, there have been several cases of wildlife consumers switching to replacement products, whether as a result of a targeted campaign or exogenous shocks. However, whether the replacement products can remain environmentally beneficial and sustainable is another question.

In sum, there is no panacea—both blanket bans and licensed supply are fraught with problems, difficulties, challenges, and downsides. The factors that determine the level of effectiveness of licensing wildlife trade in particular species are to a large extent the same as those that determine the effectiveness of bans: the level and quality of law enforcement; the elasticity of demand; the ability to supply licensed products cheaply and on a large scale; the strength of non-price-driven effects on consumer preferences, such as caring for biodiversity preservation; the property rights regimes in place; the timing of the scheme; and the value of non-consumption uses, such as ecotourism. Biological factors—such as species' reproductive rates and the ease with which they can be bred in captivity—obviously also play a crucial role.

Effective law enforcement is also crucial for the good functioning of legal markets—whether in wildlife or drugs. If legal markets are not well policed, they will merely serve to hide illegal supplies and encourage both poaching and an undesirable growth in demand.

Local Community Buy-Ins, Alternative Livelihoods, Community-Based Natural Resource Management, and Conditional Cash Transfers

Despite current characterizations of poaching networks as highly organized criminal enterprises, many poachers are members of marginalized and desperately poor local communities. Sometimes their poaching activities are fully separate from wildlife trafficking. At other times, these local poachers supply global trafficking networks or work for

CONCLUSIONS AND POLICY RECOMMENDATIONS

them as hunters, carriers, trackers, and spotters, as corrupt park rangers also sometimes do. Focusing on finding legal livelihoods for them can be an important component of policies to reduce the illegal trade in wildlife and incentivize communities to resist wildlife trafficking. Although creating economic incentives for communities to support conservation does not address the problem of high-level wildlife smugglers and organized criminal groups, it can simplify and focus law-enforcement efforts. It can encourage community cooperation with law-enforcement bodies, enhance the political sustainability of restrictions on wildlife trade, and reduce political conflict.

Generating economic incentives for the poor to support conservation policies is also important normatively because the marginalized communities dependent on hunting for basic livelihoods have often suffered greatly as a result of environmental conservation. They have been forced off their land in areas designated as protected, and their livelihoods and human security have been compromised. Laws and regulations are easiest to enforce if the vast majority of people accept them as legitimate and internalize them. The devolution of decision-making power to those who have been poor, marginalized, and without rights may not only be politically and economically beneficial, it can also be psychologically rewarding and enabling.

Yet many alternative livelihood schemes have not been effective. While frequently prescribed as the mechanism for conservation, ecotourism has unfortunately failed to be a consistent remedy against poaching and wildlife trafficking or a reliable tool of economic growth. It rarely generates sufficient income and jobs, as a result of both its own internal limitations and resource capture by elites. Thus the objective that communities earn enough to maintain prior subsistence levels, let alone achieve economic and social advancement, from ecotourism, other alternative livelihoods, or the limited hunting of wildlife, while crucial, is often elusive.

There is a large variation of outcomes concerning the effectiveness of alternative livelihoods and local community involvement in managing habitats and wildlife, some of which reveal the detrimental effects of allowing or not allowing local communities to exist in national parks. Factors such as a community's short-term versus long-term economic horizons, income and employment levels, their consumption attitudes or values-based attitudes toward nature, community cohesion and leadership structures, as well as property rights and their enforcement, also determine whether local community ownership or allowing local people to live within protected areas results in desirable environmental protection. Alternative livelihoods that address all of the structural drivers of illicit economies have the highest chance of being effective.

CBNRM schemes often go beyond alternative livelihoods, ecotourism, compensation, or allowing limited hunting. They seek to transfer rights to local communities and achieve three objectives: political empowerment, poverty alleviation and local economic development, and environmental protection. They can be based around trophy hunting, ecotourism, or other alternative livelihoods. Sometimes they have worked spectacularly well, but the outcomes have not been uniform, and for all the successes there are deficiencies and outright failures. Beyond good implementation, successes are dependent on a steady and large flow of tourists, trophy hunters, and customers, and sufficiently low densities of people compared to wildlife. Thus they have worked better in arid areas, where agriculture is not profitable but where iconic wildlife species that tourists want to see are present, than in either fertile areas, where converting land to agriculture is profitable, or tropical forests, where animals are difficult to see and where industrial logging brings far greater revenues than ecotourism. In some cases, communities became richer as a result of CBRNM policies, but then intensified unsustainable hunting and logging to further augment their economic resources at the expense of environmental conservation.

CBNRM advocates maintain that poor communities are not environmentally unfriendly. Indeed, many poor people do have strong interests and values that promote environmental conservation. So do some well-off people. But many among the poor may want to maximize their current (not just long-term) economic benefits and make money, even if that means causing major environmental degradation—just as middle-class and rich people and businesses have been and continue to be responsible for environmental degradation. Being "local" and poor hardly determines whether one's actions will be benevolent or detrimental to the environment.

Moreover, whether the CBNRM tenet that a local community should have the right to local land and its natural resources can be put

CONCLUSIONS AND POLICY RECOMMENDATIONS

into practice will depend on, among other things, whether local communities are sufficiently cohesive and whether eligibility can be established. In areas where there are too many competing claims over land rights, other approaches, including the establishment of exclusive protected areas, may be necessary.

Technical assistance to local communities and financial support by governmental and NGO donors is often necessary for the success of CBNRM projects. This assistance may have to include building leadership and accountability structures, managerial and business training, the diversification of income sources, and the restoration of customary patterns of natural-resource use. Advice on the management of ecotourism projects may be needed so that the local community can compete with large eco-businesses, but also so that any resulting tourist influx does not cause greater environmental degradation. In all these scenarios, the local community should have a strong voice. Nonetheless, outside proponents of CBNRM approaches often become deeply and lengthily enmeshed in the complexities of local community politics. Just as conservation does not escape global politics, it cannot escape local political contexts.

Conditional money transfers to poor local populations—such as paying people not to poach and to collaborate in implementing wildlife regulations—is an underutilized tool in conservation efforts, though more widely adopted in other policy domains. Its effective design requires extensive monitoring and enforcement assets. Such schemes are often fraught with moral hazard and messaging problems. Any such scheme must carefully evaluate the size of the pool of potential beneficiaries and claimants, eligibility cutoffs, and sufficient levels of compensation.

Anti-Money-Laundering Efforts

Adding AML measures to the toolkit of wildlife protection brings benefits. The designation of wildlife trafficking as a predicate offense—a basis for deploying financial intelligence tools and seizing assets—can be useful. Some potential customers may be deterred from poaching as a result, and many are likely to be deterred from using banks as a mechanism of money laundering. Financial intelligence can help identify networks and enhance physical interdiction and prosecution.

Nonetheless, the effectiveness of AML measures is highly contingent on the specific susceptibilities of the target. Financial interdiction, just like physical interdiction, is very difficult. Beyond banking, many money-laundering and evasion mechanisms exist. What is often required for success is coupling financial intelligence with other streams and sources of intelligence. But, as has been found with drug trafficking, AML mechanisms are unlikely to deprive poaching and wildlife-trafficking groups of significant money flows.

Demand Reduction

Neither bans combined with intensified law enforcement, nor licensing, nor the establishment of alternative livelihoods are sufficient to stem illegal trade in drugs or wildlife on their own. Directly tackling demand for illegally sourced wildlife is crucial.

The demand for wildlife is very diverse and includes many different drivers of consumption and many different types of users. Since target populations and the markets that supply them are highly heterogenous, policy compliance or demand-reduction efforts need to be designed specifically for each subgroup. Having a nuanced understanding of each subgroup and of the variation in demand for wildlife products across space and time is critical for policy effectiveness. Reducing demand for bushmeat by ensuring that there are acceptable alternatives available to meet people's subsistence and protein needs requires different policies from those needed to reducing demand for luxury exotic meat from endangered species.

There are important instances of successful campaigns to reduce the consumption of wildlife products. Even in entrenched markets with some of the most widespread and deeply engrained wildlife use, such as in East Asia and China, there have been modest changes in attitude and a reduction in the taste for the products of the wild.

Experience from environmental campaigns as well as others aimed at drug users and cigarette smokers shows that solely relying on consumers' altruism—whether toward other humans or other creatures—is often not very effective. This is particularly the case when problems and costs are diffuse and delayed. Instead, some of the most effective campaigns have been those in which people perceive the behavior that the campaign wishes to change as a threat to themselves personally in

CONCLUSIONS AND POLICY RECOMMENDATIONS

the short term—for example, that health or sex life will suffer negative effects within a relatively short period of time.

Anti-use messaging must be maintained over time and adapted to new subgroups of potential consumers, otherwise even previous accomplishments may wither. A campaign that resonates with one subgroup at a specific time may not resonate with another subgroup five years later.

Perhaps the most effective method of discouraging drug use and problematic alcohol use are programs based on mild but swift and certain punishments. Applying such a regime should be explored for dealing with wildlife law violations, even if detection and identification problems can create far larger obstacles to enforcement than is the case with drugs and alcohol. Compliance rates are often low when noncompliance is difficult to monitor and when the target to be monitored is widely dispersed. However, even less-than-perfect and certain enforcement increases compliance with a new norm or law. Consistent and mild punishments are far preferable from both human rights and resource-demand perspectives and are more politically sustainable than arresting the users of illegal wildlife products for lengthy periods. Establishing mild punishment for low-level buyers, escalating with the intensity and frequency of violations, will be more effective in discouraging demand than not punishing users at all.

The most effective drug-use prevention strategies, though not very effective overall, appear to be those which focus on early-age intervention (with children as young as five or six), and these are not specific to drugs but seek to build self-confidence and resistance to peer pressure. Focusing on demand-reduction measures for wildlife products among young children and teenagers so that they do not become heavy users, and so that they become transmitters of new norms toward adults in their vicinity, makes good sense.

Demand-reduction efforts to reduce the consumption of illegally sourced wildlife need to be carefully designed so as not to eliminate all demand for wildlife products when conservation policy is built on economic incentives for conservation. But fine tuning policies and programs to boost the consumption of licensed green-certified products without increasing overall demand, including for illegally sourced products, is very complex and dependent on a host of pre-existing market conditions, including whether demand for a particular commodity has already reached a plateau or continues to rise.

Ultimately, it is not the number of seizures or arrests, the amount of money delivered to poor communities to incentivize their commitment to conservation, or the intensity of adverts to stop the consumption of illegal wildlife products that are the true indicators of success. If the primary goal remains environmental conservation (as opposed to community empowerment), the real measures of policy effectiveness are an increasing or stable size of wild animal populations and a decreasing number of animals poached.

Exemplary Success Stories

All of the policy tools analyzed in this book resulted in some successes and many failures. As they are few and far between, the successes deserve special highlighting. Among these are the bans on trade in rhino and elephant products during the 1990s, after several decades of failure. Both were enabled by law-enforcement successes in range countries, and crucially by a decline in demand for ivory and rhino horn in the United States and Europe, as well as ultimately in Japan, Yemen, and South Korea. The cumulative success, however, did not withstand the rise in demand in China, Vietnam, and Southeast Asia over the past two decades, which affected both elephants (in whose ivory there is domestic legal trade in several countries and one-off international sales were authorized twice since the ban came into effect) and rhinos (whose products have been subject to a total ban). Bali's ban on collecting marine turtles for local ceremonies, combined with an effective campaign to reduce local Balinese demand, is another example of an effective policy. The ban on hunting Tibetan antelope in the India–China border region is another success, crucially enabled by a reduction in demand for the *shahtoosh* shawl and the involvement of the local community in law enforcement.

Perhaps the greatest success of allowing legal supply in order to preserve a species has been that of the legal farming of crocodilians, which dramatically reduced poaching by satisfying existing demand. The fact that crocodilians are easy to breed on farms crucially enabled the policy's success. However, with renewed demand for crocodilian

CONCLUSIONS AND POLICY RECOMMENDATIONS

products recently emerging, there also seems to have been a rise in poaching, at least in Latin America. Another (partial) success story concerns parrots, with supplies from breeding facilities having helped reduce the poaching of them in the wild. Once again, the ease of captive breeding combined with the diligent monitoring of breeding facilities and law enforcement against poaching in the wild was crucial for success. However, if captive breeding is difficult and the monitoring of breeding facilities and protected areas poor, poaching and laundering tend to persist even when there is a legal supply. Therefore, as in the case of parrots, poaching can increase over time despite the existence of a legal supply from captive-breeding facilities.

The conservation of the vicuña in Peru—based on giving local communities a set of economic incentives to preserve the species, including the managed hunting and marketing of vicuña pelts—is one of the exemplary success stories regarding creating local support for conservation. In other cases, allowing locals to hunt antelopes and zebras—species that easily breed in the wild—such as in Laikipia Conservancy in Kenya, have similarly greatly enhanced conservation efforts and are another prominent example of successful CBNRM.

Namibia is yet another example. With highly auspicious circumstances in terms of the relative densities of people and wildlife, and the prevalence of arid land unsuitable for agriculture, the establishment of communal conservancies based on CBNRM models led to the recovery of black rhinos, elephants, Hartmann's zebra, and other wildlife species in large parts of the country's north between the 1980s and 2010s. Moreover, revenues obtained have been shared equitably within communities. But even here, global wildlife trafficking pressures are undermining the success, and poaching, notably of rhinos, has increased.

Finally, examples of effective demand-reduction campaigns include the anti-fur campaign in Europe, the United States, and Canada during the 1980s and early 1990s, which led to significant decreases in the hunting of several species for pelts, including seals.

But these successes do not point to one single effective policy template. Even as many species are rapidly running out of time, there is no panacea. Thus, in designing environmental regulations and choosing among bans, legal supply, CBNRM, alternative livelihoods, and compensation strategies, and among specific designs of AML measures or demand reduction, anti-poaching and conservation policies need to pay attention to local institutional and cultural contexts, as well as the biological characteristics of species. Different localities have varied enforcement capacities and different susceptibilities to rule and norm internalization. Corruption can kill the best regulatory design. As with drugs, context specificity critically determines what regulations work in which settings and accounts for a great variety of outcomes. In the case of environmental regulation, context specificity is even more badly needed since the unique characteristics of particular species further increase the variability of policy effects.

Experimentation with policy designs, coupled with effective and constant policy evaluation and monitoring, is thus essential. Changing policies is hard, however, particularly when vested interests, such as powerful industries, government or law-enforcement agencies, scientific lobbies, or environmental NGOs, manage to capture policy and seek to derail policy adjustments.

That said, in the current context of vastly expanded demand for unsustainable wildlife products and the pervasive laundering of illegally sourced products through legal supply chains, legal sales of ivory and rhino horn should not be permitted. China should be encouraged to implement its promise to close down its ivory market as soon as possible, with forethought given now as to how to design interdiction and law-enforcement policies. I provide some ideas for such an interdiction and enforcement design below. Appropriately at the September 2016 Conference of the Parties to CITES, Swaziland was denied a license to sell its stock of rhino horn and Namibia and Zimbabwe were denied licenses to sell their stocks of ivory. These and other countries should not be given licenses to sell either ivory or rhino horn until both demand and poaching significantly decline and better enforcement structures are in place in countries where there is supply and/or demand. Such outcomes are unlikely to materialize for many years. In the current environment, one-off sales will not cause a price crash in the market; suppliers will wait them out, and too much leakage of illegally sourced product into the legal supply chain will take place.

However, if conditions change—if demand levels off, and law enforcement is tightened so that there is a lot less leakage of illegally sourced products—then serious consideration should again be given to

CONCLUSIONS AND POLICY RECOMMENDATIONS

permitting a tightly controlled legal supply. The environmental advocacy community will do little good for wildlife by doggedly sticking to one policy regardless of how the context has changed. Providing economic incentives for conservation to as many stakeholders as possible greatly facilitates the enforcement of wildlife regulations. Thus the US decision not to permit the importation of hunting trophies from countries such as Zimbabwe, where legal hunting and wildlife trade are pervaded by corruption and where regulations and enforcement do not guarantee wildlife sustainability, is correct—not because all hunting should be prohibited but because the regulations in those countries are not adequately enforced. But it is equally important and appropriate that the United States continues to promote such economic incentives, including trophy hunting, in countries with well-designed and robustly enforced policies, and help the problematic ones to improve so they too can access economic incentives for conservation.[1]

Lessons from Wildlife Policies for Dealing with the Drug Trade

Although this book is about what can be learned from the field of drug policies and applied to that of combating wildlife trafficking, the drug policy community can also learn a lot from global efforts to combat poaching and wildlife trafficking. Much of drug-policy advocacy has centered on the simplistic assumption that legal trade in addictive drugs will *ipso facto* end drug-related violence and eliminate drug trafficking. Such outcomes are in fact highly contingent on policy settings. As the experience of permitting legal wildlife trade shows, a legal economy can be equally pervaded by corruption, criminality, and the laundering of illegally sourced goods, as well as associated with large increases in the trafficking of illegally sourced commodities.

Moreover, the legal wildlife business—whether trophy hunting or photographic ecotourism—is often captured by local, national, or global businesses and elites, and the benefits do not trickle down to poor, local populations. Only with a lot of regulatory design and concerted efforts to empower local communities and provide them with technical, business, and managerial skills can those communities expect to compete effectively against large eco-businesses. The lessons for those who believe that poor, marginalized farmers who grow illegal crops will easily and immediately benefit from drug legalization is clear. The odds are very high that they will not. A lot of effort will be needed to increase equity and restructure social contracts in places like Colombia or Afghanistan, and the political, economic, and technical empowerment of poor farmers must be a core feature of any regulatory package. Otherwise, even if the local security situation permits legalization, the odds are that big businesses (including out-of-country ones) will snatch drug production from the *cocaleros* and opium farmers, who may end up worse off than if drug production continued to be illegal. Clearly delineated land rights are a crucial element of success. So is involving local communities in decision-making processes, empowering them, and taking into account their preferences in policy designs.

That does not mean that one should never allow legal trade in wildlife or drugs. At least in the case of wildlife, the predisposition should be toward legal trade. But the outcomes will depend on local institutional and cultural settings, as well as the inherent characteristics of the traded commodity.

It is equally crucial that efforts to mitigate the illegal drug trade and drug cultivation look beyond the effects on drugs and also examine the interactive effects with other illicit economies, including poaching, wildlife trafficking, and illegal logging. All too often policies to suppress or mitigate drug cultivation, such as in Colombia or Burma, have led to increased destruction of natural ecosystems and the increased slaughter of species, with logging and wildlife trafficking replacing the illegal drug trade.

Policy Guidelines and Recommendations for Addressing Poaching and Wildlife Trafficking

In designing environmental regulations and their enforcement as well as demand-reduction strategies, the policy community should consider the following guidelines and recommendations.

Basic Guidelines

Regulatory approaches, whether bans or licensing schemes, should be tailored carefully to fit local institutional and cultural settings, and

CONCLUSIONS AND POLICY RECOMMENDATIONS

species' local ecological requirements based on careful data-based analysis of such settings. Designs for policy interventions against the illegal wildlife trade simply cannot follow a general template, given highly variable outcomes of policies. This may well mean that a species will be licensed for farming or trade in one area, but not in another. Case specificity—informed by lessons from other cases, no doubt—is a critical ingredient in designing an effective regulatory framework.

When determining whether to ban or license managed trade in a species in a particular locality, careful consideration needs to be given to: the level and quality of law enforcement in that locality; the degree and patterns of corruption; the elasticity of demand for the species; the ability to supply licensed valuable products cheaply and on a large scale; the strength of non-price-driven effects on consumer preferences; the property rights regimes in place; the timing of the licensing scheme or ban; and the value of non-consumption uses, such as photographic ecotourism.

To promote compliance with regulations concerning supply, transshipment, and demand elements of the trade, policies need to mitigate the direct and opportunity costs of compliance in terms of monetary value, income loss, time and effort, prestige loss, and the cultural dissonance suffered. That may mean distributing pharmaceuticals or other legal medication to customers who use rhino horn to cure cancer and lack access to alternatives, or providing poor, marginalized communities with compensation and CBMRM mechanisms. Although it makes good sense to prioritize the most important barrier to compliance, prioritization judgments may need to be adjusted over time as preferences and market conditions change.

Adopted policy designs need to be diligently monitored. There needs to be a willingness and expectation to alter, even radically, policy designs that do not produce effective outcomes in particular locales. This may mean extending or removing licenses, intensifying alternative livelihood efforts in one area and abolishing them in another, or increasing law enforcement. To effectively assess policy effectiveness, systemic and repeated surveys of wildlife populations must be regularly undertaken.

Law Enforcement and Interdiction

In large parts of the world, there is an urgent need to increase resources for wildlife law enforcement. Equally, in many parts of the world, it is crucial to enforce existing wildlife laws far more diligently. Policy designers should seek to reduce trade-offs and competition, and increase complementarities with other law-enforcement priorities. Thus, checking for particular illicit commodities may easily include and incorporate searching for other illegal contraband. For example, anti-piracy flotillas off Somalia can and should also act against illegal fishing. But while boosting the focus given to wildlife enforcement is necessary, it is important to avoid exaggerating the links between poaching and terrorism, militancy, and organized crime.

Scarce wildlife enforcement resources need to be concentrated rather than thinly dispersed. Law enforcement should thus focus on critical smuggling hubs. However, such an approach needs to be complemented by ensuring sufficient law-enforcement capacity to detect the emergence of a more covert black market.

Interdiction patterns to reduce poaching need to be tailored to local patterns and networks. A careful analysis of existing local structures and patterns of wildlife poaching networks is crucial for designing effective interdiction policies. Generalized assumptions about the vertical integration of smuggling networks or the involvement of militant groups may often be incorrect.

That said, focusing interdiction efforts on the middle operational layer of wildlife smuggling networks, not merely on high-level traders, will greatly augment their effectiveness. Ideally, as much as possible, the middle operational layer of a smuggling ring should be brought down in one enforcement action as possible, as opposed to in a piecemeal staggered fashion, to complicate the group's replacement capacity.

In situ law enforcement is by far the most effective and important form of law enforcement as the goal is to minimize the number of animals illegally removed from the wild. Seizures are not a good measure of enforcement effectiveness and *en route* interdiction should not be overemphasized since it can result in greater numbers of animals poached to supply the retail market. In that case, *en route* interdiction efforts would be directly counterproductive to the fundamental objec-

CONCLUSIONS AND POLICY RECOMMENDATIONS

tive of conservation efforts. But even though *in situ* law enforcement is crucial, care must be taken to avoid human rights violations. Indeed, the effectiveness of wildlife law-enforcement units needs to be judged not only in terms of whether poaching is being reduced and/or environmental conservation enhanced, but also in terms of their human rights record and how they affect human security. Policies to shoot poachers on sight should never be adopted—they are morally reprehensible and they are not effective. Wildlife law-enforcement units must respect human rights, and should not employ force beyond what is necessary to deter violators and to protect themselves.

Enforcement also needs to be significantly augmented in retail markets, with sellers of illegal wildlife products meaningfully punished within local contexts. That does not mean prison terms for all violators. In the case of illegal luxury wildlife products—such as traded ivory, rhino horn, or tiger products as opposed to subsistence bushmeat from non-endangered species—retailers who egregiously violate regulations should be shut down. Traders who sell bushmeat from poached non-endangered antelopes in local subsistence markets should face much milder penalties and perhaps be dealt with through nonpunitive strategies to steer them toward lawful behavior.

Law-enforcement strategies need to anticipate problematic side effects and build mitigation mechanisms into policy designs. That means anticipating how law enforcement will displace poaching to new areas and transshipment routes, and perhaps even to new species that poachers will target. As much as possible, wildlife enforcement assets should be prepositioned in those new smuggling domains and attempt to deter them.

Given the low baseline of punishment for wildlife crime, increasing penalties for high-level smugglers and middle-level operators makes sense. But anti-poaching efforts need to avoid flooding prisons with low-level wildlife offenders. Such an outcome is counterproductive for many reasons: ethical (as many poachers are often very poor, marginalized, and dependent on hunting for their livelihood, even if they supply global wildlife trafficking networks), economic (the cost of prisons), law enforcement (prisons tend to become schools for criminals, not places of rehabilitation), social (the violence and devastation imprisonment causes to the families of the imprisoned), and political (the likelihood that communities will mobilize against all enforcement). By contrast, mild but swift and certain punishment for low-level poachers and assistants, such as trackers and spotters, can actually augment policy effectiveness.

Unless corruption among rangers, prosecutors, and courts dealing with wildlife crime is reduced, even perfectly conceived interdiction policies will fail. Thus intensifying training and anti-corruption efforts among park rangers makes good sense. This must include rewarding effective and committed rangers and law-enforcement officials who dedicate themselves to wildlife: they should be paid well, promoted, and celebrated. But such capacity-building measures need to be coupled with demands for accountability, merit-based appointments (even within patronage networks) and the constant vetting of rangers and other involved officials. Corrupt rangers need to be punished and removed from the force. Community supervision may sometimes be a useful tool for monitoring rangers fired for corruption.

It is equally essential to monitor special interdiction units, which may themselves become the most potent poachers and traffickers of wildlife. If pervasive corruption persists among wildlife law-enforcement bodies, particularly at the highest level of government in source countries, cutting off at least some foreign aid will be appropriate and necessary.

Providing Economic Incentives for Conservation and Designing and Implementing Licensing Systems

Developing economic and non-economic stakes beyond law enforcement for governments, interest groups, and local communities is crucial for the effective preservation of wildlife and natural ecosystems. Indeed, the predisposition should be toward permitting legal and licensed trade. It is equally necessary to remove or minimize economic incentives for the conversion of land to agricultural use; tools for this include differential taxes on land use, conservation subsidies, and communal property rights schemes. Appropriate property rights systems must be in place and dispute resolution mechanisms available.

The licensing and regulation of legal wildlife trade must be carefully designed and monitored. This requires developing databases of legal stocks and supplies and undertaking systematic registration within all sectors of the wildlife trade sector. Also, breeding facilities should be

CONCLUSIONS AND POLICY RECOMMENDATIONS

required to install CCTV with live feeds to wildlife law-enforcement agencies so as to minimize the chances of wildlife laundering. If wildlife farming facilities turn out to be economically unviable because they do not produce sufficient numbers of animals at a relatively cheap price and thus launder poached animals, they need to be shut down swiftly. They should also be shut down promptly if the farmed animals are treated cruelly and are kept in inhumane deplorable conditions. Such monitoring requires conducting frequent and extensive inspections along the entire supply chain and diligently punishing violators of regulations.

To improve the effectiveness of certification, it is necessary to have certifiers who are fully independent and not paid by the businesses or governments seeking to obtain the particular legal, environmental, or socially-desirable-conduct certification.

Alternative Livelihoods, Community-Based Natural Resource Management, and Conditional Cash Transfers for Low-Level Poachers and Marginalized Populations Dependent on Natural Resource Extraction

Providing alternative livelihoods for marginalized populations dependent on wildlife use for their basic livelihood is necessary for both human rights reasons and the effectiveness of policies to conserve wildlife. But such programs must be designed to stimulate broad rural and social development, and they must include policies that specifically target marginalized communities that are dependent on natural resources and are responsive to their needs. General poverty reduction measures may not be accessible to forest-dependent communities and may not be effective at weaning them off the over-utilization of wildlife. Ecotourism that also significantly benefits poor, local communities—not just any ecotourism—needs to be promoted.

Alternative livelihoods and other socio-economic measures to benefit poor local communities should not be limited to generating income, crucial as that is. They can and should also include infrastructure improvements (hopefully in an environmentally sound manner, even though roads often cause major environmental degradation by enabling the influx of new people to an area). They should also include the delivery of medical and educational services, and the development of local safety nets. Promoting agricultural productivity that increases yields without adding pressure on the environment, and improving soil fertility and water resource management, are also important. Consulting the local community and listening to its preferences will be critical to success.

In sum, alternative livelihood efforts need to address all the structural drivers of the illegal wildlife economy. Design elements must encompass the generation of sufficient employment opportunities (including off-farm and off-park employment), infrastructure building, the distribution of new technologies, marketing help and the development of value-added chains, the facilitation of local microcredit facilities, establishment of access to land without the need to participate in the illicit economy, and functioning and enforced property rights—to name a few of the most prominent components. They should not be designed as discrete and isolated handouts. But the ecological impact of such livelihood efforts must be built into policy assessments. As legal alternative livelihoods become available, law enforcement against the community's participation in poaching should intensify.

Whenever possible, CBNRM approaches should be adopted. Local communities should be given rights to conservation land and any proceeds from sustainable wildlife utilization. Communities should be assisted by NGOs to secure their rights and to have a strong voice in determinations of land use, and they should have protection to achieve environmental equity and sustainability.

As part of creating buffer zones and areas beyond them, and sometimes even in core parts of protected areas, it may make good sense to permit the limited hunting of non-endangered species and the limited sustainable extraction of natural resources to mitigate food insecurity and income loss among poor, local communities. With good monitoring and the enforcement of sustainable practices, trophy hunting managed by local communities should also be permitted, at least for species not experiencing extraordinary poaching rates such as elephants and rhinos.

Thus it is crucial to make case-by-case assessments of what type of hunting is sustainable at what levels. Careful monitoring and reassessments need to be conducted regularly and repeatedly to reassess whether wildlife populations and the ecosystem are bearing up well under limited exploitation, as the level of impact may change over

CONCLUSIONS AND POLICY RECOMMENDATIONS

time—whether as a result of increased human population density or the arrival of external poachers.

The rights conferred should be limited, with certain restrictions employed to preserve biodiversity. This does not preclude sustainable logging or the hunting of non-endangered species, either for subsistence or trophy hunting, nor does it preclude limited grazing in protected areas. If necessary global hunting bans or bans on particular wildlife products constrain economic resources for local communities, those communities should be given adequate compensation, and the onus must be on governments and environmental NGOs to raise money for such compensation. Obviously an eligibility cutoff will have to be established, over which compensation will no longer be provided.

Such conditional rights should also include the community's entitlement to 100 percent of the resources derived from sustainable wildlife management; however, the resources should be taxed. That way government support for a local community's efforts can be established, and the state may have fewer incentives to collude in deforestation or poaching by non-locals; it will also be less likely to leave the community high and dry when other actors, such as external wildlife traffickers or logging industries, threaten local nature.

If allowing people to settle in or exploit environmentally sensitive ecosystems results in those ecosystems being degraded, then consultative and compensated resettlement may well be necessary. At a minimum, compensation must ensure that the local community is economically no worse off than when it lived in the protected area, and ideally the level of compensation should also reduce poverty. That means, however, that proponents of exclusive protected areas must help in raising money for use-restriction and resettlement programs.

Providing conditional money transfers to poor local communities who may be negatively affected by conservation should be considered. To prevent wildlife poaching, small-scale illegal logging, or illegal grazing by the poor, such conditional money transfers would require effective, extensive, and corruption-free monitoring systems and the application of penalties. If a family persisted in poaching, its monthly payments would be suspended. If violations still persisted, additional short and swift penalties could be added. Crucially, law-enforcement agents would have to be able to provide evidence of poaching by specific offenders.

A cutoff of eligible beneficiaries would have to be established so that people do not start moving into an area in order to get on the payroll. Initially, not just low-level poachers and other violators, but also local residents who do not violate environmental rules should be included in compensation schemes to prevent moral hazard and for the sake of basic equity.

Sufficient funds to cover a large enough segment of the local population and to make a meaningful difference to household income would have to be ensured. Beyond cash payments, other incentives may have to be included when other costs and drivers motivate wildlife crimes. As in the case of alternative livelihoods, these may include security threats, lack of access to microcredit, lack of value-added chains, and cultural habits. Involving local communities in all elements of the compensation scheme design will be crucial.

Anti-Money-Laundering Efforts

AML methods should be included in the toolkit of wildlife protection, particularly to facilitate intelligence gathering and interdiction. But it is imperative not to overestimate the size and scope of the deterrent effect of this, and the capacity of AML measures to block money flows to poaching and wildlife trafficking networks.

Demand Reduction

A strong focus on demand reduction is crucial. Demand-reduction strategies need to be tailored to the specific sources and causes of demand as well as specific user subgroups. For example, if poaching is driven by protein needs, appropriate demand-reduction strategies may include allowing sustainable hunting, dispersing non-endangered prey to zones where hunting is permitted, or providing alternative sources of protein. If demand is driven by a desire for luxury goods and status-driven display, punishing retailers and customers and driving demand underground make good sense. If the high cost and limited availability of painkillers and cancer medication drive customers to consume rhino horn and other kinds of Traditional Chinese Medicine (TCM), widely distributing cheap, non-TCM medication can dry up the undesired

CONCLUSIONS AND POLICY RECOMMENDATIONS

demand (assuming someone is willing to pay for the distribution of the substitute drugs and legal distribution channels can be established). Of course, handing out a limited number of pills is easier than establishing accessible medical care, though the latter might be more important than the availability of medication.

Messaging equally needs to be carefully tailored to particular consumer subgroups, based on the issues that matter to them. The campaigns should not rely on general awareness programs or altruism-based messages. To reduce demand for many TCM products, messaging strategies should emphasize the negative effects to health and sex resulting from wildlife consumption. Messaging campaigns can also emphasize that consuming illegally sourced wildlife undermines attempts at "being cool" and "saving face" with respect to peer groups that matter to consumers. Mobilizing local opinion makers—picked for specific user subgroups—enhances messaging effectiveness. Outsiders such as foreign NGOs are often less effective as the principal messengers of demand-reduction measures. It is necessary to maintain anti-use messaging over time, but the content of messaging needs to be adapted to new entrants to the market. It is useful to prioritize early-age intervention before consumers become heavy users, and preferably before they become users at all.

Coercive approaches to reduce demand for wildlife products can also play an important role. Sanctioning low-level wildlife product users with mild but swift and certain punishment should be tried—for example, shaming them through sentencing them to undertake public works if caught buying small amounts of ivory or rhino horn. They should also be subject to increasing fines. Mild, but escalating, swift, and certain punishment can be meted out to restaurant owners who sell exotic meat from endangered species and to TCM traders who sell potions from endangered species. Identifying violations among the latter group will be easier than identifying customers who order illegal products.

If regulatory designs are critically based on economic incentives for conservation, demand-reduction strategies need to be designed with nuance so as to reduce demand for illegally sourced animals, plants, or wildlife products, but not to reduce overall demand for all wildlife products, including the licensed ones.

As with all policy tools, it is crucial to diligently monitor demand-reduction measures. Evaluations cannot be based merely on potential customer and user surveys. They must also include surveys of sellers' perceptions of changing demand and be accompanied by robust qualitative, including undercover, assessments. Likely demand-reduction policies will need to change over time as markets change—whether as a result of policies, internal market fads, or exogenous factors.

Overarching Lessons and Guidelines

There are perhaps two major policy lessons. The first is that policy designs need to keep in mind the frequent tension between legality and legitimacy that many restrictive conservation policies feature. They should be designed not only to maximize biodiversity and species preservation but also to minimize the cost and harm they generate, particularly to poor and marginalized populations. Sometimes there will be a happy overlap of local communities' political and economic empowerment and species preservation. But sometimes there will be a trade-off. Hunting will not always be sustainable, or a local community's presence in a protected area may not be consistent with environmental conservation. In those cases, the cost to the community of conservation measures, such as bans on hunting or of resettlement, needs to be justly and adequately compensated. If laws and regulations are rejected by most in the community and not internalized, law enforcement becomes far more challenging and difficult. The goal of policy should be to incentivize as many people as possible to support regulatory designs that maximize conservation. Offsetting the costs that environmental conservation imposes on the poor and marginalized is not only normatively desirable and necessary, it also facilitates law enforcement. At minimum, it reduces the chances that members of local communities will be recruited as poachers by wildlife trafficking networks; under auspicious circumstances, it can actively encourage the participation of local communities in vital *in situ* law enforcement.

The second most important policy implication is that given the serious trade-offs, downsides, and limitations of each policy tool, as well as their highly different outcomes, policy adjustments and flexibility are necessary. What this implies is that policy designs will need to vary not only from place to place and species to species, but also over time. Exogenous shocks may eviscerate a policy that has worked

CONCLUSIONS AND POLICY RECOMMENDATIONS

very well in a particular locale. Sometimes that may mean merely adjusting some element of that policy, such as redesigning demand-reduction strategies to fit new entrants, or strengthening controls on trophy hunting to reduce corruption and other kinds of misbehavior—so that while biodiversity is preserved and conservation is enhanced, poor and marginalized populations, as well as other stakeholders, are not deprived of vital means of income. At other times, exogenous shocks and internal policy deficiencies may mean that a regulatory design needs to be completely overhauled, such as when a disease so reduces the size of a previously abundant species that further hunting becomes unsustainable. Conversely, there can be positive perturbations—for example, if the legal trade in rhino horn or ivory declines and stabilizes, there may well come a time when it is possible to permit legal trade in previously prohibited commodities to generate resources for poor, local populations and materially incentivize other stakeholders to support conservation.

But the need to review policies on a regular basis and to be prepared to significantly alter them is not easy since stakeholders tend to develop vested interests in the preservation of existing policies from which they benefit materially or to which they are ideologically wedded. Nonetheless, as difficult as such reevaluations of policy are politically and psychologically, their necessity is perhaps the most important lesson of this book. The predisposition should be toward permitting sustainable legal trade and toward CBNRM programs as long as they are consistent with biodiversity preservation. Only when they stop being so should bans and law enforcement be intensified, thus necessitating compensation for the poor and marginalized who unfairly bear the burdens of biodiversity preservation and conservation. But a few years down the road, such as when demand for a particular species declines and stabilizes and when licensed trade builds in better regulatory controls and monitoring, greater sustainable legal trade should be allowed again.

Given these general policy recommendations, what would a specific policy design, tailored to a particular area, look like? What are the policy implications, for example, for the upcoming ivory bans in China and Hong Kong, arguably two of the most important new environmental regulations regarding wildlife trafficking?

THE EXTINCTION MARKET

How to Design a Ban on the Ivory Trade in China and Hong Kong

At the end of 2016, the governments of China and Hong Kong made substantial commitments toward combating wildlife trafficking. Most significantly, China promised to outlaw and shut down its domestic market in ivory by the end of 2017, while Hong Kong promised to do so by 2021. At the September 2016 Conference of the Parties to CITES, China strongly endorsed ending all domestic markets in ivory (though insisting that domestic markets in products from farmed tigers should remain legal and were not the business of CITES). Both China and Hong Kong should swiftly implement their promises regarding ivory bans. The longer they delay implementing the signaled prohibition of legal ivory markets, the more poaching will increase as traders anticipate the ban. Moreover, the more time lapses before implementation takes place, the greater opportunities traders (legal and illegal) have to prepare for operating in a more clandestine manner, and the more they will be able to develop evasion and corruption strategies.

Until 2017, much of the policy focus in China was on assessing the size of legal stocks and setting levels of compensation for the thirty-four officially registered traders who will lose significant income and capital as a result of the prohibition—some $600 million. China is believed to have spent about $100 billion on various environmental compensation and support projects, so the compensation scheme for ivory traders is financially feasible.[2] China also plans to encourage master ivory carvers who will become unemployed to work in museums repairing ivory artifacts. The scheme is important, but to ultimately overcome laundering and clandestine supply, China needs to seize all the ivory that it pays compensation for and as much of illegal ivory in circulation as it can, and needs to destroy it. China can also provide a limited window of opportunity (for example, thirty days) during which illegal traders could hand over their illegal ivory stocks without fear of prosecution. Similarly, Hong Kong should not wait until legal stocks run out, given how poor its registry and licensing systems are, as they will continue to be resupplied with illegal ivory. Destroying the ivory, as opposed to merely storing it in a secure place, will send a signal to current ivory traders and launderers that they cannot simply ride out the ban.

CONCLUSIONS AND POLICY RECOMMENDATIONS

Chinese and Hong Kong law-enforcement officials also need to map out the distribution networks of as many legal and illegal traders as possible and inform them that their activities will be diligently monitored, and that any violations of the ban will be strictly punished with escalating penalties. Because the ivory trade is concentrated in the hands of relatively few investors and speculators, identification should not be difficult, and measures aimed at deterring them from breaching the regulations can be delivered fairly easily. But the enforcement of laws concerning ivory and wildlife should be handed over to regular police departments and removed from any environmental agency that may have more interest in deriving economic benefit from wildlife utilization than in wildlife protection. At minimum, a special enforcement branch within an environmental agency, insulated as much as possible from political pressures, should be set up to enforce the new regulations.

Over the past few years, China has toughened the enforcement of regulations concerning the legal ivory market to some extent. In December 2013, for example, a Chinese man was sentenced to ten years in prison in Beijing for ordering two whole tusks and 163 carved ivory figures from illegal sources. In addition to robust penalties for traders, Chinese and Hong Kong authorities should also adopt the use of mild, swift, and certain penalties for low-level buyers attempting to acquire ivory after the ban is implemented. Such penalties could include escalating fines as well as compulsory public works, such as sweeping the streets for a day, and could also involve shaming, such as publishing lists of violators. Prison sentences should not be lengthy, perhaps even as short as a day or two unless repeated violations take place.

Implementing such a scheme in China (unlike, for example, in Tanzania, where corruption eviscerates the deterrent capacity of law enforcement) has many advantages, particularly given the current crackdown on crime in China. Since President Xi Jinping came to power, Chinese law-enforcement authorities have undertaken extensive raids and arrested numerous people involved in criminal activities and corruption in China, with some officials in very high posts having been given severe sentences for a variety of offences. The deterrent effects of law enforcement in China are currently at a peak.

The Chinese and Hong Kong governments, as well as environmental NGOs, also need to continue improving their demand-reduction messages. Structuring campaigns around immediate material costs, such as costs to one's prestige and dating prospects, is often highly effective. So how about campaigns showing attractive women rejecting suitors who display ivory trinkets?

Chinese and Hong Kong authorities also need to anticipate—and as much as possible deter—traders attempting to switch from ivory to other illegal wildlife commodities, such as rhino horn. Undercover investigations or even formal interviews of traders by law-enforcement agents and environmental NGOs, including conversations with traders, about possible replacement commodities, can provide valuable information on evasive maneuvers and adaptation strategies.

Both enforcement and demand-reduction strategies will have to be carefully evaluated, including by independent evaluators. Policy refinements and adjustments will have to take place to fit the Chinese context, and programs will have to be allowed to evolve in the light of experience of implementing the regulations.

If in the future, such as a decade from now, poaching elephants has sufficiently abated and the feverish level of demand in China has cooled off, with demand stabilizing at a much lower level, there may well be good reasons to permit trade in ivory again. At that point, ivory obtained as a result of natural deaths will allow local, poor populations to offset their losses from not poaching or degrading protected areas, and national governments and wildlife management authorities to maintain protected areas.

The international community needs to continue working with China to further motivate it to embrace measures to combat poaching and wildlife trafficking. Policy actions in the Chinese market and among Chinese diaspora communities will have pronounced effects on illegal wildlife markets since China and Chinese communities are the world's leading consumers of wildlife products, including illegal ones. Should China and Hong Kong strongly embrace their commitments to wildlife, their resolve will also have important positive effects on other wildlife smuggling and markets in countries such as Myanmar, Laos, and Indonesia. China could become the world's leader in combating wildlife trafficking.

This example of how to design the ban on ivory trading in China and Hong Kong is just one specific case. It is neither set in stone, nor the

CONCLUSIONS AND POLICY RECOMMENDATIONS

model for how to deal with all cases of poaching and wildlife trafficking. Setting up a ban on local bushmeat trade, if it features endangered species, might look very different, involving compensation to poor populations whose protein needs would be compromised, or the provision of alternative sources of protein. In other cases, licensing should be allowed, but with better monitoring, such as installing CCTV throughout reptile-breeding facilities in Indonesia with live feeds to wildlife law-enforcement officials, rather than banning the trade altogether.

In all wildlife regulation, particularly in the context of current poaching and trafficking crises and the escalating demand for wildlife products, experimentation in program design and restructuring will be frequently needed. Accordingly, honest, frequent, and comprehensive evaluations of policy effectiveness are equally necessary. Allowing for policy failure is key. If monitoring, including self-monitoring, reveals that an environmental regulation and its enforcement are not working as planned, its designers and implementers should be given a chance to redesign the program to address its deficiencies instead of being simply denied funding. Otherwise, there will be a strong tendency to obfuscate assessment measures and wildlife will continue to disappear irretrievably.

The Road Ahead

As if in a desolate Edward Hopper landscape, the orangutan was clinging to one last tree that stood next to the river in Kutai National Park in eastern Kalimantan, Indonesia. My joy at seeing this magnificent primate was spoiled by his destroyed habitat—under normal circumstances, the orangutan would never venture so far from trees. But here he was, in an extremely degraded and marginal habitat, probably looking for food he could no longer find inside the forest. Although once a jewel of biodiversity in Indonesia, teeming with Sumatran rhinoceros and *banteng* (a species of Southeast Asian wild cattle), and long portrayed as one of the greatest wildernesses areas left on the Indonesian side of Borneo, much of Kutai today looks like a disaster zone. Kilometers deep into its boundaries, the park has been stripped of trees. Despite the fact that the park is nominally a protected area, the hardwood trees have been logged for their timber and to clear land for the cultivation of African oil palms. The park was also badly affected by extensive fires several years ago. The big dipterocarp trees that are the essence of a Southeast Asian rainforest—on which many animal species depend for survival, but whose timber is unfortunately very valuable—have been all but eliminated from vast tracts of the protected area. The one last standing dipterocarp a kilometer deep into the forest has become a tourist attraction. Consequently, few hornbills are left in much of the park. Many areas designated as protected in Indonesia, even national parks, have already been commercially logged out, with their biodiversity degraded. With commercially valuable trees logged-out in the park, many artisanal loggers have turned to hunting in the park for the animals still left there. Some had previously been imprisoned for illegal logging, but could not make ends meet. Besides, they were intrigued by stories of poachers entering the park with big weapons and making a lot of money from poaching.[3]

Leaving Kutai several days later, we stopped at a roadside shack to photograph the destroyed forest. A poor local Dayak woman was selling various wares. Eagerly she tried to talk us into buying parts of animals her father killed in the park, such as hornbill feathers. She excitedly told us that many men were now coming into the park to hunt hornbills or buy them from local people. These beautiful and enigmatic birds were now bringing in more money than ever, and local and non-local hunters were after them. For the family, that meant bigger profits than ever per bird killed, but also serious competition from other hunters, so the family's income had not significantly increased. The result of all this is that the birds were becoming scarcer and scarcer, making hunting more difficult and tragically reducing the population size of these majestic birds, so iconic of the forests of Kalimantan and Southeast Asia.

But it is not their feathers that enticed the new round of poaching. Rather, the bird's bills have emerged as a new fad in the global TCM market, the poor man's rhino horn, presumably used to cure impotence and promote vitality and longevity. In Kalimantan, the bills would fetch 2 million Indonesian rupiahs (approximately $200); in the markets of China, Singapore, Macau, and Hong Kong, they would bring far more. The presence of well-heeled Chinese coal and timber companies in Kalimantan facilitated the trade, and the companies were already paying off the Indonesian police, military, navy, and coastguard

CONCLUSIONS AND POLICY RECOMMENDATIONS

for turning a blind eye to their illegal activities. Even without extensive bribes, stopping the traffic in hornbills would be of far lower priority for Indonesian law-enforcement agencies than interdicting illegal artisanal mining, which big mining companies have an interest in stopping to reduce competition and divert law-enforcement focus from their own illegal activities.

Although disappointed that she could not entice us to buy her wildlife goods, the Dayak woman told us that she frequently sees orangutans cross the paved highway. On either side of the road, there was little forest left—just palms as far as the eye could see. It was not clear to us where the orangutans would be going as there did not seem to be any forest left across the road, only palm plantations. Possibly there was so little food left in the forest that even here, in a national park, the orangutans are forced to eat the insides of nuts from African oil palms, a poor dietary substitute. This coping mechanism frequently puts them in conflict with plantation owners and gets them killed. While I was looking at the highway and the destroyed forest, a paraphrase of the famous line in Cormac McCarthy's post-apocalyptic novel *The Road* ran though my head: Borrowed time and borrowed world and whose eyes with which to sorrow it?[4]

Three years later, in the summer of 2015, the international outcry that followed the killing of Cecil the Lion in Zimbabwe drove home to me that, more than ever, there were the world's eyes to sorrow poaching and natural ecosystem destruction. Moreover, the world had not just eyes to sorrow, but also voices to challenge wildlife crimes, to demand better environmental policies and their enforcement. More than ever there is mobilization across the globe to halt the ruinous effects of wildlife trafficking, illegal logging and mining, and habitat destruction. But commitments to act continue to be highly varied from place to place, and in many places sadly deficient.

Moreover, effective conservation is not merely about commitment and passion, it is equally about nuanced, hard-nosed, flexible, and sometimes uncomfortable policies. There are no black and white solutions, no silver bullets with which to end poaching—and certainly no generic blueprint for conservation. Bans and law enforcement come with constrains and serious negative downsides, as do policies that allow legal hunting and trade. Trophy hunting often brings important economic benefits to national governments and incentivizes them to support conservation. Although it is vital to clean up the corruption that has permeated trophy hunting in much of Africa, to suspend it indefinitely will hurt, not advance conservation.

Bans and enforcement often also compromise human security and the basic livelihoods of poor and marginalized populations dependent on hunting and resource extraction for their basic income. Many have been victimized by exclusionary and restrictive wildlife policies for decades, and hence reject them. Legacies of colonialism and racial discrimination—of privileged wealthy white men hunting for fun while forcibly preventing poor black people from hunting for subsistence—still run deep. Indeed, the outcry in the United States over Cecil the Lion mostly did not resonate in Zimbabwe. Instead, it reinforced the view that rich whites value animals more than they value poor black people.[5] But of course, Asian wildlife traffickers and black political and economic elites in Africa also capture wildlife revenues to the exclusion of the poor and marginalized.

Efforts to redress these past wrongs and empower local communities politically and economically, as well as to give them incentives to engage in conservation, have often featured trophy hunting. Indeed, the most successful CBNRM efforts, such as Zimbabwe's CAMPFIRE program or others in Botswana and Namibia, have derived far more income and development resources for poor local people from trophy hunting than from other forms of ecotourism. Imposing indefinitely long blanket bans on trophy hunting greatly hurts local communities economically, and it undermines their support for conservation.

Effective policies will need to avail themselves of many tools and make highly context-specific judgments as to which tool or combination of tools to pick at what time and in what place. Sometimes that will mean bans on hunting and legal trade. For other times and places, allowing hunting and legal trade is what will save a species. In order to deflate the current craze for ivory and rhino horn, efforts must be made to reduce the out-of-control demand for them; efforts must also be made to halt the poaching, trafficking, and laundering that permeate trade in these items. It currently makes good sense to ban ivory sales in China, intensify enforcement efforts against trade in ivory and rhino horn in East Asia and elsewhere, and suspend further trading licenses for sup-

CONCLUSIONS AND POLICY RECOMMENDATIONS

plier countries. But looking ahead, there may well be a time to permit trade again. In fact, the predisposition should be toward permitting legal trade and toward giving local communities incentives to support conservation through CBNRM efforts. Only when CBNRM schemes are failing should other policies, with just compensation, be adopted.

But policy complexities and numerous challenges—even the sad recognition that for each policy tool there are more failures than successes, and that each comes with severe downsides and limitations—should not lead to apathy. Both to mitigate the multiple threats to humans that wildlife trafficking poses, and for basic moral reasons, it is necessary to do all that is possible to preserve species and biodiversity, and do so in ways that do not unjustly hurt poor and marginalized populations. We need to persevere, recognizing that policy effectiveness and policy outcomes are contingent on local settings, and that shifts in policy, even radical ones; may be needed over time. Rather than latching onto one particular policy tool and hanging onto it forever, the road to conservation means putting one foot ahead of the other, constantly assessing, revising, and overhauling policies and trying to find a particular design that works in a particular place at a particular time.

NOTES

1. INTRODUCTION

1. In 2015, the White House requested an $8 million increase in the total budget for US wildlife enforcement agencies, up from $75.4 million in 2014. See, for example, Ron Nixon, "Obama Administration Plans to Aggressively Target Wildlife Trafficking," *New York Times*, February 11, 2015.
2. Prices are given in US dollars throughout, unless otherwise specified.
3. Author's research in Indonesia's wildlife markets, September to November 2012.
4. Author's interviews with representatives of environmental NGOs in Indonesia, September 2012.
5. Author's research on illegal wildlife trafficking and fishing, Flores and Komodo, Indonesia, November 2012.
6. Paul Jepson, Richard Ladle, and Sujatnika, "Assessing Market-Based Conservation Governance Approaches: A Socio-Economic Profile of Indonesian Markets for Wild Birds," *Oryx*, 45(4), October 2011: 482–91; and Robert Cribb, "Conservation in Colonial Indonesia," *Interventions*, 9(1), 2007: 49–61.
7. Author's research on illegal wildlife trafficking in Myanmar, winter 2006.
8. Vincent Nijman, "An Overview of International Wildlife Trade from Southeast Asia," *Biodiversity Conservation*, 19 (4), 2010: 1102.
9. Ben Davis, "Black Market: Inside the Endangered Species Trade in Asia", (San Rafael: Earth Aware Editions, 2005) and Tom Milliken, et al., "Monitoring of Illegal Trade in Ivory and Other Elephant Specimens," (Bangkok: TRAFFIC, 2015) (available at: www.cites.org/sites/default/files/eng/cop/16/doc/E-CoP16-53-02-02.pdf).
10. Fiona Underwood, Robert Burn, and Tom Milliken, "Dissecting the

Illegal Ivory Trade: An Analysis of Ivory Seizure Data," *PLOS One*, October 18, 2013.

11. World Wildlife Fund, "Souvenir Alert Highlights Deadly Trade in Endangered Species" (available at: www.wwf.org.uk/news/scotland/n_0000000409.asp). In 1986, WWF changed its name to the World Wide Fund for Nature, but in the United States and Canada kept the original acronym. See also the Association of Southeast Asian Nations Wildlife Enforcement Network, "Illegal Wildlife Trade in Southeast Asia Factsheet", March 5, 2009 (available at www.asean-wen.org/index-php?option=com_docman&task=doc_details&gid=5&Itemid=80).

12. Duan Biggs, Franck Courchamp, Rowan Martin, and Hugh Possingham, "Legal Trade of Africa's Rhino Horn," *Science* 339(6123), 2013: 1038–39; Daniel Challender, "Asian Pangolin: Increasing Affluence Driving Hunting Pressure," *TRAFFIC Bulletin*, 23(1), October 2011: 92–93.

13. Yang Qing et al., "Trade of Wild Animals and Plants in China–Laos Border Areas Status and Suggestion for Effective Management," *Biodiversity Science*, 8(3), 2000: 284–56; Hanneke Nooren and Gordon Claridge, *Wildlife Trade in Laos: The End of the Game*, Amsterdam: Netherlands Committee for IUCN, 2001; and William Duckworth et al., "Why South-East Asia Should Be the World's Priority for Averting Imminent Species Extinctions, and a Call to Join a Developing Cross-Institutional Programme to Tackle This Urgent Issue," *Sapiens*, 5(2), 2012: 1–19.

14. Mark Auliya, "Hot Trade in Cool Creatures: A Review of the Live Reptile Trade in the European Union in the 1990s," *TRAFFIC*, 2003 (available at: www.traffic.org/reptiles-amphibians/).

15. United Nations Environmental Programme, "Elephants in the Dust: The African Elephant Crisis," March 2013 (available at: www.unep.org/pdf/RRAivory_draft7.pdf); and International Fund for Animal Welfare, "Bidding against Survival: The Elephant Poaching Crisis and the Role of Auctions in the US Ivory Market," August 2014 (available at: www.ifaw.org/sites/default/files/IFAW-Ivory-Auctions-bidding-against-survival-aug-2014_0.pdf).

16. See, for example, Rob Barnett (ed.), "Food for Thought: Utilization of Wild Meat in Eastern and Southern Africa" (Nairobi: TRAFFIC, 2000), (available at: www.traffic.org/general-reports/traffic_pub_gen7.pdf).

17. David Fennell and David Weaver, "The Ecotourism Concept and Tourism-Conservation Symbiosis," *Journal of Sustainable Tourism*, 13(4), 2005: 373–90; and Jeffrey McNeeley et al., "Conservation Biology in

Asia: The Major Policy Challenges," *Conservation Biology*, 23(4), 2009: 805–10.
18. Elizabeth Bennett et al., "Hunting the World's Wildlife into Extinction," *Oryx*, 36(4), 2002: 328–9; David Wilkie and Julia Carpenter, "Bushmeat Hunting in the Congo Basin: An Assessment of Impacts and Options for Mitigation," *Biodiversity and Conservation*, 8(7), 1999: 927–55; David Wilkie et al., "Defaunation, Not Deforestation: Commercial Logging and Market Hunting in Northern Congo," in Robert Fimbel, Alejandro Grahal, and John Robinson (eds), *The Cutting Edge: Conserving Wildlife in Logged Tropical Forests* (New York: Columbia University Press, 2001): 375–99; John Robinson and Kent Redford (eds), *Neotropical Wildlife Use and Conservation* (Chicago: University of Chicago Press, 1991); and Elizabeth L. Bennett and John G. Robinson, "Hunting of Wildlife in Tropical Forests: Implications for Biodiversity and Forest Peoples" (Washington, D.C.: World Bank, 2000).
19. Author's research with Abdi Jama into wildlife trafficking, charcoal production, and other criminality in Somaliland, March and April 2013.
20. Author's interview with a water station guard, Somaliland, March 2013.
21. United Nations Office on Drugs and Crime puts the number of elephants poached during this period at 92,000. See, United Nations Office on Drugs and Crime, "World Wildlife Crime Report: Trafficking in Protected Species," May 2016 (available at: www.unodc.org/documents/data-and-analysis/wildlife/World_Wildlife_Crime_Report_2016_final.pdf: 43); "The Elephants Fight Back," *Economist*, November 21, 2015; and Samuel Wasser et al., "Genetic Assignment of Large Seizures of Elephant Ivory Reveals Africa's Major Poaching Hotspots," *Science*, 349(6243), 2015: 1–7.
22. International Union for the Conservation of Nature (IUCN), "Poaching Behind Worst African Elephant Losses in 25 Years—IUCN Report," 2016 (available at: www.iucn.org/news/poaching-behind-worst-african-elephant-losses-25-years-%E2%80%93-iucn-report).
23. The total number of poached elephants might, however, be higher as not all slaughtered elephants are found. See Convention on International Trade in Endangered Species of Wild Fauna and Flora (CITES), "Monitoring the Illegal Killing of Elephants: Update on Elephant Poaching Trends in Africa to 31 December 2014" (available at: https://cites.org/sites/default/files/i/news/2015/WWD-PR-Annex_MIKE_trend_update_2014_new.pdf).
24. George Wittemyer et al., "Illegal Killing for Ivory Drives Global

Decline in African Elephants," *Proceedings of National Academy of Sciences*, 111(36), 2014: 13117–21.

25. Great Elephant Census, "Better Data for a Crisis: Second Tanzania Count Part of Ongoing Population Monitoring," *Great Elephant Census*. September, 9 2015. (available at: www.greatelephantcensus.com/blog/2015/9/8/better-data-for-a-crisis-second-tanzania-count-part-of-ongoing-population-monitoring); and "Government of Mozambique Announces Major Decline in National Elephant Population," World Conservation Society (WCS) Newsroom, May 26, 2015 (available at: http://newsroom.wcs.org/News-Releases/articleType/ArticleView/articleId/6760/Govt-of-Mozambique-announces-major-decline-in-national-elephant-population.aspx).
26. Wildlife Protection Society of India, "Elephant Poaching 2009–2014" (available at: www.wpsi-india.org/crime_maps/elephant_poaching.php).
27. Reese Halter, "Insatiable Demand for African Rhino Horn Spells Extinction," *Huffington Post*, October 6, 2013.
28. "South African Group Reports Slight Drop in Rhino Poaching," *Associated Press*, January 2, 2016; and Rajan Amin et al., "An Overview of the Conservation Status of and Threats to Rhinoceros Species in the Wild," *International Zoo Yearbook*, 40(1), 2006: 96–117.
29. Tom Milliken, "Illegal Trade in Ivory and Rhino Horn: An Assessment Report to Improve Law Enforcement under the Wildlife TRAPS Project" (Cambridge: TRAFFIC, 2014) (available at: www.usaid.gov/sites/default/files/documents/1865/W-TRAPS-Elephant-Rhino-report.pdf).
30. Jonathan Jones, "A Picture of Loneliness," *Guardian*, May 12, 2015.
31. Wildlife Protection Society of India, "Rhino Poaching 2009–2014" (available at: www.wpsi-india.org/crime_maps/rhino_poaching.php).
32. Rachel Nuwar, "It's Official: Vietnam's Javan Rhino is Extinct. Which Species is Next?" *Take Part*, January 2, 2013 (available at: www.takepart.com/article/2012/12/12/its-official-vietnams-javan-rhino-extinct-and-other-species-will-likely-follow).
33. Tom Milliken and Jo Shaw, "The South Africa–Viet Nam Rhino Horn Trade Nexus: A Deadly Combination of Institutional Lapses. Corrupt Wildlife Industry Professionals and Asian Crime Syndicates" (Johannesburg: TRAFFIC, 2012), (available at: www.worldwildlife.org/publications/the-south-africa-viet-nam-rhino-horn-trade-nexus).
34. "South African Group Reports Slight Drop in Rhino Poaching," *Associated Press*, January 2, 2016; and Ron Nixon, "Obama Administration."
35. George Wittemyer et al., "Poaching Policy: Rising Ivory Prices Threaten Elephants," *Nature*, 476, 2011: 282–3; and Ronald Orenstein,

Ivory, Horn, and Blood: Behind the Elephant and Rhinoceros Poaching Crisis (New York: Firefly Books, 2013).

36. Wildlife Protection Society of India, "WPSI's Tiger Poaching Statistics" (available at: www.wpsi-india.org/statistics/index.php); and Kristin Nowell, "Asian Big Cat Conservation and Trade Control in Selected Range States: Evaluating Implementation and Effectiveness of CITES Recommendations," *TRAFFIC*, 2007 (available at http://www.traffic.org/species-reports/traffic_species_mammals16.pdf).

37. World Bank, "Global Tiger Recovery Program 2010–2012," 2012 (available at: http://globaltigerinitiative.org/news-blog/by-tag/global-tiger-recovery-program/).

38. Shibao Wu et al., "Assessment of Threatened Status of Chinese Pangolin (*Manis Pentadactyla*)," *Chinese Journal of Applied Environmental Biology*, 10(4), 2004: 456–61; and Daniel Challendar et al., "*Manis pentadactyla*," IUCN Red List of Threatened Species 2014 (available at: http://dx.doi.org/10.2305/IUCN.UK.2014–2.RLTS.T12764A45222544.en); see also Erica Goode, "A Struggle to Save the Scaly Pangolin," *New York Times*, March 30, 2015.

39. WildAid, "Pangolins: On the Brink," 2016 (available at: www.wildaid.org/sites/default/files/resources/WildAid-Pangolins%20on%20the%20Brink-2016.pdf).

40. Audrey Garric, "Pangolins under Threat as Black Market Grows," *Guardian*, March 12, 2013; and Ella Davies, "'Shocking' Scale of Pangolin Smuggling Revealed," *BBC News*, March 13, 2014 (available at: www.bbc.co.uk/nature/26549963).

42. Rosaleen Duffy, *Nature Crime: How We're Getting Conservation Wrong* (New Haven: Yale University Press, 2010).

43. Michael 't Sas-Rolfes, "Assessing CITES: Four Case Studies," in Jon Hutton and Barnabas Dickson (eds), *Endangered Species, Threatened Convention: The Past, Present, and Future of CITES* (London: Earthscan, 2000): 69–87.

44. Masayuki Sakamoto, "Black and Grey: Illegal Ivory in Japanese Markets," (Japan Wildlife Conservation 2004), (available at: www.ifaw.org/sites/default/files/Black%20and%20Grey%20Illegal%20ivory%20in%20Japan%20Markets%20-%202004.pdf).

45. Rachael Bale, "US–China Deal to Ban Ivory Trade is Good News for Elephants," *National Geographic Magazine*, September 25, 2015 (available at: http://news.nationalgeographic.com/2015/09/150925-ivory-elephants-us-china-obama-xi-poaching/).

46. Simon Denyer, "China Pledges to End Ivory Trading—But Says the US Should, Too," *Washington Post*, June 5, 2015.

47. Nick Davies and Oliver Holmes, "China Accused of Defying its Own Ban on Breeding Tigers to Profit from Body Parts," *Guardian*, September 27, 2016.
48. Vanessa Piao and Cherie Chan, "Beijing Destroys Confiscated Ivory and Vows to End Trade," *New York Times*, May 29, 2015.
49. Dan Levin, "China Bans Import of Ivory Carvings for One Year," *New York Times*, February 27, 2015.
50. Simon Denyer, "China's Vow to Shut Down Its Ivory Trade by the End of 2017 is a 'Game Changer' for Elephants," *Washington Post*, December 30, 2016.
51. Peter Baker and Jada Smith, "Obama Administration Targets Trade in African Elephant Ivory," *New York Times*, July 25, 2015.
52. Bale, "US–China Deal."
53. Daniel Stiles, "Elephant Ivory Trafficking in California, USA," (Natural Resources Defense Council, 2015), (available at: http://docs.nrdc.org/wildlife/files/wil_15010601a.pdf).
54. Denyer, "China Pledges to End Ivory Trading."
55. Nigel Leader-Williams, "Regulation and Protection: Successes and Failures in Rhinoceros Conservation," in Sara Oldfield (ed.), *The Trade in Wildlife* (London: Earthscan, 2003): 89–99.
56. Andrea Crosta, Kimberly Sutherland, and Mike Beckner, "Blending Ivory: China's Old Loopholes, New Hopes," (Los Angeles: Elephant Action League, 2015) (available at: http://elephantleague.org/wp-content/uploads/2015/12/EAL_BLENDING_IVORY_CHINA_Report-Dec2015.pdf).
57. Ed Lavandera, "Winner of Black Rhino Hunting Auction: My $350,000 Will Help Save the Species," *CNN*, January 17, 2014 (available at: www.cnn.com/2014/01/16/us/black-rhino-hunting-permit/).
58. AFP, "South African Judge Lifts Domestic Ban on Rhino Horn Trade," *Guardian*, November 26, 2015. (Available at http://www.theguardian.com/environment/2015/nov/26/south-african-judge-lifts-domestic-ban-on-rhino-horn-trade).
59. Department for the Environment, Food and Rural Affairs, "Declaration: London Conference on the Illegal Wildlife Trade" February 12–13, 2014 (available at: http://www.gov.uk/government/uploads/system/uploads/attachment_data/file/281289/london-wildlife-conference-declaration-140213.pdf); and United Nations Convention against Transnational Organized Crime (available at: www.unodc.org/documents/treaties/UNTOC/Publications/TOC%20Convention/TOCebook-e.pdf).
60. White House, "National Strategy for Combating Wildlife Trafficking"

(available at: www.whitehouse.gov/sites/default/files/docs/national-strategywildlifetrafficking.pdf).
61. Colman O'Criodain, "CITES and Community-Based Conservation: Where Do We Go from Here?" in Max Abensperg-Traun et al. (eds.), *CITES and CBNRM Proceedings of an International Symposium on the Relevance of CBNRM to the Conservation and Sustainable Use of CITES-Listed Species in Exporting Countries, Vienna, Austria, May 18–20, 2011* (London: International Institute for Environment and Development, 2011), (available at: http://pubs.iied.org/pdfs/14616IIED.pdf).
62. Born Free Foundation, "Inconvenient but True: The Unrelenting Global Trade in Elephant Ivory" (report prepared for the 14th Meeting of the Conference of the Parties to CITES, The Hague, June 3–15, 2007);—,"Bloody Ivory: Stop Elephant Poaching and Ivory Trade," December 12, 2012 (available at: www.bloodyivory.org); Environmental Investigation Agency, *Stop Stimulating Demand! Let Wildlife-Trade Bans Work* (London/Washington: EIA, 2013); Daniel Stiles, "The Ivory Trade and Elephant Conservation," *Environmental Conservation*, 31(4), 2004: 309–21; Erica Thorson and Chris Wold, "Back to Basics: An Analysis of the Object and Purposes of CITES and a Blueprint for Implementation," (Portland: International Environmental Law Project, 2010), (available at: www.lclark.edu/live/files/4620); and Rosaleen Duffy, "Global Environmental Governance and North–South Dynamics: The Case of CITES," *Environment and Planning C: Government and Policy*, 31(2), 2013: 222–39.

2. THE WILDLIFE TRADE AND THE DRUG TRADE

1. See also Nigel South and Tanya Wyatt, "Comparing Illicit Trades in Wildlife and Drugs: An Exploratory Study," *Deviant Behavior*, 32(6), June 2011: 538–61.
2. Chris Huxley, "CITES: The Vision," in Jon Hutton and Barnabas Dickson (eds), *Endangered Species, Threatened Convention: The Past, Present, and Future of CITES, the Convention on International Trade in Endangered Species of Wild Fauna and Flora* (London: Earthscan, 2000): 3–12.
3. José Octavio Velásquez Gomar and Lindsay Stringer, "Moving toward Sustainability? An Analysis of CITES' Conservation Policies," *Journal of Environmental Policy*, 21(4), August 2011: 240–58.
4. Martin Jelsma, "UNGASS 2016: Prospects for Treaty Reform and UN System-Wide Coherence on Drug Policy," (Washington, D.C.: Brookings Institution, 2015), (available at: www.brookings.edu/~/media/Research/Files/Papers/2015/04/global-drug-policy/Jelsma—United-Nations-final.pdf?la=en).

5. Christopher Hallam at al., "Scheduling in the International Drug Control System," Series on Legislative Reform of Drug Policies, No 25 (Amsterdam: Transnational Institute, June 2014), (available at: www.tni.org/files/download/dlr25_0.pdf).
6. See, for example, Vanda Felbab-Brown, "Deterring Non-State Actors," in Steven Pifer et al. (eds.), *US Nuclear and Extended Deterrence. Considerations and Challenges*, Brookings Arms Control Series, Paper No. 9 (Washington D.C.: Brookings Institution, 2010), (available at: www.brookings.edu/~/media/research/files/papers/2010/6/nuclear-deterrence/06_nuclear_deterrence.pdf).
7. Vanda Felbab-Brown and Harold Trinkunas, "UNGASS 2016 in Comparative Perspective: Improving the Prospects for Success," (Washington, D.C.: Brookings Institution, 2015), (available at: www.brookings.edu/~/media/Research/Files/Papers/2015/04/global-drug-policy/FelbabBrown-TrinkunasUNGASS-2016-final-2.pdf?la=en).
8. See, for example, "Jailhouse Nation: Justice in America," *Economist*, June 20, 2015.
9. Jonathan Caulkins, "The Real Dangers of Marijuana," *National Affairs*, 26, Winter 2016: 21–34(available at: http://www.nationalaffairs.com/publications/detail/the-real-dangers-of-marijuana)
10. William Adams, *Against Extinction: A Story of Conservation* (London: Earthscan, 2004).
11. Philip Burnham, *Indian Country, God's Country: Native Americans and National Parks* (Island Press: Washington: 2000).
12. James Suzman, "Etosha Dreams: An Historical Account of the Hai//Om Predicament," *Journal of Modern African Studies*, 42(2), June 2004: 221–38; Terence Ranger, *Voices from the Rocks: Nature, Culture and History in the Matopos Hills of Zimbabwe* (Oxford: James Currey, 1999); and John MacKenzie, *Empire of Nature: Hunting Conservation and British Imperialism* (Manchester: Manchester University Press, 1988).
13. Jane Carruthers, *The Kruger National Park: A Social and Political History* (Pietermaritzburg: Natal University Press, 1995).
14. Roderick Neumann, "Moral and Discursive Geographies in the War for Biodiversity in Africa," *Political Geography*, 23(7), 2004: 813–37.
15. Jane Carruthers, "'Police Boys' and Poachers: Africans, Wildlife Protection and National Parks, the Transvaal 1902 to 1950," *Koedoe*, 36(2), 1993: 11–22.
16. Roderick Neumann, "Africa's 'Last Wilderness': Reordering Space for Political and Economic Control in Tanzania," *Africa*, 71(4), November 2001: 641–65.
17. See, for example, Michael Cernea, "Restriction of Access is Displacement: A Broader Concept and Policy," *Forced Migration Review*, 23, May

2005: 48–49; Simon Schama, *Landscape and Memory* (London: Fontana Press, 1996); and Dawn Chatty and Marcus Colchester (eds), *Conservation and Mobile Indigenous Peoples: Displacement, Forced Settlement and Sustainable Development* (New York: Berghahn Books, 2002).

18. See, for example, Nancy Lee Peluso and Peter Vandergeest, "Political Ecologies of War and Forests: Counterinsurgency and the Making of Natural Natures," *Annals of the Association of American Geographers*, 101(3), March 2011: 587–608.

19. Daniel Brockington and James Igoe, "Eviction for Conservation: A Global Overview," *Conservation and Society*, 4(3), September 2006: 424–70; and Michael Casimir, "Of Lions, Herders and Conservationists: Brief Notes on the Gir Forest National Park in Gujarat (Western India)," *Nomadic Peoples*, 5(2), 2001: 154–61.

20. Charles Geisler, "A New Kind of Trouble: Evictions in Eden," *International Social Science Journal*, 55(175), March 2003: 69–78.

21. Robert Nelson, "Environmental Colonialism: 'Saving' Africa from Africans," *Independent Review*, 8(1), 2003: 65–87; and Monique Borgerhoff-Mulder and Peter Coppolillo, *Conservation: Linking Ecology, Economics and Culture* (Princeton: Princeton University Press, 2005).

22. Kai Schmidt-Soltau and Daniel Brockington, "Protected Areas and Resettlement: What Scope for Voluntary Relocation," *World Development*, 35(12), November 2007: 2182–202.

23. Daniel Brockington, *Fortress Conservation: The Preservation of the Mkomazi Game Reserve, Tanzania* (Bloomington: Indiana University Press, 2002).

24. Arun Agrawal, *Community in Conservation: Beyond Enchantment and Disenchantment* (Gainesville: Conservation and Development Forum, 1997); see also William Adams, *Future Nature: A Vision of Conservation* (London: 1996).

25. Marshall Murphree, "Strategic Pillars of Communal Natural Resource Management: Benefit, Empowerment, and Conservation," *Biodiversity and Conservation*, 18(10), September 2009: 2551–62.

26. Elinor Ostrom, *Governing the Commons: The Evolution of Institutions for Collective Action* (Cambridge: Cambridge University Press, 1990).

27. William Adams and David Hulme, "Conservation and Communities: Changing Narratives, Policies and Practices in African Conservation," in David Hulme and Marshall Murphree (eds), *African Wildlife and African Livelihoods: The Promise and Performance of Community Conservation* (Oxford: James Currey, 2001): 9–23.

28. See, for example, David Western, "Amboseli National Park: Enlisting Land Owners to Conserve Migrating Wildlife," *Ambio*, 11(5), 1982: 302–8.

29. Clive Spinage, "Social Change and Conservation Misrepresentation in

Africa," *Oryx*, 32(4), 1998: 265–76; John Terborgh, *Requiem for Nature* (Washington, D.C.: Island Press, 1999); John Oates, "The Dangers of Conservation by Rural Development: A Case Study from the Forests of Nigeria," *Oryx*, 29(2), April 1995: 115–22; and David Wilkie et al., "Parks and People: Assessing Human Welfare Effects of Establishing Protected Areas for Biodiversity Conservation," *Conservation Biology*, 20(1), February 2006: 247–9.

30. Krystyna Swiderska et al., "The Governance of Nature and the Nature of Governance: Policy that Works Biodiversity and Livelihoods" London: International Institute for Environment and Development, 2008), (available at: http://pubs.iied.org/pdfs/14564IIED.pdf).

31. See, for example, Paula Miraglia, "Drugs and Drug Trafficking in Brazil: Trends and Policies" (Washington D.C.: Brookings Institution, 2015), (available at: www.brookings.edu/~/media/Research/Files/Papers/2015/04/global-drug-policy/Miraglia—Brazil-final.pdf?la=en).

32. See, for example, Vanda Felbab-Brown, "Enabling War and Peace: Drugs, Logs, and Wildlife in Thailand and Burma" (Washington D.C.: Brookings Institution, 2015), (available at: www.brookings.edu/~/media/research/files/papers/2015/12/thailand-burma-drugs-wildlife-felbabbrown/enabling_war_and_peace_final.pdf).

33. See, for example, Jason Breslow, "A Staggering Toll of Mexico's Drug War," *PBS Frontline*, July 27, 2015 (available at: www.pbs.org/wgbh/frontline/article/the-staggering-death-toll-of-mexicos-drug-war/).

34. Vanda Felbab-Brown, "Calderón's Caldron: Lessons from Mexico's Battle Against Organized Crime and Drug Trafficking in Tijuana, Ciudad Juárez, and Michoacán," *Brookings Latin America Initiative Paper Series*, (Washington, D.C.: Brookings Institution, 2011), (available at: www.brookings.edu/~/media/research/files/papers/2011/9/calderon-felbab-brown/09_calderon_felbab_brown.pdf); and Vanda Felbab-Brown, *Narco Noir: Mexico's Cartels, Cops, and Corruption* (Washington D.C.: Brookings Institution Press, forthcoming).

35. Author's communication with professor of drug policy, February 2016.

36. See IFAW, "Elephants Sent into Safari Slavery from Zimbabwe's World Famous Hwange National Park," November 8, 2006 (available at: www.ifaw.org/united-states/node/11256); IFAW, "South Africa Slams the Door on Elephant Tourism, NGOs Rejoice," February 25, 2008 (available at: www.ifaw.org/international/node/25661).

3. THE CHARACTER AND SCALE OF THE ILLEGAL WILDLIFE ECONOMY

1. Dilys Roe et al., "Making a Killing or Making a Living? Wildlife Trade,

Trade Controls, and Rural Livelihoods," *Biodiversity and Livelihoods*, 6, 2002 (available at: www.traffic.org/general-reports/traffic_pub_trade4.pdf).

2. On protein needs: Georgius Koppert et al., "Consommation alimentaire dans trois populations forestières de la région côtière du Cameroun: Yassa, Mvae et Bakola," in Claude Marcel Hladik et al. (eds), *L'Alimentation en forêt tropicale: interactions bioculturelles et perspectives de développement* (Paris: UNESCO, 1996): 477–96; on household income: Malcolm Starkey, "Commerce and Subsistence: The Hunting, Sale and Consumption of Bushmeat in Gabon," PhD thesis (Cambridge: Cambridge University, 2004). For comparisons with the Amazon basin, see also Robert Nasi et al., "Empty Forests, Empty Stomachs? Bushmeat and Livelihoods in the Congo and Amazon Basins," *International Forestry Review*, 13(3), January 2011: 355–68.

3. Li Zhang et al., "Wildlife Trade, Consumption, and Conservation Awareness in Southwest China," *Biodiversity and Conservation*, 17(6), 2008: 1513.

4. "Police Suspect Killing of 10,000 Tibetan Antelopes," *Xinhua News Agency*, February 21, 2013 (available at: www.news.xinhuanet.com/english/world/2013–02/21/c_132183480.htm); Camille Leclerc et al., "Overcoming Extinction: Understanding Processes of Recovery of the Tibetan Antelope," *Ecosphere*, 6(9), September 2015: 1–14; and Mao Da and Mei Xueqin, "Protecting the Tibetan Antelope: A Historical Narrative and Missing Stories," in Marco Armiero and Lise Sedrez (eds.), *A History of Environmentalism: Local Struggles, Global Histories* (London: Bloomsbury, 2014): 83–104.

5. Jonathan Franzen, "Last Song for Migrating Birds," *National Geographic Magazine*, 224(1), July 2013 (available at: http://ngm.nationalgeographic.com/2013/07/songbird-migration/franzen-text).

6. Author's interviews with representatives of environmental NGOs in Nairobi, and with park rangers, wardens, and local community members with knowledge of local poaching, Tsavo West National Park, Maasai Mara, Kenya, February and May 2013.

7. Michael 't Sas-Rolfes, "The Rhino Poaching Crisis: A Market Analysis," February 2012 (available at: www.rhinoresourcecenter.com/pdf_files/133/1331370813.pdf); and Rebecca Drury, "Identifying and Understanding Consumers of Wild Consumer Animal Products in Hanoi, Vietnam: Implications for Conservation Management," (PhD diss., University College London, 2009).

8. Ronald Orenstein, *Ivory, Horn, and Blood: Behind the Elephant and Rhinoceros Poaching Crisis* (New York: Firefly Books, 2013).

9. Deanna Donovan, "Cultural Underpinnings of the Wildlife Trade in

Southeast Asia," in John Knight (ed.), *Wildlife in Asia: Cultural Perspectives* (London: Routledge, 2004): 88–111.

10. Chinese medical assessments tend to ascribe far higher curative properties to TCM than Western ones, which mostly do not find any. See Kristin Nowell, "Species Trade and Conservation Rhinoceroses; Assessment of Rhino Horn as a Traditional Medicine," TRAFFIC, 2012 (available at: https://cites.org/sites/default/files/eng/com/sc/62/E62-47-02-A.pdf); and Helen Laburn and Duncan Mitchell, "Extracts of Rhinoceros Horn are not Antipyretic in Rabbits," *Journal of Basic and Clinical Physiology and Pharmacology*, 8(1/2), 1997: 1–11.

11. Karl Amman, "The Rhino and the Bling," 2013 (available at: www.karlammann.com/pdf/rhino-bling.pdf); Peter Gwin, "Rhino Wars," *National Geographic Magazine*, 221(3), March 2012 (available at: http://ngm.nationalgeographic.com/2012/03/rhino-wars/gwin-text); and Annette Michaela Hübschle, "A Game of Horns: Transnational Flows of Rhino Horn," (PhD thesis) (International Max Planck Research School, 2016), 161.

12. Xu Hongfa and Craig Kirkpatrick, eds, "The State of Wildlife Trade in China: Information on the Trade in Wild Animals and Plants in China 2007" (Beijing: TRAFFIC, 2008) 9, (available at: http://www.trafficj.org/publication/08-State_of_Wildlife_China.pdf_ 9

13. David Jolly, "Whaling Talks in Morocco Fail to Produce Reductions," *New York Times*, June 23, 2010; Martin Fackler, "Uncertainty Buffets Japan's Whaling Fleet," *New York Times*, May 15, 2010; and David Jolly and John Broder, "UN Rejects Export Ban on Atlantic Bluefin Tuna," *New York Times*, March 18, 2010.

14. Vicki Crook, "Trade in Anguilla Species, with a Focus on Recent Trade in European Eel A. Anguilla," TRAFFIC, 2010 (available at: https://portals.iucn.org/library/sites/library/files/documents/Traf-114.pdf).

15. Daan van Uhm, "Illegal Wildlife Trade to the EU and Harms to the World," in Rob White et al (eds), *Environmental Crime and the World* (London: Ashgate, 2014): 43–66.

16. Ganapathiraju Pramod et al., "Estimates of Illegal and Unreported Fish in Seafood Imports to the USA," *Marine Policy*, 48(C), September 2014: 102–13.

17. Li Zhang et al., "Wildlife Trade," 1494; and Peter Li and Gareth Davey, "Culture, Reform Politics, and Future Directions," *Society and Animals*, 21(1), 2013: 34–54.

18. Brian Jones, "Policy Lessons from the Evolution of a Community-Based Approach to Wildlife Management, Kunene Region, Namibia," *Journal of International Development*, 11(2), March/April 1999: 295–304.

19. Li Zhang et al., "Wildlife Trade," 1493, 1515–16.

20. Ibid.: 1512.
21. United Nations International Drug Control Program (UNDCP), *The Social and Economic Impact of Drug Abuse and Control* (Vienna: UNDCP, 1994): 29. For a devastating exposé of the looseness of drug estimates, see Peter Reuter, "The Mismeasurement of Illegal Drug Markets: The Implications of Its Irrelevance," in Susan Pozo (ed.), *Exploring the Underground Economy* (Kalamazoo: W.E. Upjohn Institute, 1996): 63–80.
22. United Nations Office on Drugs and Crime (UNODC), 'World Drug Report 2005' (available at: www.unodc.org/pdf/WDR_2005/volume_1_web.pdf): 17.
23. Ibid., ix.
24. UNEP, *Year Book 2014 Emerging Issues Update: Illegal Trade in Wildlife*, 2014 (available at: www.unep.org/yearbook/2014/PDF/chapt4.pdf): 25; see also UNODC, "Transnational Organized Crime in the Fishing Industry," 2011 (available at: www.unodc.org/documents/human-trafficking/Issue_Paper_-_TOC_in_the_Fishing_Industry.pdf); and Darren Calley, *Market Denial and International Fisheries Regulation* (Leiden: Martinus Nijhoff, 2011).
25. UNEP-INTERPOL, "The Rise of Environmental Crime," 2016, http://pfbc-cbfp.org/news_en/items/unep-interpol.enen.html
26. Zia Morales, "Fighting Wildlife Crime to End Extreme Poverty and Boost Shared Prosperity," UN Chronicle, 51(2), September 2014: 23–24.
27. International Fund for Animal Welfare, *Fighting Illicit Wildlife Trafficking: A Consultation with Governments*, 2012 (available at: http://wwf.panda.org/about_our_earth/species/problems/illegal_trade/wildlife_trade_campaign/wildlife_trafficking_report/).
28. Liana Sun Wyler and Pervaze A. Sheikh, "International Illegal Trade in Wildlife: Threats and US Policy," Washington D.C.: Congressional Research Service, 2008 (available at: www.fas.org/sgp/crs/misc/RL34395.pdf): 2.
29. Defenders of Wildlife, "Combating Wildlife Trafficking from Latin America: The Illegal Trade from Mexico, the Caribbean, Central America and South America and What Can We Do to Address It," 2015 (available at: www.defenders.org/sites/default/files/publications/combating-wildlife-trafficking-from-latin-america-to-the-united-states-and-what-we-can-do-to-address-it.pdf.)
30. Nguyen Van Song, "Tracking the Trade: Vietnam's Illegal Wildlife Business" (Hanoi: Hanoi Agricultural University, 2003) (available at: www.idrc.ca/eepsea/ev-47045-201-1-DO_TOPIC.html).
31. Tom Milliken and Jo Shaw, "The South Africa–Viet Nam Rhino Horn Trade Nexus: A Deadly Combination of Institutional Lapses, Corrupt

Wildlife Industry Professionals and Asian Crime Syndicates," (Johannesburg: TRAFFIC, 2012), (available at: http://www.npr.org/documents/2013/may/traffic_species_mammals.pdf).
32. Rebecca Drury, "Hungry for Success: Urban Consumer Demand for Wild Animal Products in Vietnam," *Conservation and Society*, 9(3), 2011: 247–57.
33. Samuel Wasser et al., "Elephants, Ivory, and Trade," *Science*, 327(5971), 2010: 1331–32.
34. Peter Reuter, "The Continued Vitality of Mythical Numbers," *National Affairs*, 75, Spring 1984: 135–47.
35. World Wildlife Fund, *Fighting Illicit Wildlife Trafficking*.
36. United Nations Office on Drugs and Crime, "World Wildlife Crime Report: Trafficking in Protected Species" (Vienna: UNODC, 2016), (available at: www.unodc.org/documents/data-and-analysis/wildlife/World_Wildlife_Crime_Report_2016_final.pdf).
37. John Robinson and Elizabeth Bennett, "Will Alleviating Poverty Solve the Bushmeat Crisis?" *Oryx*, 36(4), October 2002: 332; John Fa and Carlos Peres, "Hunting in Tropical Forests," in John Reynolds et al. (eds.), *Conservation of Exploited Species* (Cambridge: Cambridge University Press, 2001): 203–41; and Elizabeth Bennett, "Is There a Link between Wild Meat and Food Security," *Conservation Biology*, 16(3), October 2002: 590–92.
38. Elizabeth Bennett et al., "Saving Borneo's Bacon: The Sustainability of Hunting in Sarawak and Sabah," in John Robinson and Elizabeth Bennett (eds), *Hunting for Sustainability in Tropical Forests* (New York: Columbia University Press, 2000): 305–24.
39. Vincent Nijman, "An Overview of International Wildlife Trade from Southeast Asia," *Biodiversity Conservation*, 19, 2010: 1102.
40. Ibid.
41. Ibid.; see also, Edmund Green and Francis Shirley, *The Global Trade in Corals* (Cambridge: World Conservation Monitoring Centre, 1999).
42. Nijman, "Overview."
43. Arthur Blundell and Michael Mascia, "Discrepancies in Reported Levels of International Wildlife Trade," *Conservation Biology*, 19(6), 2005: 2020–25.
44. Dennis Gray, "Asia's Wildlife Hunted for China's Appetite," *Associated Press*, April 6, 2004 (available at: www.nbcnews.com/id/4585068/ns/us_news-environment/t/asias-wildlife-hunted-chinas-appetite/#.V_6XEiSD6M8).
45. "Peru Police Seize Thousands of Dried Seahorses," BBC News, August 24, 2012 (available at: www.bbc.com/news/world-latin-america-19364702).

46. Sabine Schoppe, "Status, Trade Dynamics, and Management of the Southeast Asian Box Turtle in Indonesia," (Kuala Lumpur: TRAFFIC, 2009), (available at: www.traffic.org/species-reports/traffic_species_reptiles19.pdf).
47. Nijman, "Overview": 1109.
48. Chris Shepherd, "Export of Live Freshwater Turtles and Tortoises from Northern Sumatra and Riau, Indonesia: A Case Study" in Peter Paul Van Dijk et al. (eds.), *Asian Turtle Trade: Proceedings of a Workshop on Conservation and Trade of Freshwater Turtles and Tortoises in Asia* (Lunenburg, MA: Chelonian Research Foundation, 2000): 112–19.
49. Peter Paul Van Dijk et al. (eds), *Asian Turtle Trade: Proceedings of a Workshop on Conservation and Trade of Freshwater Turtles and Tortoises in Asia, Phnom Penh, Cambodia, 1–4 December 1999* (Lunenburg: Chelonian Research Foundation, 2000).
50. TRAFFIC, "The Trade in Marine Turtle Products in Vietnam," (2004), (available at: www.traffic.org/species-reports/traffic_species_reptiles23.pdf).
51. Frances Humber et al., "So Excellent a Fishe: A Global Overview of Legal Marine Turtle Fisheries," *Diversity and Distributions*, 20(5), 2014: 579–90; and Allison Winter, "Report Finds 42,000 Turtles Harvested Each Year by Legal Fisheries," Environmental News Network, February 21, 2014 (available at: www.enn.com/top_stories/article/47074).
52. Luh De Suriyani, "Green Turtle Smuggling Continues," *Jakarta Post*, January 21, 2013.
53. "Information about Sea Turtle: Green Sea Turtle," *Sea Turtle Conservancy* (available at: www.conserveturtles.org/seaturtleinformation.php?page=green).
54. Chris Shepherd and Nolan Magnus, "Nowhere to Hide: The Trade in Sumatran Tiger" (Cambridge TRAFFIC 2004) (available at: www.traffic.org/species-reports/traffic_species_mammals15.pdf).
55. Christine Hauser, "Number of Tigers in the Wild is Rising, Wildlife Groups Say," *New York Times*, April 11, 2016.
56. Mihir Srivastava, "Tracking the Tiger Killers," *India Today*, May 28, 2010; "WII to Start Tiger Census This Month," *Times of India*, October 5, 2009; and Bill Marsh, "Fretting about the Last of the World's Biggest Cats," *New York Times*, March 6, 2010.
57. "Global Wild Tiger Population Increases, But Still a Long Way to Go," WWF (available at: www.worldwildlife.org/press-releases/global-wild-tiger-population-increases-but-still-a-long-way-to-go-3).
58. Sarah Stoner and Natalia Pervushina, "Reduced to Skin and Bones Revisited: An Updated Analysis of Tiger Seizures from 12 Tiger Range

Countries (2000–2012)" (Cambridge TRAFFIC, 2004), 2013 (available at: www.worldwildlife.org/publications/reduced-to-skin-and-bones-revisited).
59. "Delhi Hub for Trade in Leopard Body Parts: Report," *Business Standard*, September 28, 2012.
60. David Garshelis and Robert Steinmetz, "Ursus Thibetanus," IUCN Red List of Threatened Species, 2008 (available at: http://dx.doi.org/10.2305/IUCN.UK.2008.RLTS.T22824A9391633.en).
61. Daniel Stiles et al., "Stolen Apes: The Illicit Trade in Chimpanzees, Bonobos, Gorillas, and Orangutans" (United Nations Environment Programme, 2013), (available at: https://cld.bz/bookdata/KY3u76i/basic-html/page-1.html).
62. Christina Vallianos, "Pangolins: On the Brink" (San Francisco: WildAid, 2016), (available at: www.wildaid.org/sites/default/files/resources/WildAid-Pangolins%20on%20the%20Brink-2016.pdf).
63. Dennis Gray, "Wildlife at Risk in Southeast Asia: Species Being Used for Food and Medicine," *Washington Post*, April 4, 2004; and Rachel Nuwer, "In Vietnam, Rampant Wildlife Smuggling Prompts Little Concern," *New York Times*, March 30, 2015 (available at http://www.nytimes.com/2015/03/31/science/in-vietnam-rampant-wildlife-smugglingprompts-little-concern.html
64. Bryan Stuart et al., "Homalospine Watersnakes: The Harvest and Trade from Tonle Sap, Cambodia," *TRAFFIC Bulletin*, 18(3), 2000: 115–24.
65. Rodrigo Díaz, "Crimen organizado opera tráfico ilegal de buche de totoaba," *Mexicali Digital*, August 4, 2014 (available at: http://mexicalidigital.mx/2014/opera-crimen-organizado-trafico-ilegal-de-buche-de-totoaba-19992.html).
66. Michael Lohmuller, "How China Fuels Wildlife Trafficking in Latin America," *InSight Crime*, June 10, 2015 (available at: www.insightcrime.org/news-analysis/how-china-fuels-wildlife-trafficking-latin-america).
67. Mike Gaworecki, "Time Running Out to Save World's Most Endangered Porpoise, Environmentalists Warn," *Mongabay*, September 21, 2016 (available at: www.news.mongabay.com/2016/09/time-running-out-to-save-worlds-most-endangered-porpoise-environmentalists-warn/). Eighth Meeting of the Comité Internacional para la Recuperación de la Vaquita (CIRVA-8), Southwest Fisheries Science Center, November 29–30th, 2016, La Jolla, CA, http://www.iuncn-vsg.org/wp-content/uploads/2010/03/CIRVA-8-Report-Final.pdf; and Elisabeth Malkin, "Before Vaquitas Vanish, a Desperate Bid to Save Them," *New York Times*, February 27, 2017
68. Gaworecki, "Time." See also James Bargent, "Eco-Trafficking in Latin

America: The Workings of a Billion-Dollar Business," *InSight Crime*, July 7, 2014 (available at: www.insightcrime.org/news-analysis/eco-trafficking-latin-america-billion-business); and James Bargent, "Eco-Trafficking in Latin America: The Failures of the State," *InSight Crime*, July 8, 2014 (available at: www.insightcrime.org/news-analysis/eco-trafficking-in-latin-america-the-failures-of-the-state).

69. Michael Lohmuller, "Industrial-Scale 'Shark Finning' in Ecuador," InSight Crime, May 28, 2015 (available at: www.insightcrime.org/news-briefs/industrial-scale-harvesting-shark-fins-ecuador).
70. Matt McGrath, "Shark Kills Number 100 Million Annually, Research Says," BBC News, March 1, 2013 (available at: www.bbc.com/news/science-environment-21629173).

4. WHY THE ILLEGAL WILDLIFE AND DRUG ECONOMIES MATTER

1. Mancur Olson, *The Logic of Collective Action* (Cambridge: Harvard University Press, 1965).
2. World Resources Institute, "Global Biodiversity Strategy: Guidelines for Action to Save, Study, and Use Earth's Biotic Wealth Sustainably and Equitably" (Washington D.C.: World Resources Institute, 1992), (available at http://pdf.wri.org/globalbiodiversitystrategy_intro.pdf)
3. Association of Southeast Asian Nations Wildlife Enforcement Network *ASEAN-WEN*, March 5, 2009 "Illegal Wildlife Trade in Southeast Asia: Factsheet"(available at: http://asean-wen.net/doc_details&gid=5&temid=80).
4. Robert T. Paine, "A Conversation on Refining the Concept of Keystone Species," *Conservation Biology*, 9(4), 1995: 962–64.
5. Simon Denyer, "China Tried to Drive a Furry Mammal to Extinction. Maybe That Wasn't Such a Good Idea," *Washington Post*, July 22, 2016; and George Schaller, *Tibet Wild* (Washington D.C.: Island Press, 2012).
6. Robert Nasi and Tony Cunningham, "Sustainable Management of Non-Timber Forest Resources: A Review with Recommendations for the SBSTTA" (Montreal: Secretariat to the Convention on Biological Diversity, 2001), (available www.srs.fs.usda.gov/pubs/VT_Publications/01t37.pdf)
7. Michael Cavendish, *The Economics of Natural Resource Utilisation by Communal Area Farmers of Zimbabwe* (Oxford: Oxford University Press, 1997).
8. James Mayers and Sonja Vermeulen, "Power from Trees: How Good Forest Governance Can Help Reduce Poverty" (London: International Institute for Environment and Development, 2002), (available at: http://pubs.iied.org/pdfs/11027IIED.pdf).
9. Jitendra Kumar Das and Om Prakash, "Measuring Market Channel

Efficiency and Strategy to Improve Income to Local Communities Dependent on Tropical Forests," *Journal of Sustainable Forestry*, 15(4), 2002: 28; and Ajay Kumar Mahapatra et al., "The Impact of NTFP Sales on Rural Households' Cash Income in India's Dry Deciduous Forest," *Environment Management*, 35(3), 2005: 258–65.

10. Sophie Chao, "Forest Peoples: Numbers Across the World," Forest Peoples Programme, 2012 (available at: www.forestpeoples.org/sites/fpp/files/publication/2012/05/forest-peoples-numbers-across-world-final_0.pdf).
11. David Kaimowitz and Douglas Sheil, "Conserving What and For Whom? Why Conservation Should Help Meet Basic Human Needs in the Tropics," *Biotropica*, 39(6), 2007: S12–S17.
12. Cavendish, *Economics of Natural Resource Utilisation*.
13. Pippa Trench et al., "Beyond Any Drought: Root Causes of Chronic Vulnerability in the Sahel" (Sahel Working Group, 2007), (available at: www.livestock-emergency.net/userfiles/file/assessment-review/Trench-et-al-2007.pdf).
14. Darrell Posey and Graham Dutfield, "Beyond Intellectual Property Rights: Toward Traditional Resource Rights for Indigenous Peoples and Local Communities" (Ottawa: International Development Research Centre, 1996); and Michel Pimbert, "SwedBio and Sida strategic workshop: The role of biodiversity for ecosystem services and its importance for poor people and local livelihoods in developing countries—priorities for the future" in Swedish International Biodiversity Program and Swedish Biodiversity Centre (eds.) "The Role of Biodiversity for Ecosystem Services and its Importance for People and Local Livelihoods in Developing Countries: Priorities for the Future," (Uppsala: Swedish International Biodiversity Programme, 2003), (available at http://citeseerx.ist.psu.edu/viewdoc/download?rep=rep1&type=pdf&doi=10.1.1.136.5346
15. United Nations Environment Program, "République Démocratique du Congo: Evaluation Environnementale Post-Conflit" (2012) (available at: http://postconflict.unep.ch/publications/UNEP_DRC_PCEA_full_FR.pdf).
16. Barney Dickson, "What is the Goal of Regulating Wildlife Trade? Is Regulation a Good Way to Achieve This Goal," in Sara Oldfield (ed.), *Trade in Wildlife: Regulation for Conservation* (London: Earthscan, 2003): 23–32.
17. Catrina MacKenzie et al., "Spatial Patterns of Illegal Resource Extraction in Kibale National Park, Uganda," *Environmental Conservation*, 39(1), 2011: 38–50.

18. United Nations Food and Agriculture Organization, "The State of World Fisheries and Aquaculture" (2016), (available at: www.fao.org/3/a-i5555e.pdf).
19. World Bank, "Turning the Tide—Saving Fish and Fishers: Building Sustainable and Equitable Fisheries and Governance," (2005), (available at: http://siteresources.worldbank.org/ESSDNETWORK/Publications/20631963/seaweb_FINAL_pt.1.pdf).
20. Millennium Ecosystem Assessment, *Ecosystems and Well-Being: Synthesis for Decision-Makers* (Washington D.C.: Island Press, 2006).
21. Dilys Roe et al., "Which Components or Attributes of Biodiversity Influence Which Dimensions of Poverty?" *Environmental Evidence*, 3(3), 2014: 1–15.
22. Laura Tangley, "The Sustainable Extraction of Rainforest Products in Guatemala," *US News and World Report*, 124(15), 1998: 40–41 and 44.
23. WRI, "Global Biodiversity Strategy."
24. Dan Simon, "Mexican Cartels Running Pot Farms in US National Forest," *CNN*, August 8, 2008 (available at: www.cnn.com/2008/CRIME/08/08/pot.eradication/).
25. Kieran Cooke, "California is Left High and Dry by Cannabis Growers," Climate News Network, February 4, 2015 (available at: http://climatenewsnetwork.net/california-is-left-high-and-dry-by-cannabis-growers/).
26. Mohammed Jamjoom and Gena Somra, "Qat Crops Threaten to Drain Yemen Dry," *CNN*, December 13, 2010 (available at: www.cnn.com/2010/WORLD/meast/12/02/yemen.water.crisis/).
27. UNEP, "République Démocratique du Congo."
28. World Travel Tourism Council, "Travel and Tourism, Economic Impact: Tanzania" (available at: www.wttc.org/-/media/files/reports/economic%20impact%20research/countries%202015/tanzania2015.pdf): 3.
29. Batswana authorities cite the higher percentage of 15 percent. See "Deadly Borders: 30 Namibians Killed through Bostwana's Shoot-to-Kill Policy" *The Namibian*, March 9, 2016. The World Travel Tourism Council gives the much lower estimate of 8.6 percent; see World Travel Tourism Council, "Travel and Tourism": 3.
30. United Nations Office on Drugs and Crime (UNODC), "Afghanistan Opium Survey 2013" December 2013 (available at: www.unodc.org/documents/crop-monitoring/Afghanistan/Afghan_Opium_survey_2013_web_small.pdf): 12.
31. William Byrd and David Mansfield, "Afghanistan's Opium Economy: An Agricultural, Livelihoods, and Governance Perspective," report pre-

pared for the World Bank Agriculture Sector Review, June 2014 (copy in author's possession).

32. UNODC, "Opium Amounts to Half of Afghanistan's GDP in 2007, Reports UNODC," November 16, 2007 (available at: www.unodc.org/india/afghanistan_gdp_report.html); and Christopher Ward and William Byrd, "Afghanistan's Opium Drug Economy" *World Bank Report No. SASPR-5* (Washington D.C.: World Bank, 2004), (available at: http://siteresources.worldbank.org/INTAFGHANISTAN/Publications-Resoucres/20325060/AFOpium-Drug-Economy-WP.pdf.pdf).

33. Mara Zimmerman, "The Black Market for Wildlife: Combatting Transnational Organized Crime in the Illegal Wildlife Trade," *Vanderbilt Journal of Transnational Law*, 36, 2003: 1656–89; and Julie Ayling, "What Sustains Wildlife Crime? Rhino Horn Trading and the Resilience of Criminal Networks," *Journal of International Wildlife Law and Policy*, 16, 2013: 57–80.

34. Francisco Thoumi, *Illegal Drugs, Economy, and Society in the Andes* (Baltimore: Johns Hopkins University Press, 2004); and Mauricio Reina, "Drug Trafficking and the National Economy," in Charles Berquist et al. (eds), *Violence in Colombia 1990–2000: Waging War and Negotiating Peace* (Wilmington: Scholarly Resources, 2001): 75–94.

35. David Pimentel et al., "Environmental and Economic Costs of Nonindigenous Species in the United States," *BioScience*, 50(1), January 2000: 53–65.

36. Ian Sample and John Gittings, "In China, the Civet Is a Delicacy—And May Have Caused SARS," *Guardian*, May 23, 2003; and "Out of the Shadows: The Origin of SARS," *Economist*, November 1, 2013.

37. See, for example, William Karesh, "Wildlife Trade and Global Disease Emergence," *Emerging Infectious Diseases*, 11(7), 2005: 1000–2; and William Karesh et al., "Implications of Wildlife Trade on the Movement of Avian Influenza and Other Infectious Diseases," *Journal of Wildlife Diseases*, 43(3), 2007: 55–9.

38. "An Ounce of Prevention," *Economist*, March 19, 2016.

39. Ibid.

40. Amadou Sy and Amy Copley, "Understanding the Economic Effects of the 2014 Ebola Outbreak in West Africa," Brookings Institution, 2014 (available at: www.brookings.edu/blogs/africa-in-focus/posts/2014/10/01-ebola-outbreak-west-africa-sy-copley).

41. Greg Warchol, "The Transnational Illegal Wildlife Trade," *Criminal Justice Studies* 17(1), March 2004: 57–73; Jolen Lin, "Tackling Southeast Asia Wildlife Trade," *Singapore Yearbook of International Law*, 9, 2005: 191–208; Nigel South et al., "The Illegal Wildlife Trade: A Case Study Report on the Illegal Wildlife Trade in the United Kingdom, Norway,

Colombia and Brazil," Study in the Framework of EFFACE Research Project, University of Oslo and University of South Wales, 2015 (available at: http://efface.eu/sites/default/files/EFFACE_Illegal%20 Wildlife%20Trade_revised.pdf).

42. Camilo Mejia Giraldo and James Bargent, "Are Mexican Narcos Moving into Lucrative Fish Bladder Market?" InSight Crime, August 6, 2014 (available at: www.insightcrime.org/news-briefs/mexico-narcos-fish-bladder-market).

43. Nick Davies and Oliver Holmes, "The Crime Family at the Centre of Asia's Animal Trafficking Network," *Guardian*, September 26, 2016 (available at: www.theguardian.com/environment/2016/sep/26/bach-brothers-elephant-ivory-asias-animal-trafficking-network).

44. Ibid.

45. Jonny Steinberg, "The Illicit Abalone Trade in South Africa," Institute for Security Studies, 2005 (available at: www.dlist.org/sites/default/files/doclib/Abalone%20Trade%20in%20South%20Africa.pdf).

46. Vanda Felbab-Brown, "Calderón's Caldron: Lessons from Mexico's Battle Against Organized Crime and Drug Trafficking in Tijuana, Ciudad Juarez, and Michoacán," Latin America Initiative Paper Series, Brookings Institution, 2011 (available at: www.brookings.edu/~/media/research/files/papers/2011/9/calderon-felbab-brown/09_calderon_felbab_brown.pdf); and Vanda Felbab-Brown, *Narco Noir: Mexico's Cartels, Cops, and Corruption* (Washington: Brookings Institution Press, forthcoming).

47. Stephan Sina et al., "Wildlife Crime," Study for the ENVI Committee, Directorate-General for Internal Policies, Policy Department A: Economic and Scientific Policy, European Parliament, 2016 (available at: www.europarl.europa.eu/RegData/etudes/STUD/2016/570008/IPOL_STU(2016)570008_EN.pdf): 9; see also Donald Liddick, *Crimes Against Nature: Illegal Industries and the Global Environment* (Oxford: Praeger, 2011); and Erica Alacs and Arthur Georges, "Wildlife Across Our Borders: A Review of the Illegal Trade in Australia," *Australian Journal of Forensic Sciences*, 40(2), 2008: 147–60.

48. See, for example, Dee Cook et al., "The International Wildlife Trade and Organized Crime: A Review of the Evidence and the Role of the UK," 2002 (available at: www.ibrarian.net/navon/paper/The_International_Wildlife_Trade_and_Organised_Cr.pdf?paperid=1569954); Robin Thomas Naylor, "The Underworld of Ivory," *Crime, Law, and Social Change*, 42(4/5), 2004: 261–95; and Jacqueline Schneider, *Sold Into Extinction: The Global Trade in Endangered Species* (Santa Barbara: Praeger, 2012).

49. Mauricio Rubio, "Violence, Organized Crime, and the Criminal Justice System in Colombia," *Journal of Economic Issues*, 32(2): 605–10.
50. Coletta A. Youngers and Eileen Rosin (eds.), *Drugs and Democracy in Latin America* (Boulder: Lynne Rienner, 2005); and Jamie Fellner, "Punishment and Prejudice: Racial Disparities in the War on Drugs," Human Rights Watch, May 2000 (available at: www.precaution.org/lib/hrw_war_on_drugs.2000.pdf).
51. Vanda Felbab-Brown, Shooting Up: *Counterinsurgency and the War on Drugs* (Washington: Brookings Institution, 2010).
52. Thor Hanson et al., "Warfare in Biodiversity Hostpots," *Conservation Biology*, 23(3), 2009: 578–87.
53. Paula Kahumbu and Andrew Halliday, "Case Proven: Ivory Trafficking Funds Terrorism," *Guardian*, August 30, 2015; Aislinn Laing, "LRA Warlord Joseph Kony Uses Ivory Trade to Buy Arms, *Daily Telegraph*, January 12, 2016; and Louisa Lombard, "Ivory Wars," *New York Times*, September 20, 2012.
54. Jeffrey Gettleman, "Africa's Elephants Are Being Slaughtered in a Poaching Frenzy," *New York Times*, September 3, 2012; and Michael Marshall, "Elephant Ivory Could Be Bankrolling Terrorist Groups," *New Scientist*, October 2, 2013.
55. Natascha White, "The 'White Gold of Jihad': Violence, Legitimisation and Contestation in Anti-Poaching Strategies," *Journal of Political Ecology*, 21, 2014: 452–74.
56. Robin Thomas Naylor, *Patriots and Profiteers: Economic Warfare, Sanctions, Embargo Busting, and Their Human Cost* (Montreal: McGill-Queen's University Press, 1999): 178.
57. Sophia Benz and Judith Benz-Schwarzburg, "Great Apes and New Wars," *Civil Wars*, 12(4), 2011: 395–430; Jasper Humphreys and M.L.R. Smith, "War and Wildlife: The Clausewitz Connection," *International Affairs*, 87(1), 2011: 121–42; and Linda Norgrove and David Hulme, "Confronting Conservation at Mount Elgon, Uganda," *Development and Change*, 37(5), 2006: 1093–116.
58. Sharon Begley, "Big Business: Wildlife Trafficking," *Newsweek*, 151(10), March 1, 2008; Naylor, *Patriots and Profiteers*; Naylor, "The Underworld of Ivory"; Greg L. Warchol et al., "Transnational Criminality: An Analysis of the Illegal Wildlife Market in Southern Africa," *International Criminal Justice Review*, 13(1), 2003: 1–27; and Vanda Felbab-Brown, "The Political Economy of Illegal Domains in India and China," *International Lawyer*, 43(4), 2009: 1411–28.
59. Author's interviews with Afghan powerbrokers involved in smuggling during the 1990s and US officials responding to the involvement of

NATO soldiers in Afghanistan in wildlife smuggling from the country, Kabul, Afghanistan, fall 2009.
60. George Jambiya et al., "Night Time Spinach: Conservation and Livelihood Implications of Wild Meat Use in Refugee Situations in North-Western Tanzania," TRAFFIC East-Southern Africa, 2007 (available at: https://portals.iucn.org/library/sites/library/files/documents/Traf-101.pdf).
61. Nabin Baral and Joel Heinen, "The Maoists' People's War and Conservation in Nepal," *Politics and the Life Sciences*, 24 (1/2), 2006: 2–11.
62. Daniel Winkler, "The Mushrooming Fungi Market in Tibet Exemplified by *Cordyceps sinensis* and *Tricholoma matsutake*," *Journal of the International Association of Tibetan Studies*, 4, 2008: 1–47.
63. Nabil Baral and Joel Heinen, "Resources Use, Conservation Attitudes, Management Intervention and Park–People Relations in the Western Terai Landscape of Nepal," *Environmental Conservation*, 34(1), 2007: 64–72.
64. "Nepal Winning Battle against One-Horned Rhino Poachers," *New York Daily News*, May 11, 2012.
65. Vanda Felbab-Brown, "The Impact of Organized Crime on Governance: The Case Study of Nepal," in Camino Kavanagh (ed.), *Impact of Organized Crime on Governance*, Center for International Cooperation, New York University, 2013 (available at: http://cic.nyu.edu/sites/default/files/kavanagh_crime_developing_countries_nepal_study.pdf).
66. Stephen Ellis, "Of Elephants and Men: Politics and Nature Conservation in South Africa," *Journal of Southern African Studies* 20(1), 1994: 53–69; and Roz Reeve and Stephen Ellis, "An Insider's Account of the South African Security Force's Role in the Ivory Trade," *Journal of Contemporary African Studies*, 13(2), 1995: 222–43.
67. Kristof Titeca, "Central Africa: Ivory Beyond the LRA – Why a Broader Focus Is Needed in Studying Poaching," All Africa, September 17, 2013 (available at: http://allafrica.com/stories/201309170982.html); and Taylor Toeka Kakala, "Soldiers Trade in Illegal Ivory," *Inter Press Service*, July 25, 2013 (available at: www.ipsnews.net/2013/07/soldiers-trade-in-illegal-ivory/).
68. "Sudan Army Accused of Ivory Trade," Al Jazeera, March 14, 2005 (available at: www.aljazeera.com/archive/2005/03/200841092645497609.html).
69. Environmental Investigative Agency (EIA), "Under Fire: Elephants in the Front Line," January 31, 1992 (available at: https://eia-international.org/report/under-fire-elephants-in-the-front-line).
70. Vanda Felbab-Brown, "Enabling War *and* Peace: Drugs, Logs, and

Wildlife in Thailand and Burma," Brookings Institution, 2015 (available at: www.brookings.edu/~/media/research/files/papers/2015/12/thailand-burma-drugs-wildlife-felbabbrown/enabling_war_and_peace_final.pdf)

71. See, for example, Christopher Anthony Loperena, "Conservation by Radicalized Dispossession: The Making of an Eco-Destination on Honduras's North Coast," *Geoforum*, 69, 2016: 184–93.
72. Louisa Lombard, "Threat Economies and Armed Conservation in Northeastern Central African Republic," *Geoforum*, 69, 2016: 218–26.
73. Felbab-Brown, "Enabling War *and* Peace."
74. Rosaleen Duffy, "War, By Conservation," *Geoforum*, 69, 2016: 238–48; Megan Ybarra, "'Blind Passes' and the Production of Green Security through Violence on the Guatemalan Border," *Geoforum*, 69, 2016: 194–206.
75. Elizabeth Lunstrum, "Green Militarization: Anti-Poaching Efforts and the Spatial Contours of Kruger National Park," *Annals of the Association of American Geographers*, 104(4), 2014: 816–32; Rosaleen Duffy, "Waging A War to Save Biodiversity: The Rise of Militarized Conservation," *International Affairs*, 90(4), 2014: 819–34; and Alice Kelly and Megan Ybarra, "Introduction to the Themed Issue: 'Green Security in Protected Areas'," *Geoforum*, 69, 2016: 171–75.
76. Michael Dwyere et al., "The Security Exception: Development and Militarization in Laos's Protected Areas," *Geoforum*, 69, 2016: 207–17.
77. Loperena, "Conservation by Radicalized Dispossession."
78. Diana Bocarejo and Diana Ojeda, "Violence and Conservation: Beyond Unintended Consequences and Unfortunate Coincidences," *Geoforum*, 69, 2016: 176–83.
79. Author's interviews in Bogotá, Nariño, Catatumbo, and middle Magdalena Valley, fall 2008 and 2009, and spring 2015.
80. Author's interviews in Kenya, February to May 2013.
81. Nir Kalron and Andrea Crosta, "Africa's Wild Gold of Jihad: Al Shabaab and Conflict Ivory," Elephant Action League (available at: www.elephantleague.org/project/africas-white-gold-of-jihad-al-shabaab-and-conflict-ivory/); and Johan Bergenas et al., "Killing Lions, Buying Bombs," *New York Times*, August 9, 2013.
82. Tom Maguire and Cathy Haenlein, "An Illusion of Complicity: Terrorism and the Illegal Ivory Trade in East Africa," Royal United Services Institute (RUSI), Occasional Papers No. 21, 2015 (available at: https://rusi.org/sites/default/files/201509_an_illusion_of_complicity_0.pdf).
83. Vanda Felbab-Brown, "Little to Gloat About," Cipher Brief, 2016 (avail-

able at: www.thecipherbrief.com/article/africa/little-gloat-about-1089).
84. Felbab-Brown, *Shooting Up*.
85. Ibid.
86. Vanda Felbab-Brown, "No Easy Exit: Drugs and Counternarcotics Policies in Afghanistan," Brookings Institution, 2015 (available at: www.brookings.edu/~/media/Research/Files/Papers/2015/04/global-drug-policy/FelbabBrown–Afghanistan-final.pdf?la=en).
87. Journalists for Justice, "Black and White: Kenya's Criminal Racket in Somalia," 2015 (available at: www.jfjustice.net/userfiles/file/Research/Black%20and%20White%20Kenya's%20Criminal%20Racket%20in%20Somalia.pdf).
88. Author's interviews with environmental groups operating in Nagaland, India (May to June 2008).
89. Author's interviews in *dacoit* communities in Uttar Pradesh, India, July 2007.
90. Lombard, "Threat Economies."
91. Felbab-Brown, *Shooting Up*: 1–33.
92. Vanda Felbab-Brown, "Human Security and Crime in Latin America: The Political Capital and Political Impact of Criminal Groups and Belligerents Involved in Illicit Economies," FIU/WHEMSAC, 2011 (available at: www.brookings.edu/research/articles/2011/09/latin-america-crime-felbab-brown).
93. Cynthia McClintock, *Revolutionary Movements in Latin America: El Salvador's FMLN and Peru's Shining Path* (Washington: United States Institute of Peace Press, 1998).
94. Annette Michaela Hübschle, "A Game of Horns: Transnational Flows of Rhino Horn," PhD thesis (Cologne: International Max Planck Research School, 2016): 312 and 333.
95. Ibid.: 310 and 321.
96. Ibid.: 308.
97. Maria Hauck and Neville A. Sweijd, "A Case Study of Abalone Poaching in South Africa and Its Impact on Fisheries Management," *ICES Journal of Marine Species: Journal du Conseil*, 56(6), 1999: 1024–32.
98. These factors are explored in Felbab-Brown, *Shooting Up*.

5. THE ACTORS

1. Erica von Essen and Michael Allen, "Reconsidering Illegal Hunting as Crime of Dissent: Implication for Justice and Deliberative Uptake," *Criminal Law and Philosophy*, 2015: 1–16; and Sandra Bell et al., "The

Political Culture of Poaching: A Case Study from Northern Greece," *Biodiversity and Conservation*, 16(2), 2007: 399–418.

2. "Pardhi Tribe Termed the Biggest Threat to Wildlife," *Express India*, January 7, 2008; Uday Mahurkar, "King in Shackles," *India Today*, April 23, 2007; and "If They Were Crooks, Wouldn't They Be Richer?" *Economist*, April 22, 2010.

3. See, for example, Pillenahalli Basavarajappa Tejaswi, "Non-Timber Forest Products (NTFPs) for Food and Livelihood Security: An Economic Study of Tribal Economy in the Western Ghats of Karnataka, India," MSc thesis (Ghent: University of Ghent, 2008): 2.

4. Izabella Koziell, "The Links between Biodiversity and Poverty" (United Nations Development Programme, 2000), (available at: https://ec.europa.eu/europeaid/sites/devco/files/publication-biodiversity-in-development-brief-12001_en.pdf).

5. Evan Bowen-Jones, "Bushmeat: Traditional Regulation or Adaptation to Market Forces," in Sara Oldfield (ed.), *The Trade in Wildlife* (London: Earthscan, 2003): 121–31.

6. Elizabeth Bennett and John Robinson (eds), *Hunting for Sustainability in Tropical Rainforests* (New York: Columbia University Press, 2000).

7. Author's interviews with United Nations Office on Drugs and Crime officials in Myanmar and poppy farming communities, December 2005.

8. Vanda Felbab-Brown, "Asia's Role in the Illicit Trade of Wildlife," *Boston Globe*, March 20, 2006.

9. Madhu Rao et al., "Hunting Patterns in Tropical Forests Adjoining the Hkakaborazi National Park, North Myanmar," *Oryx*, 39(3), 2005: 292–300; Chris Shepherd and Vincent Nijman, "The Wild Cat Trade in Myanmar," TRAFFIC Malaysia, 2008 (available at: www.traffic.org/species-reports/traffic_species_mammals40.pdf); Chris Shepherd and Vincent Nijman, "The Trade in Bear Parts from Myanmar: An Illustration of the Ineffectiveness of the Enforcement of International Wildlife Trade Regulations," *Biodiversity Conservation*, 17(1), 2008: 35–42.

10. World Bank, "Going, Going, Gone … The Illegal Trade in Wildlife in East and Southeast Asia," Discussion Paper, 2005 (available at: www-wds.worldbank.org/external/default/WDSContentServer/WDSP/IB/2005/09/08/000160016_20050908161459/Rendered/PDF/334670PAPER0Going1going1gone.pdf): 6.

11. Dennis Gray, "Asia's Wildlife Hunted for China's Appetite," *Associated Press*, April 6, 2004 (available at: www.nbcnews.com/id/4585068/ns/us_news-environment/t/asias-wildlife-hunted-chinas-appetite/#.V_6XEiSD6M8).

12. Bryan Christy, "Asia's Wildlife Trade," *National Geographic*, 217(1), January 2010 (available at: http://ngm.nationalgeographic.com/2010/01/asian-wildlife/christy-text).
13. George Rushby, *No More the Tusker* (London: W.H. Allen, 1965).
14. Annette Michaela Hübschle, "A Game of Horns: Transnational Flows of Rhino Horn." (PhD thesis) (Cologne: International Max Planck Research School, 2016): 321.
15. Bryan Christy, *The Lizard King* (Guilford: Lyons Press, 2010).
16. "Sansar Chand, Notorious Tiger Poacher, Dead," *Times of India*, March 19, 2014 (available at: www.timesofindia.indiatimes.com/city/jaipur/Sansar-Chand-notorious-tiger-poacher-dead/articleshow/32261903.cms).
17. Jo Hastie et al., "Back in Business: Elephant Poaching and the Ivory Black Markets of Asia," *Environmental Investigation Agency*, 2002 (available at: www.eia-international.org/wp-content/uploads/Back-in-Business-2002.pdf).
18. Julian Rademeyer, *Killing for Profit* (Cape Town: Zebra Press, 2012); Tom Milliken and Jo Shaw, "The South Africa–Viet Nam Rhino Horn Trade Nexus: A Deadly Combination of Institutional Lapses, Corrupt Wildlife Industry Professionals and Asian Crime Syndicates," (Cambridge: TRAFFIC, 2012), (available at: www.npr.org/documents/2013/may/traffic_species_mammals.pdf).
19. Thomas Fuller, "US Offers Reward in Wildlife-Trade Fight," *New York Times*, November 13, 2013.
20. Nick Davies and Oliver Holmes, "The Crime Family at the Centre of Asia's Animal Trafficking Network," *Guardian*, September 26, 2016 (available at: www.theguardian.com/environment/2016/sep/26/bach-brothers-elephant-ivory-asias-animal-trafficking-network).
21. Oliver Holmes and Nick Davies, "Revealed: The Criminals Making Millions from Illegal Wildlife Trafficking," *Guardian*, September 26, 2016 (available at: https://www.theguardian.com/environment/2016/sep/26/revealed-the-criminals-making-millions-from-illegal-wildlife-trafficking).
22. Author's interviews with US and Mexican environmental NGOs focusing on totoaba poaching and smuggling and their negative effects on the conservation of the critically endangered porpoise vaquita at a San Francisco workshop on these issues, October 14, 2016. See also Camilo Mejia Giraldo and James Bargent, "Are Mexican Narcos Moving into Lucrative Fish Bladder Market?" *InSight Crime*, August 6, 2014 (available at: www.insightcrime.org/news-briefs/mexico-narcos-fish-bladder-market).

23. Fiona Underwood et al., "Dissecting the Illegal Ivory Trade: An Analysis of Ivory Seizure Data," *PLOS One*, October 18, 2013.
24. Varun Vira et al., "Out of Africa: Mapping the Global Trade in Illicit Elephant Ivory," *C4ADS*, 2014 (available at: www.wwf.se/source.php/1578610/out%20of%20africa.pdf); see also Varun Vira and Thomas Ewing, "Ivory's Curse: The Militarization and Professionalization of Poaching in Africa," *C4ADS*, 2014 (available at: www.rhinoresourcecenter.com/pdf_files/139/1398477046.pdf?view).
25. Lucy Vigne and Esmond Martin, "Increasing Rhino Awareness in Yemen and a Decline in the Rhino Horn Trade," *Pachyderm*, 53, 2013: 51–8.
26. Ukesh Raj Bhuju et al., "Report on the Facts and Issues on Poaching of Mega Species and Illegal Trade in Their Parts in Nepal" (Amsterdam: Transparency International Nepal, 2009), (available at: www.rhinoresourcecenter.com/pdf_files/131/1315014970.pdf).
27. Ko-Lin Chin and Sheldon Zhang, *The Chinese Heroin Trade: Cross-Border Drug Trafficking in Southeast Asia and Beyond* (New York: New York University Press, 2015).
28. Patrick L. Clawson and Rensselaer W. Lee III, *The Andean Cocaine Industry* (New York: St. Martin's, 1996): 40–41; Mark Bowden, *Killing Pablo* (New York: Atlantic Monthly, 2001); and George Grayson, *The Cartels* (New York: Praeger, 2014).
29. Peter Reuter, *Disorganized Crime: Economics of the Visible Hand* (Cambridge: MIT Press, 1983).
30. Davies and Holmes, "Crime Family."
31. Ibid.
32. Rachel Bale, "The World Is Finally Getting Serious about Tiger Farms," *National Geographic*, September 29, 2016 (available at: www.news.nationalgeographic.com/2016/09/wildlife-watch-tiger-farms-cites-protections/).
33. Debbie Banks and Julian Newman, "The Tiger Skin Trail" (Washington D.C.: Environmental Investigation Agency, 2004), (available at: www.eia-international.org/wp-content/uploads/TheTigerSkinTrail-Low-Res.pdf).
34. Author's interviews with border patrol and national park officials and local community members, Solukhumbu district, and with a leading conservationist, Kathmandu, May 2012.
35. "Dalai Lama Campaigns for Wildlife," *BBC News*, April 6, 2005 (available at: www.news.bbc.co.uk/2/hi/science/nature/4415929.stm).
36. World Bank, "Going, Going, Gone": 10.
37. Xu Hongfa and Craig Kinspatrick, eds, "The State of Wildlife Trade in China: Information on the Trade in Wild Animals and Plants in China

2007" (Beijing: TRAFFIC, 2008), 9 (available at: http://www.trafficj.org/publication/08-State_of_Wildlife_China.pdf 13.

38. Author's interviews with officials of Kenya Wildlife Society, Nairobi, February 2013; and "South African Group Reports Slight Drop in Rhino Poaching," *Associated Press*, January 2, 2016 (available at: http://www.bostonherald.com/news/international/2016/01/south_african_group_reports_slight_drop_in_rhino_poaching).
39. Hübschle, "Game of Horns": 302–3.
40. Ibid: 19.
41. Author's research and interviews in Mong La, Myanmar, January 2006.
42. Daniel Stiles, "Elephant and Ivory Trade in Thailand" (Petaling Jaya: TRAFFIC, 2009), (available at: www.traffic.org/species-reports/traffic_species_mammals50.pdf).
43. Xiyun Yang, "Tiger Deaths Raise Alarms About Chinese Zoos," *New York Times*, March 18, 2010.
44. Tanya Wyatt, "Exploring the Organization of Russia Far East's Illegal Wildlife Trade: Two Case Studies of the Illegal Fur and Illegal Falcon Trades," *Global Crime*, 10(1/2), March 2009: 144–54.
45. Louisa Lombard, "Threat Economies and Armed Conservation in Northeastern Central African Republic," *Geoforum*, 69, 2016: 219.
46. Conor Joseph Cavanagh et al., "Securitizing REDD+: Problematizing the Emerging Illegal Timber Trade and Forest Carbon Interface in East Africa, *Geoforum*, 60, February 2015: 72–82; Jennifer Devine, "Counterinsurgency Ecotourism in Guatemala's Maya Biosphere Reserve," *Environment and Planning D: Society and Space*, 32(6), December 2014: 981–1001; and Francis Massé and Elizabeth Lunstrum, "Accumulation by Securitization: Commercial Poaching, Neoliberal Conservation, and the Creation of New Wildlife Frontiers," *Geoforum*, 69, February 2016: 227–37.

6. POLICY RESPONSE I: BANS AND LAW ENFORCEMENT

1. See, for example, Robert MacCoun and Peter Reuter, *Drug War Heresies: Learning from Other Vices, Times, and Places* (Cambridge: Cambridge University Press, 2001).
2. John Walsh and Geoff Ramsey, "Uruguay's Drug Policy: Major Innovations, Major Challenges." The Brookings Institution, April 2015 (https://www.brookings.edu/wp-content/uploads/2016/07/Walsh-Uruguay-final.pdf). Mark Kleiman, "Legal Commercial Cannabis Sales in Colorado and Washington: What Can We Learn?" Brookings Institution, 2015 (available at: http://www.brookings.edu/~/media/

Research/Files/Papers/2015/04/global-drug-policy/Kleiman—Wash-and-Co-final.pdf?la=en).
3. Cyril Lombard and Pierre du Plessis, "The Impact of the Proposal to List Devil's Claw on Appendix II of CITES," in Sara Oldfield (ed.), *The Trade in Wildlife* (London: Earthscan, 2003): 132–16.
4. "Call of the Wild," *Economist*, March 8, 2008.
5. Vanda Felbab-Brown, "Indonesia Field Report IV: The Last Twitch? Wildlife Trafficking, Illegal Fishing, and Lessons from Anti-Piracy Efforts," Brookings Institution, March 2013 (available at: www.brookings.edu/research/reports/2013/03/25-indonesia-wildlife-trafficking-felbabbrown).
6. Philippe Rivalan et al., "Can Bans Stimulate Wildlife Trade?" *Nature* 477(7144), 2007: 529–30.
7. John Frederick Walker, *Ivory Ghosts: The White Gold of History and the Fate of the Elephants* (New York: Grove Press, 2010).
8. Author's interviews about illegal trade in wildlife in the lower Amazon and Pantalan, Brazil, December 2009 and January 2010. Conservation awareness in Brazil continues to be severely lacking.
9. Author's interviews with conservation biologists, Nairobi, Kenya, February 2012.
10. Michael 't Sas-Rolfes, "Assessing CITES: Four Case Studies," in Jon Hutton and Barnabas Dickson (eds.), *Endangered Species, Threatened Convention: The Past, Present, and Future of CITES* (London: Earthscan, 2000): 69–87.
11. For an excellent discussion, see Nigel Leader-Williams, "Regulation and Protection: Successes and Failures in Rhinoceros Conservation," in Sara Oldfield (ed.), *The Trade in Wildlife* (London: Earthscan, 2003): 89–99.
12. Aaron Bruner et al., "Effectiveness of Parks in Protecting Tropical Biodiversity," *Science*, 291(5501), January 2001: 125–8.
13. Dan Ashe, Testimony before the US House of Representatives, Committee on Foreign Affairs, Regarding the National Strategy for Combatting Wildlife Trafficking, February 26, 2014. *U.S. Fish and Wildlife Service, Department of Interior.* (available at: www.fws.gov/international/pdf/wildlife-trafficking-national-strategy-testimony.pdf); and Sharon Guynup, "US A Major Destination for Trafficked Latin American Wildlife," *Mongabay*, November 25, 2015 (available at: http://news.mongabay.com/2015/11/u-s-a-major-destination-for-trafficked-latin-american-wildlife/).
14. Claire Salisbury, "Communities and Cutting-Edge Tech Keep Cambodia's Gibbons Singing," Mongabay, February 22, 2016 (available at: www.

google.com/?gws_rd=ssl#q=Claire+Salisbury%2C+%E2%80%9CCommunities+and+Cutting-Edge+Tech+Keep+Cambodia%E2%80%99s+Gibbons+Singing%2C%E2%80%9D+Mongabay.com%2C+).

15. Hugo Jachmann, "Elephant Poaching and Resource Allocation for Law Enforcement," in Sara Oldfield (ed.), *The Trade in Wildlife* (London: Earthscan, 2003): 100–7.
16. Author's interviews with representatives of Profauna and law enforcement officials in Bali, September 2012.
17. Author's interviews with the biologist from Sulawesi's Pacific Institute and local environmental NGO representatives, Manado, Sulawesi, October 2012.
18. George Wittemyer et al., "Poaching Policy: Rising Ivory Prices Threaten Elephants," *Nature*, 476, 2011: 282–3.
19. Kristen Conrad, "Trade Bans: A Perfect Storm for Poaching?" *Tropical Conservation Science*, 5(3), 2012: 245–54.
20. Simon Bloch, "Czech Police Charge 16 for Trading in Rhino Horn," *Business Day*, December 18, 2014.
21. "Poachers Using Science Papers to Target Newly Discovered Species," *Guardian*, January 1, 2016 (available at: www.theguardian.com/environment/2016/jan/01/poachers-using-science-papers-to-target-newly-discovered-species).
22. Vanda Felbab-Brown, "Counterinsurgency, Counternarcotics, and Illicit Economies in Afghanistan: Lessons for State-Building," in Jacqueline Brewer and Michael Miklaucic (eds), *Convergence: Illicit Networks and National Security in the Age of Globalization* (Washington D.C.: National Defense University Press, 2013): 189–212.
23. Adam Pain, "Opium Trading Systems in Helmand and Ghor" (Kabul: Afghanistan Research and Evaluation Unit, 2006), (available at https://www.yumpu.com/en/document/view/42972723/opium-trading-systems-in-helamdn-and-ghor-the-afghanistan-).
24. Environmental Affairs Department of the Republic of South Africa, "Poaching Statistics" (available at: www.environment.gov.za/projectprogrammes/rhinodialogues/poaching_statistics); and "South African Group Reports Slight Drop in Rhino Poaching," *Associated Press*, January 2, 2016 (available at: http://www.bostonherald.com/news/international/2016/01/south_african_group_reports_slight_drop_in_rhino_poaching).
25. Willy Lowry, "Ring of Elephant Poachers Broken up by Tanzanian Authorities," *New York Times*, February 8, 2016.
26. Vanda Felbab-Brown, "Despite its Siren Song, High-Value Targeting Doesn't Fit All: Matching Interdiction Patterns to Specific Narcoter-

rorism and Organized-Crime Contexts," *Brookings Institution* 2013 (available at: www.brookings.edu/~/media/research/files/papers/2013/10/01-matching-interdiction-patterns-to-narcoterrorism-and-organized-crime-contexts-felbabbrown/felbabbrown—matching-interdiction-patterns-to-specific-threat-environments.pdf).

27. For such focused deterrence strategies, see, for example, Anthony Braga, "Getting Deterrence Right?" *Criminology and Public Policy*, 11(2), 2012: 201–10; David Kennedy, *Don't Shoot: One Man, A Street Fellowship, and the End of Violence in Inner-City America* (New York: Bloomsbury, 2011); and Mark Kleiman, *When Brute Force Fails: How to Have Less Crime and Less Punishment* (Princeton: Princeton University Press, 2009).

28. Nick Davies and Oliver Holmes, "The Crime Family at the Centre of Asia's Animal Trafficking Network," *Guardian*, September 26, 2016 (available at: www.theguardian.com/environment/2016/sep/26/bach-brothers-elephant-ivory-asias-animal-trafficking-network).

29. Laurel Neme, "For Rangers on the Front Lines of Anti-Poaching Wars, Daily Trauma," *National Geographic*, June 27, 2014 (available at: http://news.nationalgeographic.com/news/2014/06/140627-congo-virunga-wildlife-rangers-elephants-rhinos-poaching/).

30. Thin Green Line Foundation, "2009–2016: Ranger Roll of Honor, In Memoriam" (available at: www.europarc.org/wp-content/uploads/2016/07/2009-2016-Honour-Roll-1.pdf).

31. Stephan Sina et al., "Wildlife Crime" (Brussels: Directorate-General for Internal Policies, 2016), 9 (available at: www.europarl.europa.eu.RegData/etudes/STUD/2016/570008/IPOL_STU(2016)570008_EN.pdf)

32. John Sellar, "Policing the Trafficking of Wildlife: Is There Anything to Be Learned from Law Enforcement Responses to Drug and Firearm Trafficking?" *Global Initiatives Against Organized Crime*, February 2014 (available at: www.globalinitiative.net/download/global-initiative/Global%20Initiative%20-%20Wildlife%20Trafficking%20Law%20Enforcement%20-%20Feb%202014.pdf).

33. See Craig Packer, *Lions in the Balance: Man-Eaters, Manes, and Men with Guns* (Chicago: University of Chicago Press, 2015); and Paul Robbins, "The Rotten Institution: Corruption in Natural Resource Management," *Political Geography*, 19(4), May 2000: 423–43.

34. See, for example, Patrick Chabal and Jean-Pascal Daloz, *Africa Works: Disorder and Political Instrument* (Bloomington: Indiana University Press, 1999); and Nicolas van der Walle, *African Economies and the Politics of Permanent Political Crisis, 1979–1999* (Cambridge: Cambridge University Press, 2001).

35. Author's interviews with officials of Kenya Wildlife Society, Nairobi, February 2013.
36. Author's interviews with park rangers in the Tsavo West National Park, February 2013.
37. Author's interview with park ranger at the rhino sanctuary of the Tsavo West National Park, February 2013.
38. Louisa Lombard, "Threat Economies and Armed Conservation in Northeastern Central African Republic," *Geoforum*, 69, 2016: 218–26.
39. See, for example, Antony Lynam, "Rain-Forest Guardians," *Wildlife Conservation*, January 2005: 8–9 (available at: https://www.researchgate.net/publication/235459798_Rainforest_Guardians?fulltextDialogue=true).
40. Author's interviews with environmental NGO representatives in Nairobi, February 2013, park rangers and wardens in Tsavo West National Park, February 2013, in Maasai Mara, Kenya, May 2013, and Serengeti National Park and Ngorongoro Conservation Area, Tanzania, August 2003.
41. Author's interviews with environmental NGO representatives, officials of Kenya Wildlife Society, Kenyan journalists, and park rangers in Kenya, February, April, and May 2013.
42. Charles Hornsby, *Kenya: A History since Independence* (New York: I.B. Tauris, 2013).
43. Justin Ling, "Inside the US-Canadian Smuggling Ring for Narwhal Tusks," *Vice News*, March 18, 2016.
44. Transparency International, "Global Corruption Index 2015" (available at: www.transparency.org/cpi2015#results-table).
45. Kenya's new 2010 constitution devolved powers and changed subnational administrative structures and their boundaries.
46. Author's interviews with Brian Heath, park rangers, and managers of lodges in Maasai Mara, Kenya, May 2013.
47. See also Richard Leakey, *Wildlife Wars: My Battle to Save Kenya's Elephants* (London: Pan, 2001).
48. Corruption in Kenya's Forestry Service was widely reported during the author's interviews with wildlife officials and Kenyan and international environmental NGOs in Nairobi and around the Maasai Mara National Reserve, February 2013 and May 2013.
49. Author's interviews with Brian Heath and Mara Conservancy warden and rangers, May 2013.
50. Adam Nossiter, "US Sting that Snared Guinea-Bissau Ex-Admiral Shines Light on Drug Trade," *New York Times*, April 16, 2013. (available at http://www.nytimes.com/2013/04/16/world/africa/us-sting-that-snared-guineabissau-ex-adminral-shines-light-on-drug-trade.html)

51. For an evaluation of CICIG's effectiveness, see, for example, "La comisión internacional contra la impunidad en Guatemala: un estudio de investigación de WOLA sobre la experiencia de la CICIG," Washington Office on Latin America (WOLA), March 2015 (available at: www.wola.org/sites/default/files/CICIG%203.25.pdf).
52. Joey Berning and Moses Montesh, "Countering Corruption in South Africa: The Rise and Fall of the Scorpions and the Hawks," *South Africa Crime Quarterly*, 39, 2012: 3–9; see also, Wendell Roelf, "South Africa's Scorpion Crime Fighters to Be Disbanded," *Reuters*, February 12, 2008 (available at: www.reuters.com/article/us-safrica-scorpions-idUSL1223 114320080212).
53. Nicholas Yong, "Drop in Illegal Wildlife Trade Here," *Straits Times*, May 2, 2009.
54. DLA Piper, "Empty Threat: Does the Law Combat Illegal Wildlife Trade? An Eleven-Country Review of Legislative and Judicial Approaches," 2014 (available at: www.dlapiperprobono.com/export/sites/pro-bono/downloads/pdfs/Illegal-Wildlife-Trade-Report-2014.pdf): 131 and 194.
55. TRAFFIC, "An Assessment of the Domestic Ivory Carving Industry and Trade Controls in India," 2003 (available at: www.traffic.org/species-reports/traffic_species_mammals20.pdf).
56. DLA Piper, "Empty Threat": 266.
57. Ibid.: 71.
58. Sina et al., "Wildlife Crime": 71–85 and 93–4.
59. John Cruden, Poaching and Terrorism: A National Security Challenge Hearing, "Statement before the Committee on Foreign Affairs, Subcommittee on Terrorism, Nonproliferation, and Trade, US House of Representatives," The United States Department of Justice, April 22, 2015 (available at: www.justice.gov/opa/speech/poaching-and-terrorism-national-security-challenge-statement-assistant-attorney-general).
60. "China Ivory Smuggler Fined $1 A Piece," *News24*, March 26, 2013 (available at: www.news24.com/Africa/News/China-ivory-smuggler-fined-1-a-piece-20130326).
61. Fumbuka Ng'wanakilala, "Tanzania Court Jails Two Chinese Men for Ivory Smuggling: Media," *Reuters*, March 19, 2016 (available at: www.reuters.com/article/us-tanzania-poaching-idUSKCN0WL0NK).
62. Jeffrey Ulmer and Darrell Steffensmeier, "The Age and Crime Relationship: Social Variation, Social Explanations," in Kevin Beaver et al. (eds.), *The Nurture Versus Biosocial Debate in Criminology: On the Origins of Criminal Behavior and Criminality* (London: Sage, 2014): 377–97.

63. Author's research on wildlife trafficking in the Pantanal and the Amazon, Brazil, winter 2010.
64. "Deadly Borders: 30 Namibians Killed through Bostwana's Shoot-to-Kill Policy," *The Namibian*, March 9, 2016.
65. David Smith, "Elephant Deaths Rise in Tanzania after Shoot-to-Kill Poachers Policy is Dropped," *Guardian*, December 31, 2013. On allegations of extrajudicial killings of local community poachers in Uganda's largest national park, Murchison Falls, see "Elephants, People Fight for Life in Nwoya," *Observer* (Kampala, Uganda), February 14, 2013.
66. Rosaleen Duffy, "Waging a War to Save Biodiversity: The Rise of Militarized Conservation," *International Affairs*, 90(4), 2014: 825.
67. Roderick Neumann, "Moral and Discursive Geographies in the War for Biodiversity in Africa," *Political Geography*, 23(7), 2004: 813–37.
68. Simon Romero, "Colombia Lists Civilian Killings in Guerrilla Toll," *New York Times*, October 29, 2008.
69. Author's interviews with Brian Heath, manager of the Mara Conservancy, Kenya, May 2013.
70. E.J. Milner-Gulland and Nigel Leader-William, "A Model of Incentives for the Illegal Exploitation of Black Rhinos and Elephants: Poaching Pays in Luangwa Valley, Zambia," *Journal of Applied Ecology*, 29(2), 1992: 388–401.

7. LEGAL TRADE

1. See, for example, John Collins, "The Economics of a New Global Strategy," in LSE Ideas (compilers), *Ending the Drug Wars* (London: London School of Economics, 2014) 8–15, (available at: www.lse.ac.uk/IDEAS/publications/reports/pdf/LSE-IDEAS-DRUGS-REPORT-FINAL-WEB.pdf).
2. For the counterinsurgency and counterterrorism arguments, see Vanda Felbab-Brown, *Shooting Up: Counterinsurgency and the War on Drugs* (Washington D.C.: Brookings Institution, 2010).
3. Edward Shepard and Paul Blackley, "Medical Marijuana and Crime: Further Evidence from the Western States," *Journal of Drug Issues*, 46(2), April 2016: 122–34.
4. Colorado Department of Revenue, "2015 Annual Report: Gross Collections and Net Collections by Fund and Tax Sources, Fiscal Year 2015" (available at: www.colorado.gov/pacific/sites/default/files/2015%20Annual%20Report_1.pdf): 49.
5. Report by ArcView Market Research and New Frontier Data, cited in Christopher Ingram, "Americans Spent More on Legal Weed than on

Cheetos and Funyuns Combined Last Year," *Washington Post*, February 2, 2016.

6. Beau Kilmer, "Marijuana Legalization, Government Revenues, and Public Budgets: Ten Factors to Consider," testimony presented before the Vermont Senate Committee on Finance, January 19, 2016 (available at: www.rand.org/content/dam/rand/pubs/testimonies/CT400/CT449/RAND_CT449.pdf).

7. United Nations Environment Programme. "Year Book 2014 Emerging Issues Update: Illegal Trade in Wildlife," 2014 (available at: www.unep.org/yearbook/2014/PDF/chapt4.pdf): 25.

8. Vanda Felbab-Brown, "Organized Criminals Won't Fade Away," *World Today Magazine*, August 2012 (available at: www.brookings.edu/research/articles/2012/08/drugs-crime-felbabbrown).

9. See, for example, Vanda Felbab-Brown and Anna Newby, "How to Break Free of the Drugs–Conflict Nexus in Colombia," Brookings Institution, December 16, 2015 (available at: www.brookings.edu/blogs/order-from-chaos/posts/2015/12/16-colombia-drugs-conflict-nexus-felbabbrown-newby).

10. Jonathan Caulkins et al., *Marijuana Legalization: What Everyone Needs to Know* (Oxford: Oxford University Press, 2012).

11. Jonathan Caulkins, "The Real Dangers of Marijuana," *National Affairs*, 26, Winter 2016: 21–34 (available at http://www.nationalaffairs.com/publications/detail/the-real-dangers-of-marijuana).

12. Caroline Chatwin, "Mixed Messages from Europe on Drug Policy Reform: The Cases of Sweden and the Netherlands" (Washington, D.C.: Brookings Institution, 2015), (available at: www.brookings.edu/~/media/Research/Files/Papers/2015/04/global-drug-policy/ChatwinSwedenNetherlands-final.pdf?la=en).

13. Erwin Bulte and Richard Damania, "An Economic Assessment of Wildlife Farming and Conservation," *Conservation Biology*, 19(4), August 2005: 1222–33.

14. Timothy F. Wright et al., "Nest Poaching in Neotropical Parrots," *Conservation Biology*, 15(3), 2001: 710–20; Brendan Moyle, "Regulation, Conservation, and Incentives," in Sara Oldfield (ed.), *The Trade in Wildlife: Regulation for Conservation* (London: Earthscan, 2003): 41–51; and Jorge Rabinovich, "Parrots, Precaution and Project Ele: Management in the Face of Multiple Uncertainties," in Rosie Cooney and Barney Dickson (eds), *Biodiversity and the Precautionary Principle: Risk and Uncertainty in Conservation and Sustainable Use* (London: Earthscan, 2005): 173–88.

15. John Thorbjarnarson and Alvaro Velasco, "Venezuela's Caiman Harvest

Program: A Historical Perspective and Analysis of Its Conservation Benefits" (Washington D.C.: Wildlife Conservation Society, 1998); John Loveridge, "A Review of Crocodile Management in Zimbabwe" (Harare: University of Zimbabwe, 1996); and Henriette Kievit, "Conservation of the Nile Crocodile: Has CITES Helped or Hindered?" in Jon Hutton and Barnabas Dikson (eds.), *Endangered Species: Threatened Convention: The Past, Present, and Future of CITES* (London: Earthscan, 2000), 88–97.

16. Jon Hutton and Grahame Webb, "Crocodiles: Legal Trade Snaps Back" in Sara Oldfield (ed.), *The Trade in Wildlife* (London: Earthscan, 2003): 108–20; and Robert Jenkins et al., "Review of Crocodile Ranching Programs," (Gland: International Union for Conservation of Nature, 2006), (available at: https://cites.org/common/com/ac/22/EFS-AC22-Inf02.pdf).

17. James Perran Ross (ed.), "Crocodiles: Status Survey and Conservation Action Plan" (Gland: International Union for Conservation of Nature, 1998), (available at: http://iucncsg.org/ph1/modules.Publications/action_plan1998/plan1998a.htm).

18. Sandija O'Connell, "Crocodile Farms: Is it Cruel to Keep These Wild Creatures Captive?" *Independent*, October 4, 2006.

19. Defenders of Wildlife, "Combating Wildlife Trafficking from Latin America: The Illegal Trade from Mexico, the Caribbean, Central America and South America and What Can We Do to Address It," 2015 (available at: www.defenders.org/sites/default/files/publications/combating-wildlife-trafficking-from-latin-america-to-the-united-states-and-what-we-can-do-to-address-it.pdf).

20. Robert W.G. Jenkins, "The Significant Trade Process: Making Appendix II Work," in Jon Hutton and Barnabas Dickson (eds), *Endangered Species, Threatened Convention* (London: Earthscan, 2000): 47–56; and Rowan Martin, "When CITES Works and When It Does Not," in Jon Hutton and Barnabas Dickson (eds), *Endangered Species, Threatened Convention* (London: Earthscan, 2000): 29–37.

21. Jane Wheeler and Domingo Hoces R., "Community Participation, Sustainable Use, and Vicuña Conservation in Peru," *Mountain Research and Development*, 17(3), July 1997: 283–87; and Daniel Challender et al., "Toward Informed and Multi-Faceted Wildlife Trade Interventions," *Global Ecology and Conservation*, 3, 2015: 129–48.

22. Author's interviews with Brian Heath, Maasai Mara, May 2013.

23. Wouter van Hoven, "Private Game Parks in Southern Africa," in René van der Duim et al. (eds), *Institutional Arrangements for Conservation, Development, and Tourism in Eastern and Southern Africa* (Dordrecht: Springer, 2015): 101–18.

24. Brian Reilly, "Game Ranching in South Africa: An Alternative Model Mixing Economics and Conservation," *Wildlife Professional*, 8(4), 2014: 36–41 (available at: http://sustainability.colostate.edu/sites/sustainability.colostate.edu/files/Final%20Private%20Lands%20Package%20TWS%20Summer%202014.pdf).
25. Richard Emlsie and Martin Brooks, "Status Survey and Conservation Action Plan: African Rhino," IUCN/SSC African Rhino Specialist Group, 1999 (available at: https://portals.iucn.org/library/efiles/edocs/1999-049.pdf).
26. Maano Ramutsindela, *Parks and People in Postcolonial Societies: Experiences in Southern Africa* (New York: Kluver, 2006).
27. Nigel Leader-Williams, "Regulation and Protection: Successes and Failures in Rhinoceros Conservation," in Sara Oldfield (ed.), *The Trade in Wildlife* (London: Earthscan, 2003): 89–99; and Michael 't Sas-Rolfes, "The Rhino Poaching Crisis: A Market Analysis," Feburary 2012 (available at: www.rhinoresourcecenter.com/pdf_files/133/1331370813.pdf).
28. Robert Smith et al., "New Rhino Conservation Project in South Africa to Understand Landowner Decision-Making," *Oryx*, 47(3), 2013: 323.
29. Author's interview with a Tanzanian wildlife official, Washington, DC, June 2016.
30. John Frederick Walker, *Ivory Ghosts: The White Gold of History and the Fate of Elephants* (New York: Grove Press, 2010).
31. Timothy Swanson, "A Tale of Rent Seeking, Corruption, Stockpiling and (Even) Tragedy: Re-Telling the Tale of the Commons," *International Review of Environmental and Resource Economics*, 1(1), 2007: 111–50.
32. Rabinovich, "Parrots, Precaution and Project Ele."
33. Wright et al., "Nest Poaching."
34. Author's interviews with officials of Kenya Wildlife Society and representatives of environmental NGOs, Nairobi, Kenya, February 2013.
35. Author's interviews with park wardens at the Nechisar, Bale Mountains, and Simien National Parks, Ethiopia, March 2013.
36. Author's interviews with park rangers, wardens, and managers, Maasai Mara, May 2013.
37. Author's interviews with officials of Kenya Wildlife Society and representatives of environmental NGOs, Nairobi, Kenya, February 2013; and interviews with park management officials, rangers, and local community representatives in both Narok and the Mara Triangle parts of Maasai Mara, May 2013.
38. Author's interviews with the park warden at Lake Abiata-Shala National Park, Ethiopia, March 2013.

39. Alexander N. James et al., "Balancing the Earth's Accounts," *Nature*, 336(6199), 1999: 533–35.
40. William Adams and Mark Infield, "Who is on the Gorilla's Payroll? Claims on Tourist Revenues from a Ugandan National Park," *World Development*, 31(1), January 2003: 177–90.
41. See, for example, Michael Norton-Griffiths and Clive Southey, "The Opportunity Costs of Biodiversity Conservation in Kenya," *Ecological Economics*, 12(2), February 1995: 125–39, (available at http://mng5.com/papers/BiodivCosts.pdf
42. Vanda Felbab-Brown, "Not as Easy as Falling Off a Log: The Illegal Logging Trade in the Asia-Pacific Region and Possible Mitigation Strategies" (Washington D.C.: The Brookings Institution, 2011) (available at: www.brookings.edu/~/media/research/files/papers/2011/3/illegal-logging-felbabbrown/03_illegal_logging_felbabbrown.pdf); and Vanda Felbab-Brown, "Indonesia Field Report III: The Orangutan's Road: Illegal Logging and Mining in Indonesia," *Brookings Institution*, February 2013 (available at: www.brookings.edu/research/reports/2013/02/07-indonesia-illegal-logging-mining-felbabbrown).
43. Author's interviews in Sulawesi, Kalimantan, and Java, Indonesia, fall 2012.
44. Xu Hongfa and Craig Kirkpatrick, eds, "The State of Wildlife Trade in China: Information on the Trade in Wild Animals and Plants in China 2007" (Beijing: TRAFFIC, 2008)9 (available at http://www.trafficj.org/publication/08-State_of_Wildlife_China.pdf)
45. Li Zhang et al., "Wildlife Trade, Consumption, and Conservation Awareness in Southwest China," *Biodiversity and Conservation*, 17(6), 2008: 1506.
46. Susan M. Wells and Jonathan G. Barzdo, "The International Trade in Marine Species: Is CITES a Useful Control Mechanism?" *Coastal Management*, 19(1), January 1991: 135–54.
47. Vincent Nijman and Chris Shepherd, "Adding Up the Numbers: An Investigation into Commercial Breeding of Tokay Geckos in Indonesia," *TRAFFIC*, 2015 (available at: www.traffic.org/species-reports/traffic_species_reptiles47.pdf).
48. Jiao and Thomas, "State of Wildlife Trade in China": 13–15; Brian Gratwicke et al., "The World Can't Have Wild Tigers and Eat Them, Too," *Conservation Biology*, 22(1), February 2008: 222–23; Miranda Mockrin et al., "Wildlife Farming: A Viable Alternative to Hunting in Tropical Forests?" *World Conservation Society*, 2005 (available at: www.mtnforum.org/sites/default/files/publication/files/2005_mockrin.pdf); and Craig Kirkpatrick and Lucy Emerton, "Killing Tigers to Save

Them: Fallacies of the Farming Argument," *Conservation Biology*, 24(3), June 2010: 665–69.

49. Andrea Crosta, Kimberly Sutherland, and Mike Beckner. "Blending Ivory. China's Old Loopholes, New Hopes," (Los Angeles: Elephant Action League, 2015), (available at: http://elephantleague.org/wp content/uploads/2015/12/EAL_BLENDING_IVORY_CHINA_REPORT_2015.pdf); Cheryl Lo and Gavin Edwards, "The Hard Truth: How Hong Kong's Ivory Trade Is Fueling Africa's Poaching Crisis" (Hong Kong: World Wildlife Fund, 2015), (available at: http://awsassets.wwfhk.panda.org/downloads/wwf_ivorytrade_eng_eversion.pdf); Peter Knights et al., "The Illusion of Control: Hong Kong's 'Legal' Ivory Trade," *WildAid*, 2015 (available at: www.wildaid.org/sites/default/files/resources/The%20Illusion%20of%20Control-Full%20Report.pdf).

50. Brendan Moyle, "The Black Market in China for Tiger Products," *Global Crime* 10(1/2), February 2009: 124–43; Kristine Nowell and Xu Ling, "Taming the Tiger Trade: China's Markets for Wild and Captive Tiger Products since the 1993 Domestic Trade Ban," *TRAFFIC East Asia*, November 2007 (available at: www.traffic.org/species-reports/traffic_species_mammals16.pdf).

51. Nick Davies and Oliver Holmes, "China Accused of Defying Its Own Ban on Breeding Tigers to Profit from Body Parts," *Guardian*, September 27, 2016.

52. Dai Qiang et al., "Conservation Status of Chinese Giant Salamander (Andrias davidianus)," trans. Wang Yi (Beijing: Chinese Academy of Sciences, 2009), (available at: www.cepf.net/Documents/final_CIBCAS_giantsalamander_china.pdf).

53. Vincent Nijman, "An Overview of International Wildlife Trade from Southeast Asia," *Biodiversity Conservation*, 19, 2010: 1110.

54. Chris Shepherd, "The Bird Trade in Medan, North Sumatra: An Overview," *Birding ASIA*, 5, 2006: 16–24.

55. Craig Packer, *Lions in the Balance: Man-Eaters, Manes, and Men with Guns* (Chicago: University of Chicago Press, 2015).

56. Daan Buijs, "A Summary of the Introduction of White Rhino onto Private Land in the Republic of South Africa" (Johannesburg: Rhino and Elephant Foundation, 1987).

57. Tom Milliken and Jo Shaw, "The South Africa–Viet Nam Rhino Horn Trade Nexus: A Deadly Combination of Institutional Lapses, Corrupt Wildlife Industry Professionals and Asian Crime Syndicates," *TRAFFIC*, 2012 (available at: www.worldwildlife.org/publications/the-south-africa-viet-nam-rhino-horn-trade-nexus): 38.

58. Annette Michaela Hübschle, "A Game of Horns: Transnational Flows

of Rhino Horn," PhD thesis (Cologne: International Max Planck Research School, 2016): 282.
59. Dinesh Jha Panday et al., "Forensic Wildlife Parts and Product Identification and Individualization Using DNA Barcoding," *Open Forensic Science Journal*, 7, 2014: 6–13.
60. Pervaze A. Sheikh, "Illegal Logging: Background and Issues" (Washington D.C.: Congressional Research Service, 2008), (available at: http://nationalaglawcenter.org/wp-content/uploads/assets/crs/RL33932.pdf). Even the FSC is not infallible; see, for example, World Rainforest Movement, "Laos: FSC Certified Timber Is Illegal" (available at: www.illegal-logging.info/item_single.php?it_id=1683&it=news).
61. "Seeing the Wood," *Economist*, September 25, 2010: 13.
62. Ibid.
63. Author's interviews with representatives of logging companies in Samarinda, Kalimantan, October 2012.
64. Sam Lawson and Larry McFaul, "Illegal Logging and Related Trade: Indicators of Global Response" (Washington, D.C.: Chatham House, 2010) 77, (available at: www.chathamhouse.org/sites/files/chathamhouse/public/Research/Energy%2C%20Environment%20and%20Development/0710pr_illegallogging.pdf).
65. See, for example, Jago Wadley, et al., "Behind the Veneer: How Indonesia's Last Rainforests Are Being Felled for Flooring" (Washington D.C.: Environmental Investigation Agency, 2006), (available at: www.eia-international.org/cgi/reports/reports.cgi?t=template&a=117).
66. Lawson and MacFaul, *Illegal Logging*: 75–6.
67. Sadie Gurman, "Drug Traffickers Seek Haven Amid Legal Marijuana," *Associated Press*, January 28, 2016 (available at: www.salon.com/2016/01/28/drug_traffickers_seek_safe_haven_amid_legal_marijuana/).
68. Catherine Saint Louis, "New Challenge for Police: Finding Pot in Lollipops and Marshmallows," *New York Times*, May 16, 2015, (available at http://www.nytimes.com/2015/05/17/us/new-challenge-for-police-finding-pot-in-lollipops-and-marshmallows.html?mwrsm=Email&_r=0)
69. Rocky Mountains High Intensity Drug Trafficking Investigative Support Center, "The Legalization of Marijuana in Colorado: The Impact," (Denver: Rocky Mountains High Intensity Drug Trafficking Area, 2014), (available at: www.rmhidta.org/html/august%202014%20legalization%20of%20mj%20in%20colorado%20the%20impact.pdf).
70. "International Narcotics Strategy Control Report" US Department of

State, March 2007. (available at: www.state.gov/j/inl/rls/nrcrpt/2007/vol1/html/80858.htm).

71. David Mansfield, "An Analysis of Licit Opium Poppy Cultivation: India and Turkey," 2001 (available at: www.davidmansfield.org/data/Policy_Advice/UK/India-Turkeycultivation.doc); and Letizia Paoli et al., "The Global Diversion of Pharmaceutical Drugs, *Addiction*, 104(3), February 2009: 347–54.

72. Vanda Felbab-Brown, "Opium Licensing in Afghanistan: Its Desirability and Feasibility" (Washington D.C.: Brookings Institution, 2007), (available at: https://www.brookings.edu/wp-content/uploads/2016/06/felbab-brown200708.pdf).

73. Author's correspondence with a former narcotics commissioner of India, summer and fall 2015.

74. Mansfield, "Analysis of Licit Opium Poppy Cultivation."

75. Eduardo Castillo, "Ex-Mexico President Calls for Legalizing Drugs," *Associated Press*, August 9, 2010 (available at: www.highbeam.com/doc/1A1-D9HFMD304.html).

76. Beau Kilmer et al., "Reducing Drug Trafficking Revenues and Violence in Mexico: Would Legalizing Marijuana in California Help?" *RAND*, 2010 (available at: www.rand.org/content/dam/rand/pubs/occasional_papers/2010/RAND_OP325.pdf).

77. Vanda Felbab-Brown, "Calderón's Caldron: Lessons from Mexico's Battle Against Organized Crime and Drug Trafficking in Tijuana, Ciudad Juárez, and Michoacán" (Washington D.C.: Brookings Institution, 2011), (available at: www.brookings.edu/~/media/research/files/papers/2011/9/calderon-felbab-brown/09_calderon_felbab_brown.pdf).

78. Steve Scherer, "Mafia Thrives on Italy's Legalized Gambling Addiction," *Reuters*, March 11, 2015 (available at: www.reuters.com/article/us-italy-mafia-slots-idUSKBN0M720R20150312); and Domenico Conti, "Want Some Mafia with Your Pizza? How the Mob Is Taking Over Rome's Restaurants," *International Business Times*, February 14, 2014.

79. Jonathan P. Caulkins et al., "Considering Marijuana Legalization: Insights from Vermont and Other Jurisdictions" (Santa Monica: RAND, 2015) 135, (available at: http://www.rand.org/pubs/research_reports/RR864.html).

80. Dan Levin, "China Weighs Ban on Manta Ray Gills, Sold in Traditional Market as Modern Panacea," *New York Times*, January 7, 2016.

81. Author's research on poaching in Kalimantan, Indonesia, fall 2012.

82. Douglas MacMillan and Jeonghee Han, "Cetacean By-Catch in the Korean Peninsula: By Chance or by Design?" *Human Ecology* 39, October 2011: 519–33.

83. Annie Zak, "Synthetic Rhino Horn Maker: It's not Cloning, Exactly," *Puget Sound Business Journal*, February 6, 2015; Raveena Aulakh, "Lab-Grown Horns Could Save Rhinos," *Toronto Star*, February 24, 2015; Deborah Sullivan Brennan, "Can They Clone a Rhinoceros?" *San Diego Union Tribune*, December 20, 2014.
84. Mike Yves, "Vietnam Craves Rhino Horn; Costs More than Cocaine," *Associated Press*, April 4, 2012 (available at: https://www.yahoo.com/news/vietnam-craves-rhino-horn-costs-more-cocaine-062134928.html).
85. Alexander Britton, "'Rathkeale Rovers' Gang Members Jailed over Multi-Million Euro Rhino Horn Spree," *Independent*, April 4, 2016.
86. For such customer choices regarding bear bile, see Adam Dutton et al., "A Stated Preference Investigation into the Chinese Demand for Farmed vs. Wild Bear Bile," *PLOS One*, 6(7), 2011: 1–10.
87. Hübschle, "Game of Horns": 346.
88. Ibid.: 349.
89. Rosie Cooney and Paul Jepson, "The International Wild Bird Trade: What's Wrong with Blanket Bans?" *Oryx*, 40(1), January 2006: 18–23.

8. LOCAL COMMUNITY INVOLVEMENT

1. Dilys Roe et al. (eds), "Community Management of Natural Resources in Africa," International Institute for the Environment and Development, 2009 (available at: http://pubs.iied.org/pdfs/17503IIED.pdf): 12.
2. For some of the first formulations of CBNRM, see, for example, David Western and Michael Wright (eds), *Natural Connections: Perspectives in Community-Based Conservation* (Washington: Island Press, 1994).
3. William Adams and David Hulme, "Conservation and Communities: Changing Narratives, Policies and Practices in African Conservation," in David Hulme and Marshall Murphree (eds), *African Wildlife and African Livelihoods: The Promise and Performance of Community Conservation* (Oxford: James Currey, 2001): 9–23; and Dan Brockington et al., "Conservation, Human Rights, and Poverty Reduction," *Conservation Biology*, 20(1), 2005: 250–52.
4. Katrina Brandon and Michael Wells, "Planning for People and Parks: Design Dilemmas," *World Development*, 20(4), 1992: 557–70; and Michael Wells et al., *People and Parks: Linking Protected Areas Management with Local Communities* (Washington D.C.: World Bank, 1992).
5. Grazia Borrini-Feyerabend et al., "Sharing Power: Learning by Doing In Co-Management of Natural Resources Throughout the World," International Institute for the Environment and Development, 2004 (available at: http://cmsdata.iucn.org/downloads/sharing_power.pdf); and Musole Musumali et al., "An Impasse in Community Based Natural

Resource Management in Implementation: The Case of Zambia and Botswana," *Oryx*, 41(3), July 2007: 306–13.

6. See, for example, Elizabeth Rihoy (ed.), "The Commons Without the Tragedy? Strategies for Community Based Natural Resource Management in Southern Africa" (Lilongwe: SADC Wildlife Technical Coordination Unit, 1995); and Brian Jones and Marshall Murphree, "The Evolution of Policy on Community Conservation in Namibia and Zimbabwe," in David Hulme and Marshall Murphree (eds), *African Wildlife and African Livelihoods: The Promise and Performance of Community Conservation* (Oxford: James Currey, 2001): 38–58.

7. William Moseley, "African Evidence on Relation of Poverty, Time Preference, and the Environment," *Ecological Economies*, 38(3), 2001: 317–26.

8. See, for example, Isabella Koziell, "Diversity, Not Adversity: Sustaining Livelihoods with Biodiversity" (London: International Institute for Environment and Development, 2001), (available at: http://pubs.iied.org/pdfs/7822IIED.pdf); and Wolfram Dressler et al., "From Hope to Crisis and Back Again?" *Environmental Conservation*, 37(1), 2010: 5–15.

9. Author's interviews in the Maasai Mara Nature Reserve and surrounding areas, Kenya, May 2013.

10. See, for example, Simon Milledge et al., "Forestry, Governance and National Development: Lessons Learned from a Logging Boom in Southern Tanzania" (Cambridge: TRAFFIC, 2007), (available at: http://www.trafficj.org/publication/07_Forestry_Governance_and_national.pdf).

11. Peter Veit and Catherine Benson, "When Parks and People Collide" (New York: Carnegie Council for Ethics and International Relations, 2004) 13–14 (available at: www.carnegiecouncil.org/publications/archive/dialogue/2_11/section_2/4449.html/:pf_printable).

12. Vanda Felbab-Brown, "Enabling War and Peace: Drugs, Logs, and Wildlife in Thailand and Burma" (Washington D.C.: Brookings Institution, 2015), (available at: https://www.brookings.edu/wp-content/uploads/2016/07/Policy-paper-7-webv5-1.pdf).

13. See, for example, Katherine Homewood and W. Alan Rodgers, *Maasailand Ecology: Pastoralist Development and Wildlife Conservation in Ngorongoro, Tanzania* (Cambridge: Cambridge University Press, 1991).

14. Ferreira de Castro and Charles Duff, *Jungle: A Tale of the Amazon Rubber-Tappers* (New York: Viking Press, 1935).

15. See, for example, Elizabeth Macfie and Elizabeth Williamson, "Best Practice Guidelines for Great Ape Tourism" (Gland: International Union for Conservation of Nature, 2010), (available at: https://portals.iucn.org/library/efiles/edocs/ssc-op-038.pdf).

16. Kreg Lindberg, "Economic Aspects of Ecotourism," in Kreg Lindberg et al. (eds), *Ecotourism: A Guide for Planners and Managers, Volume II*. (North Bennington: Ecotourism Society, 1998): 87–117.
17. See, for example, Kirk Helliker, "Reproducing White Commercial Agriculture in South Africa" (Grahamstown: Sociology Department, Rhodes University, 2013); and John Hearne and Margaret McKenzie, "Compelling Reasons for Game Ranching in Maputaland," in Herbert Prins et al. (eds), *Wildlife Conservation by Sustainable Use* (Boston: Kluwer Academic Publishers, 2000): 417–38, (available at https://www.ru.ac.za/media/rhodesuniversity/content/politics/documents/White%20commercial%20farms%20K%20Helliker%20seminar%20paper.doc.
18. Author's interviews with ecolodge managers in Maasai Mara and Tsavo National Park, Kenya, May 2013, and in Bale Mountains National Park, Ethiopia, March 2013.
19. Author's interviews with lodge management staff, national park law-enforcement officials, and local community members in the Chitwan and Bardia districts, Terai, Nepal, May 2012.
20. See also Rosaleen Duffy, *A Trip Too Far: Ecotourism, Politics, and Exploitation* (London: Earthscan, 2002); and Paul Steele, "Ecotourism: An Economic Analysis," *Journal of Sustainable Tourism*, 3(1), 1995: 29–44.
21. Ibid., 44.
22. Ibid., 200.
23. William Adams and Mark Infield, "Who Is On the Gorilla's Payroll? Claims on Tourist Revenues from a Ugandan National Park," *World Development*, 31(1), January 2003: 177–90.
24. See, for example, William Newmark and John Hough, "Conserving Wildlife in Africa: Integrated Conservation and Development Projects and Beyond," *BioScience*, 50(7), 2000: 585–92.
25. Vanda Felbab-Brown, "Indonesia Field Report IV: The Last Twitch? Wildlife Trafficking, Illegal Fishing, and Lessons from Anti-Piracy Efforts," *Brookings Institution*, March 26, 2013 (available at: www.brookings.edu/research/reports/2013/03/25-indonesia-wildlife-trafficking-felbabbrown).
26. Author's interviews with lodge owners in Tsavo National Park and Maasai Mara, Kenya, May 2013, and in Serengeti National Park and Ngorongoro Conservation Area, Tanzania, summer 2003.
27. Ibid., 207.
28. Steve Johnson, "State of CBNRM Report 2009" (Gaborone: Botswana National CBNRM Forum, 2009).
29. Joseph Mbaiwa et al., "From Collaboration to Conservation: Insights

from the Okavango Delta, Bostwana," *Society and Natural Resources*, 24(4), January 2011: 400–11.
30. Joseph Mbaiwa, "The Success and Sustainability of Community-Based Natural Resource Management in the Okavango Delta, Bostwana," *South African Geographical Journal*, 86(1), 2004: 44–53.
31. See, for example, David Weaver, "Asian Ecotourism: Patterns and Themes," *Tourism Geographies*, 4(2), 2002: 153–72.
32. For details, see, "Going, Going, Gone … The Illegal Trade in Wildlife in East and Southeast Asia," (Washington D.C.: World Bank, 2005), (available at: http://www-wds.worldbank.org/external/default/WDSContentServer/WDSP/IB/2005/09/08/000160016_20050908161459/Rendered/PDF/334670PAPER0Going1going1gone.pdf.
33. Jacqueline Groves-Sunderland et al., "Impacts of Co-Management on Western Chimpanzee (Pan Troglodytes Verus) Habitat and Conservation in Nialama Classified Forest, Republic of Guinea: A Satellite Perspective," *Biodiversity and Conservation*, 20(12), July 2011: 2745–57.
34. Kent Redford et al., "Linking Conservation and Poverty Alleviation: Discussion Paper and Good and Best Practice in the Case of Great Ape Conservation," (London: International Institute for Environment and Development, 2013), (available at: http://pubs.iied.org/pdfs/G03714.pdf).
35. Michael Bernard and Kwesi Darkoh, "Indigenous Knowledge and Ecotourism-Based Livelihoods in the Okavango Delta in Botswana," *Botswana Notes and Records*, 39, 2008: 62–73.
36. TRAFFIC, "What's Driving the Wildlife Trade: A Review of Expert Opinion on Economic and Social Drivers of the Wildlife Trade and Trade Control Efforts in Cambodia, Indonesia, Lao PDR, and Vietnam" (Washington, D.C.: World Bank, 2008), (available at: http://www.trafficj.org/publication/08_what's_driving_the_wildlife_trade.pdf).
37. For the effectiveness of alternative livelihoods efforts in Peru and Colombia, see Vanda Felbab-Brown, *Shooting Up: Counterinsurgency and the War on Drugs* (Washington D.C: Brookings Institution, 2010): 35–112. For the effectiveness of alternative livelihoods efforts in Afghanistan, see Vanda Felbab-Brown, *Aspiration and Ambivalence: Strategies and Realities of Counterinsurgency and State-Building in Afghanistan* (Washington D.C: Brookings Institution, 2013).
38. See, for example, Ronald D. Renard, *Opium Reduction in Thailand, 1970–2000: A Thirty-Year Journey* (Bangkok: UNDCP Silkworm Books, 2001).
39. See, for example, Renard, *Opium Reduction in Thailand*; and James Windle, *Suppressing the Poppy: A Comparative Historical Analysis of Successful Drug Control* (London: I.B. Tauris, 2016).

40. See Renard, *Opium Reduction in Thailand*: 38; and United Nations Office on Drugs and Crime, "Opium Poppy Cultivation in Southeast Asia," (2008) 1, (available at: www.unodc.org/documents/crop-monitoring/East_Asia_Opium_report_2008.pdf).
41. Vanda Felbab-Brown, "Improving Supply Side Policies: Smarter Eradication, Interdiction, and Alternative Livelihoods and the Possibility of Licensing," (London: London School of Economics, 2014), (available at: https://www.brookings.edu/wp-content/uploads/2016/06/improvingsupplysidepoliciesfelbabbrown.pdf
42. David Mansfield, "The Economic Superiority of Illicit Drug Production: Myth and Reality—Opium Poppy Cultivation in Afghanistan," (paper prepared for the International Conference on Alternative Development in Drug Control and Cooperation, Feldafing, Munich, January 7–12, 2002).
43. James Fairhead et al., "Green Grabbing: A New Appropriation of Nature," *Journal of Peasant Studies*, 39(2), July 2012: 237–61.
44. Author's interviews with the park warden in Nechisar National Park, who used to be warden in the Simien Mountains National Park, Ethiopia, March 2013.
45. Author's interviews with park wardens and local people in Awash, Lake Abiata-Shala, Nechisar, and Simien Mountains National Parks, Ethiopia, March 2013.
46. Ibid.
47. Author's interviews with scholars, journalists, and human right campaigners, Addis Ababa, March 2013. See also "Analysis: Land Grab or Development Opportunity," *BBC News*, February 22, 2012 (available at: http://www.bbc.com/news/world-africa-17099348); Lorenzo Cotula; "Doing It My Way," *Economist*, March 2, 2013, (available at http://www.economists.com/news/special-report/21572379-ideological-competition-between-two-diametrically-opposed-economic-models-doing-it-my).
48. See, for example, Joanne McLean and Steffen Straede, "Conservation, Relocation, and the Paradigms of Park and People Management: A Case Study of Padampur Village and the Royal Chitwan National Park, Nepal," *Society and Natural Resources*, 16(6), 2003: 509–26.
49. Dij Raj Khanal and Ghana Shyam Pandey, "Illegal Logging and the Issue of Transparency in Nepal," FECOFUND, 2011 (available at: www.illegal-logging.info/uploads/Khanal100212.pdf).
50. Kiran Dongol, "Nepal's Initiatives to Reduce Illegal Logging" (available at: www.illegal-logging.info/sites/files/chlogging/uploads/Dongol100212.pdf).

51. See, for example, "Indonesia: Natural Resources and Law Enforcement," International Crisis Group (available at: www.crisisgroup.org/~/media/Files/asia/south-east-asia/indonesia/Indonesia%20Natural%20 Resources%20and%20Law%20Enforcement.pdf).
52. Author's interviews with logging companies and law-enforcement officials, Samarinda, Kalimantan, Indonesia, October 2012.
53. Author's interviews with the park warden and local community members, Lake Abiata-Shala National Park, March 2013.
54. Author's interviews with park wardens, park rangers, and local community members, Lake Abiata-Shala National Park, Bale Mountains National Park, and Simien Mountains National Park, Ethiopia, March 2013.
55. See, for example, Jonathan Adams and Thomas McShane, *The Myth of Wild Africa: Conservation without Illusion* (Berkeley: University of California Press, 1992); and William Denevan, "The Pristine Myth: The Landscape of the Americas in 1492," *Annals of the Association of American Geographers*, 82(3), 1992: 369–85.
56. Tim Forsyth and Andrew Walker, *Forest Guardians, Forest Destroyers: The Politics of Environmental Knowledge in Northern Thailand* (Seattle: University of Washington Press, 2008).
57. See also, for example, Katherine Willis et al., "How 'Virgin' Is Virgin Rainforest?" *Science*, 304(5669), 2004: 402–3.
58. Daniel Botkin, *Discordant Harmonies: A New Ecology for the Twenty-First Century* (New York: Oxford University Press, 1990).
59. See, for example, Johan T. Du Toit and David H.M. Cumming, "Functional Significance of Ungulate Diversity in African Savannas and the Ecological Implications of the Spread of Pastoralism," *Biodiversity and Conservation*, 8(12), December 1999: 1643–61; and Steven Higgins et al., "Changes in Woody Community Structure and Composition under Contrasting Landuse Systems in a Semi-Arid Savanna, South Africa," *Journal of Biogeography*, 26(3), 1999: 619–27.
60. See, for example, Carlos Peres and Barbara Zimmerman, "Perils in the Parks or Parks in Peril? Reconciling Conservation in Amazonian Reserves with and without Use," *Conservation Biology*, 15(3), 2001: 793–97; and Kent Redford and Steven Sanderson, "Extracting Humans from Nature," *Conservation Biology*, 14(5), 2000: 1362–64.
61. Paul Robbins et al., "Conservation as it is: Illicit Resource Use in a Wildlife Reserve in India," *Human Ecology*, 37, 2009: 356–82.
62. Author's interviews with two leading environmental conservation researchers in Kathmandu, Nepal, May 2012. See also Thapa Shova and Klaus Hubacek, "Drivers of Illegal Resource Extraction: An

Analysis of Bardia National Park, Nepal," *Journal of Environmental Management*, 92(1), January 2011: 156–64.

63. Nabil Baral and Joel Heinen, "Resources Use, Conservation Attitudes, Management Intervention and Park–People Relations in Western Terai Landscape of Nepal," *Environmental Conservation*, 34(1), March 2007: 64–72.
64. Author's interviews with forestry officials, wardens and enforcement officials, village committee members, and local community members, Solukhumbu, Chitwan, and Bardia districts, Nepal, May 2012.
65. Ibid.
66. Harini Nagendra et al., "Evaluating Forest Management in Nepal: Views across Space and Time," *Ecology and Society*, 10(1), 2004: 24–39.
67. Author's interviews with two leading Nepali conservationists, Chitwan National Park and Kathmandu, May 2012. One anecdotal piece of data confirming the limited illegal ivory market in Nepal is that during my research in the major tourist markets of Patan and Bhaktapur I did not find a store with ivory statues. During a previous research trip to India between June and August 2007, on the other hand, I easily located such illegal items in the country's major tourist and wildlife trading hubs.
68. For comparisons with Indonesia and other areas of illegal logging in the Asia-Pacific region, see Vanda Felbab-Brown, "Not as Easy as Falling Off a Log: The Illegal Logging Trade in the Asia-Pacific Region and Possible Mitigation Strategies" (Washington D.C.: Brookings Institution, 2011), (available at: www.brookings.edu/~/media/research/files/papers/2011/3/illegal-logging-felbabbrown/03_illegal_logging_felbab-brown.pdf).
69. Surya Thapa, "Neither Forests Nor Trees," *Nepali Times*, January 7, 2011.
70. Author's interviews with representatives of Resources Himalaya, Kathmandu, May 29, 2012.
71. Author's interviews with businessmen and law enforcement officials, Solukhumbu district, Nepal, May 2012.
72. Johan Oldekop et al., "A Global Assessment of the Social and Conservation Outcomes of Protected Areas," *Conservation Biology*, 30(1), February 2016: 133–41.
73. John Vidal, "WWF Accused of Facilitating Human Rights Abuses of Tribal People in Cameroon," *Guardian*, March 3, 2016.
74. Everson Mushava, "Lions, Elephants Used to Kick out Manzou Villagers," *Zimbabwe Standard*, January 18, 2015.
75. "Find Land for Displaced, State Urges," *Times of Zambia*, May 6, 2015.

76. "Kauihura Pays Shs 20 Million to Bukwo Victims of Police Brutality," *Observer* (Kampala, Uganda), January 18, 2016.
77. Pratch Rujivanarom, "Villagers Fight a Losing Battle in 2015," *The Nation* (Thailand), December 28, 2015.
78. "Tribespeople Illegally Evicted from 'Jungle Book' Tiger Reserve," *Survival International*, January 14, 2015 (available at: www.survivalinternational.org/news/10631); and "India: Tribes Face Harassment and Eviction for, 'Tiger Conservation'," *Survival International*, May 30, 2014 (available at: www.survivalinternational.org/news/10239).
79. Christo Fabricius and Chris de Wet, "The Influence of Forced Removals and Land Restitutions on Conservation in South Africa," in Dawn Chatty and Marcus Colchester (eds), *Conservation and Mobile Indigenous Peoples: Displacement, Forced Settlement and Sustainable Development* (New York: Berghahn Books, 2002): 142–57.
80. Krithi Karanth, "Making Resettlement Work: The Case of India's Bhadra Wildlife Sanctuary," *Biological Conservation*, 139, October 2007: 315–24.
81. For the problems of the park, see, for example, Teresa Conor, "Place, Belonging and Population Displacement: New Ecological Reserves in Mozambique and South Africa," *Development of Southern Africa*, 22(3), 2005: 365–82.
82. See, for example, Conor, "Place, Belonging and Population Displacement"; and Wolfram Dressler and Bram Büscher, "Market Triumphalism and the CBNRM 'Crises' at the South African Section of the Great Limpopo Transfrontier Park," *Geoforum*, 39, 2008: 452–65.
83. Annette Michaela Hübschle, "A Game of Horns: Transnational Flows of Rhino Horn," PhD thesis (Cologne: International Max Planck Research School, 2016): 295.
84. Author's interviews with the former warden of Simien Mountains National Park, who initiated and implemented the relocation scheme, and who then became warden at the Nechisar National Park, where I interviewed him, and with the subsequent warden of the Simien Mountains National Park, Nechisar National Park and Simien Mountains National Park, Ethiopia, March 2013.
85. Author's interviews with the park warden of the Simien Mountains National Park, March 2013; and representatives of the Frankfurt Zoological Society in Lalibela, Ethiopia, March 2013.
86. Author's research in the Simien Mountains National Park, Ethiopia, March 2013.
87. Author's interviews with the park warden of Nechisar National Park, March 2013.
88. Fred Pearce, "Laird of Africa," *New Scientist*, 186(2495), 2005: 48–50;

and Fred Pearce, "Big Game Losers," *New Scientist*, 187(2512), 2005: 21.
89. Brian Child, et al., "Final Evaluation Report," Zimbabwe Natural Resources Management Program–USAID/Zimbabwe, "Strategic Objective No. 1" (Washington, D.C.: USAID, 2003), (available at: http://pdf.usaid.gov/pdf_docs/pdact554.pdf).
90. Russell Taylor and Marshall Murphree, "Case Studies on Successful Southern African NRM Initiatives and Their Impact on Poverty and Governance: Zimbabwe–Masoka and Gairezi" (Washington, D.C.: International Resources Group, 2007), (available at: https://rmportal.net/library/content/frame/case-studies-on-successful-southern-african-nrm-initiatives-and-their-impacts-on-poverty-and-governance-zimbabwe-masoka-and-gairezi/at_download/file).
91. Ian Scoones, "Why Cecil the Lion Offers Lessons for Land Reform and the Role of Elites," *The Conversation*, August 5, 2015.
92. Roe et al., "Community Management": viii.
93. Peter Frost and Ivan Bond, "The CAMPFIRE Programme in Zimbabwe: Payments for Wildlife Services," *Ecological Economics*, 65(4), 2008: 776–87.
94. James Murombedzi, "Devolution and Stewardship in Zimbabwe's CAMPFIRE Programme," *Journal of International Development*, 11(2), April 1999: 287–93.
95. Jocelyn Alexander and JoAnne McGregor, "Wildlife and Politics: CAMPFIRE in Zimbabwe," *Development and Change*, 31(3), June 2000: 605–27.
96. Joanna Durbin et al., "Namibian Community-Based Natural Resource Management Programme: Project Evaluation 4–19 May 1997" (Windhoek: Intergrated Rural Development and Nature Conservation, 1997).
97. Roe et al., "Community Management": vii.
98. Jonathan Barnes et al., "Economic Efficiency and Incentives for Change within Namibia's Community Wildlife Use Initiatives," *World Development*, 30(4), 2002: 667–81.
99. Brian Jones, "Policy Lessons from the Evolution of a Community-Based Approach to Wildlife Management, Kunene Region, Namibia," *Journal of International Development*, 11(2), 1999: 295–304.
100. "Rhino Death Toll at 80," *The Namibian*, January 7, 2016.
101. See, for example, Martha Honey, *Ecotourism and Sustainable Development: Who Owns Paradise?* (Washington, D.C: Island Press, 1999).
102. Conrad Steenkamp and Jana Uhr, "The Makuleke Land Claim: Power Relations and Community-Based Natural Resource Management"

(London: International Institute for Environment and Development, 2000) 2, (available at: http://pubs.iied.org/pdfs/7816IIED.pdf).
103. Steven Robins and Kees van der Waal, "'Model Tribes' and Iconic Conservationists? The Makuleke Restitution Case in Kruger National Park," *Development and Change*, 39(1), April 2008: 53–72.
104. Christopher Barrett et al., "Conserving Tropical Biodiversity amid Weak Institutions," *Bioscience*, 51(6), 2001: 497–502; Edward Taylor et al., "The Economics of Ecotourism: A Galápagos Islands Economy-Wide Perspective," *Economic Development and Cultural Change*, 51(4), July 2003: 978–97; and Elinor Ostrom, "Beyond Markets and States: Polycentric Governance of Complex Economic Systems," *American Economic Review*, 100(3), June 2010: 641–72.
105. Amanda Stronza, "The Economic Promise of Ecotourism for Conservation," *Journal of Ecotourism*, 6(3), March 2009: 210–30.
106. Francis Karanja et al., "Equity in the Loita/Purko Naimina Enkiyio Forest in Kenya: Securing Maasai Rights to and Responsibilities for the Forest," *Forest and Social Perspectives in Conservation* No. 11 (2002) (available at: https://portals.iucn.org/library/sites/library/files/documents/2000-019-11.pdf).
107. Esther Mwangi, "The Puzzle of Group Ranch Subdivision in Kenya's Maasailand," *Development and Change*, 38(5), 2007: 889–910.
108. See, for example, Jaap Arntzen et al., "Rural Livelihoods, Poverty Reduction and Food Security in Southern Africa: Is CBNRM the Answer?" *United States Agency for International Development*, 2007 (available at: www.sarpn.org/documents/d0002585/Rural_livelihoods_CBNRM_Mar2007.pdf); Ivan Bond, "CAMPFIRE and the Incentives for Institutional Change," in David Hulme and Marshall Murphree (eds), *African Wildlife and African Livelihoods: The Promise and Performance of Community Conservation* (Oxford: James Currey, 2001): 227–43.
109. Mike Norton-Griffiths, "How Many Wildebeest Do You Need?" *World Economics*, 8(2), June 2007: 41–64.
110. See, for example, Michael Thompson et al., "Maasai Mara—Land Privatization and Wildlife Decline: Can Conservation Pay Its Way?" in Kathrine Homewood et al. (eds), *Staying Maasai? Livelihoods, Conservation, and Development in East African Rangelands* (New York: Springer Press, 2009): 77–110; and Hannah Reid and Stephen Turner, "The Richtersveld and Makuleke Contractual Parks in South Africa: Win–Win for Communities and Conservation?" in Christo Fabricius et al. (eds), *Rights, Resources, and Rural Development: Community-Based Natural Resource Management in Southern Africa* (London: Earthscan, 2004): 223–34.

111. See, for example, Paul Ferraro, "Global Habitat Protection: Limitations of Development Interventions and a Role for Conservation Performance Payments" *Conservation Biology*, 15(4), 2001: 990–1000; and Paul Ferraro and David Simpson, "The Cost-Effectiveness of Conservation Payments," (Washington, D.C.: Resources for the Future, 2000), (available at: www.rff.org/files/sharepointWorkImages/Download/RFF-DP-00-31.pdf).

112. Author's fieldwork on urban crime, poaching, and charcoal production in locations throughout Ethiopia, spring 2013.

113. African Elephant Specialist Group, "Technical Brief: Review of Compensation Schemes for Agricultural and Other Damage Caused by Elephants" (Gland: International Union for Conservation of Nature), (available at: https://cmsdata.iucn.org/downlaods/heccomreview.pdf).

114. See, for example, Vanda Felbab-Brown et al., "Assessment of the Implementation of the United States Government's Support for Plan Colombia's Illicit Crops Reduction Components," USAID, 2009 (available at: http://pdf.usaid.gov/pdf_docs/PDACN233.pdf).

115. See, for example, "Cutting Down on Cutting Down," *Economist*, June 7, 2014; Johan Eliasch, *Climate Change: Financing Global Forests* (London: UK Government, 2008).

116. Stefanie Engel et al., "Designing Payments for Environmental Services in Theory and Practice: An Overview of the Issues," *Ecological Economics*, 65, May 2008: 663–74; and Douglas MacMillan and Sharon Philip, "Can Economic Incentives Resolve Conservation Conflict: The Case of Wild Deer Management and Habitat Conservation in the Scottish Highlands," *Human Ecology*, 38, 2010: 485–94.

117. See, for example, Cyrus Samii et al., "Effects of Payment for Environmental Services (PES) on Deforestation and Poverty in Low and Middle Income Countries: A Systematic Review CEE 3–015b," *Collaboration for Environmental Evidence*, 2014 (available at: www.environmentalevidence.org/wp-content/uploads/2015/01/Samii_PES_Review-formatted-for-CEE.pdf).

118. Rhett Butler, "The Year in Rainforests," *Mongabay*, December 29, 2015 (available at: http://news.mongabay.com/2015/12/the-year-in-rainforests-2015/).

119. Alain Karsenty, "Financing Options to Support REDD+ Activities" (Brussels: European Commission, 2012), (available at: http://ur-foretssocietes.cirad.fr/content/download/4123/32260/version/3/file/REDD_study_CIRAD_final.pdf).

120. Edward Wilson, *The Future of Life* (New York: Vintage, 2002).

121. Ron Moreau and Sami Yousafzai, "Flowers of Destruction," *Newsweek*, 142(2), July 14, 2003: 33.
122. Author's interviews with representatives of international development NGOs and Afghan counter-narcotics officials, Kabul, September 2005.
123. Peter Oborne and Lucy Edwards, "A Victory for Pushers," *Spectator*, May 31, 2003.
124. Li Zhang, "China Must Act Decisively to Eradicate the Ivory Trade," *Nature*, 527(135), 2015.
125. Astrid Zabel and Karin Holm-Muller, "Conservation Performance Payments for Carnivore Conservation in Sweden," *Conservation Biology*, 22(2), April 2008: 247–51; Amy Dickman et al., "A Review of Financial Instruments to Pay for Predator Conservation and Encourage Human–Carnivore Coexistence," *Proceedings of the National Academy of Sciences of the United States of America*, 108, 2011: 13937–44.
126. Tom Clements et al., "An Evaluation of the Effectiveness of Direct Payment for Biodiversity Conservation: The Bird Nest Protection Program in the Northern Plains of Cambodia," *Conservation Biology*, 157, January 2013: 50–59.

9. ANTI-MONEY LAUNDERING EFFORTS

1. William Buckley and Gabriela Salazar Torres, "Mexico's Proposed Anti-Money Laundering Law," *Haynes and Boone's Newsroom*, July 5, 2011 (available at: www.haynesboone.com/mexican_anti-money_landering_law/).
2. International Consortium on Combatting Wildlife Crime, "Wildlife and Forest Crime Analytic Toolkit" (Vienna: United Nations Office on Drugs and Crime, 2012), (available at: www.unodc.org/documents/Wildlife/Toolkit_e.pdf).
3. Robert Dreher, "On Poaching and Terrorism: Testimony before the US House of Representatives, Committee on Foreign Affairs, Subcommittee on Terrorism, Nonproliferation, and Trade", Department of the Interior, April 22, 2015 (available at: http://docs.house.gov/meetings/FA/FA18/20150422/103355/HHRG-114-FA18-Wstate-DreherR-20150422.pdf).
4. Jeanne Giraldo and Harold Trinkunas, "The Political Economy of Terrorism Financing," in Jeanne Giraldo and Harold Trinkunas (eds), *Terrorism Financing and State Responses: A Comparative Perspective* (Stanford: Stanford University Press, 2007): 7–20; Juan Carlos Zarate, *Treasury's*

War: The Unleashing of a New Era of Financial Warfare (New York: Public Affairs, 2013); Mark Pieth, "Criminalizing the Financing of Terrorism," *Journal of International Criminal Justice*, 4(5), 2006: 1074–86; and Thomas Biersteker and Sue Eckert, *Countering the Financing of Terrorism* (New York: Routledge, 2008).

5. Richard Barrett, "Preventing the Financing of Terrorism," *Case Western Journal of International Law*, 44(3), 2011: 719.
6. Bill Mauer, "From Anti-Money Laundering to … What? Formal Sovereignty and Feudalism in Offshore Financial Centers," in Anne Clunnan and Harold Trinkunas (eds), *Ungoverned Spaces: Alternatives to State Authority in an Era of Softened Sovereignty* (Stanford: Stanford University Press, 2010): 215–32.
7. Michael Levi, "The Impact of Anti-Money Laundering Measures within Developing Countries against Proceeds of Corruption" (Cardiff: Cardiff School of Social Sciences, 2009), (available at: http://pp.lao.org.pk/wp-content/uploads/2013/07/Financial-Intelligence-Units-Anti-Money-Laundering-Efforts-and-The-Control-of-Grand-Corruption.pdf).
8. Stuart Levey, "How We're Tying up Terrorists' Cash" *Christian Science Monitor*, December 24, 2008. For al-Qaeda's financing from the African gem trade, see Douglas Farah, *Blood from Stones* (New York: Broadway Books, 2004). For independent assessments of the effectiveness of counterterrorism AML measures, see Terrence Halliday et al., "Global Surveillance of Dirty Money: Assessing Assessments of Regime to Control Money Laundering and Combat the Financing of Terrorism," (Champaign: University of Illinois College of Law, 2014), (available at: http://faculty.publicpolicy.umd.edu/sites/default/files/reuter/files/report_global_surveillance_of_dirty_money_release_date_30_january_2014.pdf).
9. Vanda Felbab-Brown, "The Not-So-Jolly Roger: Dealing with Piracy off the Coast of Somalia and in the Gulf of Guinea" (Washington, D.C.: Brookings Institution, 2013), (available at: www.brookings.edu/~/media/Research/Files/Reports/2014/foresight%20africa%202014/02%20foresight%20piracy%20somalia%20felbab%20brown.pdf).
10. Peter Reuter and Edwin M. Truman, *Chasing Dirty Money* (Washington, D.C.: Institute for International Economics, 2004).
11. See, for example, Robert Powis, *The Money Launderers* (Chicago: Probus, 1992).
12. Nauman Farooqi, "Curbing the Use of Hawala for Money Laundering and Terrorist Financing: Global Regulatory Response and Future Challenges," *International Journal of Business Governance and Ethnics*, 5(1/2), 2010: 64–75.
13. Ibid.

14. See, for example, Lauren Dellinger, "From Dollars to Pesos: A Comparison of the US and Colombian Anti-Money Laundering Initiatives from an International Perspective," *California Western International Law Journal*, 38(2), 2008: 419–54.
15. Francisco Thoumi and Marcela Anzola, "Extra-Legal Economy, Dirty Money, Illegal Capital Inflows and Outflows and Money Laundering in Colombia," 2009.
16. Charles Goredema, "Measuring Money Laundering in Southern Africa," *Institute for Security Studies*, 2005 (available at: www.issafrica.org/topics/organised-crime/01-dec-2005-measuring-money-laundering-in-southern-africa-charles-goredema).
17. Katy Migiro and Stella Dawson, "Kenya: Chic Nairobi Throbs to the Beat of Dirty Money," *Thompson Reuters Foundation News*, December 10, 2013 (available at: http://news.trust.org//item/20131209150854-1kirf/).
18. Felix Olick, "Govt Hunts for Cash Launderers," *The Star*, December 23, 2015 (available at: www.allafrica.com/stories/201512230155.html).
19. Piyanut Tumnukasetchai, "AMLO Follows Money Trail to Bring Down Siamese Rosewood Smuggling Cartel," *The Nation* (Thailand), July 21, 2014.
20. Nick Davies and Oliver Holmes, "The Crime Family at the Centre of Asia's Animal Trafficking Network," *Guardian*, September 26, 2016 (available at: www.theguardian.com/environment/2016/sep/26/bach-brothers-elephant-ivory-asias-animal-trafficking-network).
21. Wendy Stueck, "Convicted Canadian Narwhal Tusk Smuggler Extradited to US on Money-Laundering Charges," *Globe and Mail*, March 17, 2016.
22. Jade Saunders and Jens Hein, "EUTR, CITES and Money Laundering: A Case Study on the Challenges of Coordinated Enforcement in Tackling Illegal Logging" (London: Chatham House, 2015), (available at: http://efface.eu/sites/default/files/EFFACE_EUTR%20CITES%20and%20money%20laundering%20A%20case%20study%20on%20the%20challenges%20to%20coordinated%20enforcement%20in%20tacking%20illegal%20logging.pdf). See also ICCWC, *Wildlife and Forest Crime*.
23. James Tillen and Laura Billings, "Anti-Money Laundering 2015: In 23 Jurisdictions Worldwide; Getting the Deal Through," 2015 (available at: www.amt-law.com/res/news_2015_pdf/150625_4396.pdf).
24. Christie Smythe, "HSBC Judge Approves $1.9B Drug-Money Laundering Accord," *Bloomberg*, July 3, 2013 (available at: www.bloom-

berg.com/news/articles/2013-07-02/hsbc-judge-approves-1-9b-drug-money-laundering-accord).
25. Ibid.
26. "Whoops Apocalypse," *Economist*, February 27, 2016.
27. Anne Clunan, "US and International Responses to Terrorist Financing," in Jeanne Giraldo and Harold Trinkunas (eds), *Terrorism Financing and State Responses: A Comparative Perspective* (Stanford: Stanford University Press, 2007): 260–81.
28. Koh Gui Qing, "Exclusive: Risk Ranking: China Revamps Anti-Money Laundering Rules—Sources," *Reuters*, April 17, 2013 (available at: www.reuters.com/article/us-china-laundering-risk-ratings-idUSBRE 93G15A20130417).
29. "Somalia Famine 'Killed 260,000 People'," *BBC News*, May 2, 2013 (available at: www.bbc.com/news/world-africa-22380352).

10. DEMAND REDUCTION

1. See, for example, Diego Veríssimo et al., "Wildlife Trade in Asia: Start with the Consumer," *Asian Journal of Conservation Biology*, 1(2), 2012: 49–50.
2. Susan M. Wells and Jonathan G. Barzdo, "The International Trade in Marine Species: Is CITES a Useful Control Mechanism?" *Coastal Management*, 19(1), January 1991: 145.
3. "Canada: Harp Seal Hunt Begins amid Lower Demand," *New York Times*, April 8, 2010, (available at http://www.nytimes.com/2010/04/09 world/americas/09briefs-canadahunt.html?_r=0)
4. Debbie Banks et al., "Skinning the Cat: Crime and Politics of the Big Cat Skin Trade" (London: Environmental Investigation Agency, 2006) 10, (available at: http://www.wpsi-india.org/images/EIA-WPSI_ Skinning_The_Cat.pdf).
5. Matt McGrath, "'Indiana Jones' Shark Gains Protection at CITES Meeting," *BBC News*, October 3, 2016 (available at: www.bbc.com/news/science-environment-37547103).
6. Shelley Clarke et al., "Global Estimates of Shark Catches Using Trade Records from Commercial Markets," *Ecology Letters*, 9(10), October 2006: 1115–26.
7. Juliet Eilperin, "Distaste Widening for Shark Fin Soup," *Washington Post*, June 5, 2011; and Bettina Wassener, "Environmental Cost of Shark Finning Is Getting Attention in Hong Kong," *New York Times*, June 20, 2010, (available https://www.washingtonpost.com/national/distaste-widening-for-sharks-fin-soup/2011/05/18/AG3txgJH_story.html.).

8. "The Elephants Fight Back," *Economist*, November 21, 2015.
9. Sarah Karacs, "Why Appetite for Shark Fin Soup Continues to Grow Despite Efforts to Stem the Slaughter," *South China Morning Post*, September 20, 2016 (available at: www.scmp.com/news/hong-kong/health-environment/article/2020985/why-politics-stubborn-tastes-and-resilient).
10. "Sympathy for the Misunderstood Shark," Bloom Hong Kong, April 20, 2015 (available at: www.bloomassociation.org/en/bloom-hong-kong-sympathy-for-the-misunderstood-shark/).
11. Austin Ramzy, "Officials Accused of Dining on Endangered Salamander," *New York Times*, January 26, 2015, (available at http://sinosphere.blogs.nytimes.com/2015/03/31science/in-vietnam-rampant-wildlife-smugglingprompts-little-concern.html).
12. Sarah Karacs, "Shark Fin Soup Still Served at 98 per cent of Hong Kong Restaurants as Restaurants Choose Money over Environmentally Friendly Practices," *South China Morning Post*, January 25, 2016 (available at: www.scmp.com/news/hong-kong/health-environment/article/1905257/shark-fin-soup-still-served-98-cent-hong-kong).
13. Rachel Nuwer, "In Vietnam, Rampant Wildlife Smuggling Prompts Little Concern," *New York Times*, March 30, 2015, (available at: http://www.nytimes.com/2015/03/31/science/in-vietnam-rampant-wildlife-smugglingprompts-little-concern.html).
14. Li Zhang et al., "Wildlife Trade, Consumption, and Conservation Awareness in Southwest China," *Biodiversity Conservation*, 17(6), 2008: 1511.
15. Xu Hongfa and Craig Kirkpatrick, eds, "The State of Wildlife Trade in China: Information on the Trade in Wild Animals and Plants in China 2007" (Beijing: TRAFFIC, 2008), 9 (available at: http://www.trafficj.org/publication/08-State_of_Wildlife_China.pdf).
16. Stephen Nash, "Sold for a Song: The Trade in Southeast Asian non-CITES Birds" (Cambridge: TRAFFIC, 1994), (available at: www.traffic.org/species-reports/traffic_species_birds5.pdf).
17. "Ivory Demand in Hong Kong" (San Francisco: WildAid, 2015), (available at: www.wildaid.org/sites/default/files/resources/WildAid-Hong%20Kong%20Ivory%20Survey,%202015.pdf).
18. "Reducing Demand for Ivory: An International Study," *National Geographic*, August 2015 (available at: http://press.nationalgeographic.com/files/2015/09/NGS2015_Final-August-11-RGB.pdf).
19. Li Zhang and Feng Yi, "Wildlife Consumption and Conservation Awareness in China: A Long Way to Go," *Biodiversity and Conservation*, 23(9), 2014: 2371–81.

20. Merilee Grindle, "Good Enough Governance Revisited," *Development Policy Review*, 25(5), September 2007: 553–74.
21. Miriam Wells, "Don't Forget That Your Weekend Fun is Killing People, Cocaine Users Told," *Vice News*, December 2, 2015 (available at: https://news.vice.com/article/dont-forget-that-your-weekend-fun-is-killing-people-cocaine-users-told).
22. Author's conversation with top-level Colombian officials who were promoting the campaign, Bogotá, Colombia, fall 2008.
23. Melanie Wakefield et al., "Use of Mass Media Campaigns to Change Health Behavior," *Lancet*, 376(9748), October 2010: 1261–71.
24. Lisa Goldman and Stanton Glantz, "Evaluation of Antismoking Advertising Campaigns," *Journal of American Medical Science Association*, 279(10), 1998: 772–77.
25. Dan Levin, "China Weighs Ban on Manta Ray Gills, Sold in Traditional Market as Modern Panacea," *New York Times*, January 7, 2016.
26. Kim Witte et al., "A Theoretically-Based Evaluation of HIV/AIDS Prevention Campaigns along the Trans-Africa Highway in Kenya," *Journal of Health Communication*, 3(4), 1998: 345–63.
27. Li Zhang et al., "Wildlife Trade": 1513–14.
28. Bontha Babu and Shantanu Kumar Kar, "Coverage, Compliance, and Some Operational Issues of Mass Drug Administration during the Programme to Eliminate Lymphatic Filariasis in Orissa, India," *Tropical Medicine and International Health*, 9(6), June 2004: 702–9.
29. United States Department of Health and Human Services, "The Health Consequences of Smoking—50 Years of Progress: Report of the Surgeon General," 2014 (available at: www.surgeongeneral.gov/library/reports/50-years-of-progress/sgr50-chap-2.pdf): 15–41.
30. "Weight Watchers," *Economist*, April 13, 2013.
31. Rachel M. Wasser and Priscilla Bei Jiao, "Understanding Motivations: The First Step Toward Influencing China's Unsustainable Wildlife Consumption," *TRAFFIC*, 2010 (available at: www.changewildlifeconsumers.org/wp-content/uploads/2016/03/Understanding-the-Motivations.pdf).
32. Kai Schmidt-Soltau, "Conservation-Led Resettlement in Central Africa: Environmental and Social Risks," *Development and Change*, 34(3), 2003: 536.
33. Miranda Mockrin et al., "Wildlife Farming: A Viable Alternative to Hunting in Tropical Forests?" (Washington, D.C.: Wildlife Conservation Society, 2005), (available at: www.mtnforum.org/sites/default/files/publication/files/2005_mockrin.pdf).
34. Steven West and Kerry O'Neal, "Project D.A.R.E. Outcome Effectiveness Revisited," *American Journal of Public Health*, 94, 2004: 1027–29.

35. John Strang et al., "Drug Policy and the Public Good: Evidence for Effective Interventions," *Lancet*, 379(9810), January 2012: 71–83; Jonathan Caulkins et al., *An Ounce of Prevention, a Pound of Uncertainty: The Cost-Effectiveness of School-Based Drug Prevention Programs* (Santa Monica: Drug Policy Research Center, RAND, 1999); and Beau Kilmer and Rosalie Pacula, "Preventing Drug Use," in Phillip B Levine and David J. Zimmerman (eds), *Targeting Investments in Children: Fighting Poverty when Resources Are Limited* (Chicago: University of Chicago Press, 2010): 181–220.
36. Author's interviews with Chinese scholars of conservation and wildlife-trafficking, Shanghai, Beijing, and London, fall 2015.
37. The criminologist Mark Kleiman and the judge Steven Alm are the pioneers of this concept. See, for example, Mark Kleiman, *Against Excess: Drug Policy for Results* (New York: Basic Books, 1992); Mark Kleiman, "Coerced Abstinence: A Neo-Paternalistic Drug Policy Initiative," in Lawrence Mead (ed.), *The New Paternalism* (Washington, D.C.: Brookings Institution Press, 1998); and Mark Kleiman, *When Brute Force Fails: How to Have Less Crime and Less Punishment* (Princeton: Princeton University Press, 2009).
38. Beau Kilmer et al., "Efficacy of Frequent Monitoring with Swift, Certain, and Modest Sanctions for Violations: Insights from South Dakota's 24/7 Sobriety Project," *American Journal of Public Health*, 103, 2013: E37–E43.
39. Kent Weaver, "Getting People to Behave: Research Lessons for Policy Makers," *Public Administration Review*, 75(6), December 2015: 806–16.
40. Daniel Martin Varisco, "From Rhino Horns to Dagger Handles: A Deadly Business," *Animal Kingdom*, 92(3), June 1989: 44–9.
41. Lucy Vigne and Esmond Martin, "Increasing Rhino Awareness in Yemen and a Decline in the Rhino Horn Trade," *Pachyderm*, 53, 2013: 51–8.
42. Esmond Bradley Martin, Lucy Vigne, and Crawford Allan, "On a Knife's Edge: The Rhinoceros Horn Trade in Yemen" (Cambridge: TRAFFIC, 1997), (available at: http://www.google.com/url?sa=t-&rct=j&q=&esrc=s&source=web&cd=1&cad=rja&uact=8&ved=0ahUKEwjTutjxtLLOAhVp64MKHS7-CtoQFggcMAA&url=http%3A%2F%2Fstatic1.1.sqspcdn.com%2Fstatic%2Ff%2F157301%2F9113577%2F1288012352957%2Ftraffic_species_mammals58.pdf%3Ftoken%3Dg0HQoR7lebB4WHQLrvNM4bnPyfU%253D&usg=AFQjCNHP6P_JT9qi_2Bx-OeU_BZYWcSZ2w&sig2=T0yks-xPP-IFquD8rU015g).
43. Lucy Vigne et al., "Increased Demand for Rhino Horn in Yemen

Threatens Eastern Africa's Rhinos," *Pachyderm*, 43, December 2007: 73–86.
44. Ibid.
45. Jeffrey Fleishman, "Daggers Slice Yemen Tradition: Plastic Supplanting Handles Made from Rhino Horns," *Chicago Tribune*, December 18, 2009.
46. Vigne and Martin, "Increasing Rhino Awareness."
47. Lucy Vigne and Esmond Martin, "Yemen Strives to End Rhino Horn Trade," *Swara*, 16(6), 1993: 20–24.
48. Lucy Vigne and Esmond Martin, "Yemen's Attitudes toward Rhino Horn and Jambiyas," *Pachyderm*, 44, June 1993: 45–53.
49. Fred Pearce, "Rhino Rescue Plan Decimates Asian Antelopes," *New Scientist*, February 12, 2003 (available at: www.newscientist.com/article/dn3376-rhino-rescue-plan-decimates-asian-antelopes/); and Richard Ellis, *Tiger Bone and Rhino Horn: The Destruction of Wildlife for Traditional Chinese Medicine* (Washington, D.C.: Island Press, 2013).
50. Stephanie von Meibom et al., "Saiga Antelope Trade: Global Trends with a Focus on Southeast Asia," TRAFFIC-Europe, 2013 (available at: https://portals.iucn.org/library/efiles/edocs/traf-115.pdf).

11. CONCLUSIONS AND POLICY RECOMMENDATIONS

1. "Secretary of Interior: US Supports Sustainable Trophy Hunts," *Associated Press*, January 19, 2016 (available at: www.yahoo.com/news/secretary-interior-us-supports-sustainable-trophy-hunts-215748549.html?ref=gs).
2. Li Zhang, "China Must Act Decisively to Eradicate the Ivory Trade," *Nature*, 527(135), 2015.
3. Author's interviews with illegal loggers, miners, and poachers, Kutai National Park, Kalimantan, Indonesia, October 2012.
4. Cormac McCarthy, *The Road* (New York: Knopf, 2006): 130.
5. See, for example, Goodwell Nzou, "In Zimbabwe, We Don't Cry for Lions," *New York Times*, August 4, 2015.

BIBLIOGRAPHY

Adams, Jonathan and Thomas McShane, *The Myth of Wild Africa: Conservation without Illusion* (Berkeley: University of California Press, 1992)

Adams, William. *Against Extinction: A Story of Conservation*. London: Earthscan, 2004.

———. *Future Nature: A Vision of Conservation*. London: Earthscan, 1996.

Adams, William and David Hulme. "Conservation and Communities: Changing Narratives, Policies and Practices in African Conservation." In *African Wildlife and African Livelihoods: The Promise and Performance of Community Conservation* edited by David Hulme and Marshall Murphee Oxford: James Currey, 2001, 9–23.

Adams, William and Mark Infield. "Who is on the Gorilla's Payroll? Claims on Tourist Revenues from a Ugandan National Park." *World Development* 31, no. 1 (January 2003): 177–90.

Adams, Jonathan and Thomas McShane, *The Myth of Wild Africa: Conservation without Illusion* (Berkeley: University of California Press, 1992)

African Elephant Specialist Group. "Technical Brief: Review of Compensation Schemes for Agricultural and Other Damage Caused by Elephants," Technical Brief, IUCN (available at: https://cmsdata.iucn.org/downloads/heccomreview.pdf).

Agrawal, Arun. *Community in Conservation: Beyond Enchantment and Disenchantment*. Gainesville: Conservation and Development Forum, 1997.

Alacs, Erika and Arthur Georges. "Wildlife Across Our Borders: A Review of the Illegal Trade in Australia." *Australian Journal of Forensic Sciences* 40, no. 2 (December 2008): 147–60.

Alexander, Jocelyn and JoAnne McGregor. "Wildlife and Politics: CAMPFIRE in Zimbabwe." *Development and Change* 31, no. 3 (June 2000): 605–27.

Amin, Rajan, K. Thomas, Richard Emslie, Thomas J. Foose, and Nico Van Strien. "An Overview of the Conservation Status of and Threats to

BIBLIOGRAPHY

Rhinoceros Species in the Wild." *International Zoo Yearbook* 40, no. 1 (July 2006): 96–117.

Amman, Karl. "The Rhino and the Bling." 2013. http://www.karlammann.com/pdf/rhino-bling.pdf.

"An Ounce of Prevention." *Economist*. March 19, 2016.

Arntzen, Jaap, Tsehepo Setlhogile, and John Barnes. "Rural Livelihoods, Poverty Reduction and Food Security in Southern Africa: Is CBNRM the Answer?" *USAID*, 2007. http://www.sarpn.org/documents/d0002585/Rural_livelihoods_CBNRM_Mar2007.pdf.

Ashe, Dan. "Testimony before the U.S. House of Representatives, Committee on Foreign Affairs, Regarding the National Strategy for Combatting Wildlife Trafficking." U.S. Fish and Wildlife Service, Department of Interior. February 26, 2014. https://www.fws.gov/international/pdf/wildlife-trafficking-national-strategy-testimony.pdf.

Association of Southeast Asian Nations Wildlife Enforcement Network. "Illegal Wildlife Trade in Southeast Asia: Factsheet." ASEAN-WEN. March 5, 2009.

Aulakh, Raveena. "Lab-Grown Horns Could Save Rhinos." *The Toronto Star*. February 24, 2015.

Auliya, Mark. "Hot Trade in Cool Creatures: A Review of the Live Reptile Trade in the European Union in the 1990s." *TRAFFIC*, 2003. http://www.traffic.org/reptiles-amphibians/.

Ayling, Julie. "What Sustains Wildlife Crime? Rhino Horn Trading and the Resilience of Criminal Networks." *Journal of International Wildlife Law and Policy* 16 (2013): 57–80.

Babu, Bontha and Shantanu Kumar Kar. "Coverage, Compliance, and Some Operational Issues of Mass Drug Administration during the Programme to Eliminate Lymphatic Filariasis in Orissa, India." *Tropical Medicine and International Health* 9, no. 6 (June 2004): 702–9.

Baker, Peter and Jada Smith. "Obama Administration Targets Trade in African Elephant Ivory." *The New York Times*, July 25, 2015. http://www.nytimes.com/2015/07/26/world/africa/obama-administration-targets-trade-in-african-elephant-ivory.html?_r=0.

Banks, Debbie, Nitin Desai, Justin Gosling, Tito Joseph, Onkuri Majumdar, Nick Mole, Mary Rice, Belinda Wright, and Victor Wu. "Skinning the Cat: Crime and Politics of the Big Cat Skin Trade." London: Environmental Investigation Agency, 2006. http://www.wpsi-india.org/images/EIA-WPSI_Skinning_The_Cat.pdf

Bale, Rachael. "The World Is Finally Getting Serious about Tiger Farms," *National Geographic*, September 29, 2016 (available at: www.news.nationalgeographic.com/2016/09/wildlife-watch-tiger-farms-citesprotections/).

BIBLIOGRAPHY

———. "U.S.-China Deal to Ban Ivory Trade Is Good News for Elephants." *National Geographic*, September 25, 2015 (available at: http://news.nationalgeographic.com/2015/09/150925-ivory-elephants-us-china-obama-xi-poaching/.)

Banks, Debbie and Julian Newman. "The Tiger Skin Trail." Washington, D.C.: Environmental Investigation Agency, 2004. https://eia-international.org/wp-content/uploads/TheTigerSkinTrail-Low-Res.pdf.

Baral, Nabil and Joel Heinen. "Resources Use, Conservation Attitudes, Management Intervention and Park-People Relations in Western Terai Landscape of Nepal." *Environmental Conservation* 34, no. 01 (March 2007): 64–72.

"The Maoist People's War and Conservation in Nepal." *Politics and the Life Sciences* 24, nos. 1–2 (2006): 2–11.

Bargent, James. "Eco-Trafficking in *Latin* America: The Workings of a Billion-Dollar Business." InSight Crime. July 7, 2014. http://www.insightcrime.org/newsanalysis/eco-trafficking-latin-america-billion-business.

———. "Eco-Trafficking in Latin America: The Failures of the State." *InSight Crime*, July 8, 2014. http://www.insightcrime.org/news-analysis/eco-trafficking-in-latin-america-the-failures-of-the-state/

Barnes, Jonathan et al., "Economic Efficiency and Incentives for Change within Namibia's Community Wildlife Use Initiatives," *World Development*, 30(4), 2002: 667–81.

Barnett, Rob (ed.). *Food for Thought: Utilization of Wild Meat in Eastern and Southern Africa*, Nairobi: TRAFFIC East/Southern Africa, 2000 (available at: www.traffic.org/general-reports/traffic_pub_gen7.pdf).

Barrett, Christopher, Katrina Brandon, Clark Gibson, and Heidi Gjertsen. "Conserving Tropical Biodiversity amid Weak Institutions." *Bioscience* 51, no. 6 (2001): 497–502.

Barrett, Richard. "Preventing the Financing of Terrorism." *Case Western Journal of International Law* 44, no. 3 (2011): 719.

Begley, Sharon. "Big Business: Wildlife Trafficking." *Newsweek*. March 01, 2008. http://www.newsweek.com/big-business-wildlife-trafficking-83865.

Bell, Sandra, Kate Hamsphire, and Stella Topalidou. "The Political Culture of Poaching: A Case Study from Northern Greece." *Biodiversity and Conservation* 16, no. 2 (January 2007): 399–418.

Bennett, Elizabeth. "Is There a Link between Wild Meat and Food Security." *Conservation Biology* 16, no. 3 (June 2002): 590–92.

Bennett, Elizabeth L., Adrian J. Nyaoi, and Jephte Sompud. "Saving Borneo's Bacon: The Sustainability of Hunting in Sarawak and Sabah." In *Hunting for Sustainability in Tropical Forests*, edited by John Robinson and Elizabeth Bennett. New York: Columbia University Press, 2000: 305–329.

Bennett, Elizabeth, Eleanor Jane Milner-Gulland, Mohamed Bakarr, Heather

BIBLIOGRAPHY

Eves, John Robinson, and David Wilkie. "Hunting the World's Wildlife into Extinction." *Oryx* 36, no. 4 (2002): 328–9.

Bennett, Elizabeth L., and John G. Robinson. "Hunting of Wildlife in Tropical Forests: Implications for Biodiversity and Forest Peoples." Washington D.C.: World Bank, 2000.

Benz, Sophia and Judith Benz-Schwarzburg. "Great Apes and New Wars." *Civil Wars* 12 no. 4 (January 2011): 395–430.

Berning, Joey, and Moses Montesh. "Countering Corruption in South Africa: The Rise and Fall of the Scorpions and Hawks." *South African Crime Quarterly* 39 (2012): 3–10.

Bergenas, Johan, Rachel Stohl, and Ochieng Adala. "Killing Lions, Buying Bombs." *New York Times*. August 9, 2013. http://www.nytimes.com/2013/08/10/opinion/killing-lions-buying-bombs.html.

Bernard, Michael and Kwesi Darkoh, "Indigenous Knowledge and Ecotourism-Based Livelihoods in the Okavango Delta in Botswana," Botswana Notes and Records, 39, 2008: 62–73.

Bhuju, Ukesh Raj, Ravi Sharma Aryal, and Prakash Aryal. "Report on the Facts and Issues on Poaching of Mega Species and Illegal Trade in Their Parts in Nepal." Amsterdam: Transparency International, 2009 http://www.rhinoresourcecenter.com/pdf_files/131/1315014970.pdf.

Biersteker, Thomas J., and Sue E. Eckert. *Countering the Financing of Terrorism*. New York: Routledge, 2008.

Biggs, Duan, Frank Courchamp, Rowan Martin, and Hugh Possingham. "Legal Trade of Africa's Rhino Horn." *Science* 339 (2013): 1038–39.

Bloch, Simon. "Czech Police Charge 16 for Trading in Rhino Horn." *Business Day*. December 18, 2014.

Blundell, Arthur and Michael Mascia. "Discrepancies in Reported Levels of International Wildlife Trade." *Conservation Biology* 19, no. 6 (October 2005): 2020–2025.

Bocarejo, Diana and Diana Ojeda. "Violence and Conservation: Beyond Unintended Consequences and Unfortunate Coincidences." *Geoforum* 69, (November 2015): 176–183.

Bond, Ivan. "CAMPFIRE and the Incentives for Institutional Change." In *African Wildlife and African Livelihoods: The Promise and Performance of Community Conservation* edited by David Hulme and Marshall Murphree. Oxford: James Currey, 2001: 227–43.

Borgerhoff-Mulder, Monique and Peter Coppolillo, *Conservation: Linking Ecology, Economics and Culture* (Princeton: Princeton University Press, 2005).

Born Free Foundation. "Bloody Ivory." Accessed December 12, 2012, http://www.bloodyivory.org.

———. "Inconvenient but True: The Unrelenting Global Trade in Elephant

BIBLIOGRAPHY

Ivory." Report Prepared for the 14th Meeting of the Conference of the Parties to CITES, The Hague, June 3–15, 2007.

Bowden, Mark. *Killing Pablo*. New York: Atlantic Monthly, 2001.

Borrini-Feyerabend, Grazia et al., "Sharing Power: Learning by Doing In Co-Management of Natural Resources Throughout the World," International Institute for the Environment and Development, 2004 (available at: http://cmsdata.iucn.org/downloads/sharing_power.pdf)

Botkin, Daniel. Discordant Harmonies: A New Ecology for the Twenty-First Century (New York: Oxford University Press, 1990).

Bowden, Mark. *Killing* Pablo (New York: Atlantic Monthly, 2001)

Bowen-Jones, Evan. "Bushmeat: Traditional Regulation or Adaptation to Market Forces." In *The Trade in Wildlife* edited by Sara Oldfield, London: Earthscan, 2003, 121–131.

Braga, Anthony. "Getting Deterrence Right?" *Criminology and Public Policy* 11, no. 2 (2012): 201–210.

Brandon, Katrina and Michael Wells. "Planning for People and Parks: Design Dilemmas." *World Development* 20, (April 1992): 557–70.

Brennan, Deborah Sullivan. "Can They Clone a Rhinoceros?" *San Diego Union Tribune*. December 20, 2014, http://www.sandiegouniontribune.com/news/environemnt/sdut-rhino-cloning-stem-cells-san-deigo-zoo-safari-park-2014dec20-story.html

Breslow, Jason "A Staggering Toll of Mexico's Drug War," PBS Frontline, July 27, 2015: www.pbs.org/wgbh/frontline/article/the-staggering-death-toll-of-mexicos-drug-war/.

Britton, Alexander. "'Rathkeale Rovers' Gang Members Jailed over Multi-Million Euro Rhino Horn Spree." *Independent*. April 4, 2016. http://www.independent.ie/irish-news/courts/rathkeale-rovers-gang-members-jailed-over-multimillion-euro-rhino-horn-spree-34597780.html

Brockington, Daniel. *Fortress Conservation: The Preservation of the Mkomazi Game Reserve, Tanzania*. Bloomington: Indiana University Press, 2002.

Brockington, Daniel and James Igoe. "Eviction for Conservation: A Global Overview." *Conservation and Society* 4, no. 3 (September 2006): 424–70.

Brockington, Dan et al., "Conservation, Human Rights, and Poverty Reduction," *Conservation Biology*, 20(1), 2005: 250–52.

Bruner, Aaron, Raymond E. Gullison, Richard E. Rice, and Gustavo A.B. da Fonseca. "Effectiveness of Parks in Protecting Tropical Biodiversity." *Science* 291, no. 5501 (January 2001): 125–8.

Buckley, William and Gabriela Salazar Torres. "Mexico's Proposed Anti-Money Laundering Law." *Haynes and Boone's Newsroom*, July 5, 2011. http://www.haynesboone.com/mexican_anti-money_laundering_law/

Buijs, Daan. "A Summary of the Introduction of White Rhino onto Private Land in the Republic of South Africa." Johannesburg: Rhino and Elephant Foundation, 1987.

BIBLIOGRAPHY

Bulte, Erwin, and Richard Damania. "An Economic Assessment of Wildlife Farming and Conservation." *Conservation Biology* 19, (August 2005): 1222–33.

Burnham, Phillip *Indian Country, God's Country: Native Americans and National Parks* (Island Press: Washington, 2000).

Butler, Rhett. "The Year in Rainforests." *Mongabay*. December 29, 2015. http://news.mongabay.com/2015/12/the-year-in-rainforests-2015/.

Byrd, William and David Mansfield, "Afghanistan's Opium Economy: An Agricultural, Livelihoods, and Governance Perspective," report prepared for the World Bank Agriculture Sector Review, June 2014 (copy in author's possession).

"Call of the Wild," *Economist*, March 8, 2008.

Calley, Darren. *Market Denial and International Fisheries Regulation*. Leiden: Martinus Nijhoff, 2011.

"Canada: Harp Seal Hunt Begins amid Lower Demand." *New York Times*. April 8, 2010. http://www.nytimes.com/2010/04/09/world/americas/09 briefs-canadahunt.html?_r=0.

Carruthers, Jane. "'Police Boys' and Poachers: Africans, Wildlife Protection and National Parks, the Transvaal 1902 to 1950." *Koedoe* 36, no. 2 (September 1993): 11–22.

———. *The Kruger National Park: A Social and Political History*. Pietermaritzburg: Natal University Press, 1995.

Casimir, Michael. "Of Lions, Herders and Conservationists: Brief Notes on the Gir Forest National Park in Gujarat (Western India)." *Nomadic Peoples* 5, no. 2 (2001): 154–161.

Castillo, Eduardo. "Ex-Mexico President Calls for Legalizing Drugs," *Associated Press*, August 9, 2010 (www.highbeam.com/doc/1A1-D9HFMD304.html).

Caulkins, Jonathan. "The Real Dangers of Marijuana." *National Affairs* 26 (Winter 2016): 21–34. http://www.nationalaffairs.com/publications/detail/the-real-dangers-of-marijuana.

Caulkins, Jonathan, Susan S. Everingham, C. Peter Rydell, James Chiesa, and Shawn Bushway. *An Ounce of Prevention, a Pound of Uncertainty: The Cost-Effectiveness of School-Based Drug Prevention Programs*. Santa Monica: RAND, 1999.

Caulkins, Jonathan, Angela Hawken, Beau Kilmer, and Mark Kleiman. *Marijuana Legalization: What Everyone Needs to Know*. Oxford: Oxford University Press, 2012.

Caulkins, Jonathan P., Beau Kilmer, Mark A.R. Kleiman, Robert J. MacCoun, Gregory Midgette, Pat Oglesby, Rosalie Liccardo Pacula, and Peter H. Reuter. "Considering Marijuana Legalization: Insights from Vermont and Other Jurisdictions." Santa Monica: RAND, 2015: 135. http://www.rand.org/pubs/research_reports/RR864.html.

BIBLIOGRAPHY

Cavanagh, Conor Joseph, Pal Olav Vedeld, Leif Tore Traedal. "Securitizing REDD+: Problematizing the Emerging Illegal Timber Trade and Forest Carbon Interface in East Africa." *Geoforum* 60. (February 2015): 72–82.

Cavendish, Michael. *The Economics of Natural Resource Utilisation by Communal Area Farmers of Zimbabwe*. Oxford: Oxford University Press, 1997.

Cernea, Michael. "Restriction of Access Is Displacement: A Broader Concept and Policy." *Forced Migration Review* no. 23 (May 2005): 48–9.

Chabal, Patrick and Jean-Pascal Daloz, *Africa Works: Disorder and Political Instrument* (Bloomington: Indiana University Press, 1999).

Challender, Daniel. "Asian Pangolin: Increasing Affluence Driving Hunting Pressure." *TRAFFIC Bulletin* 23, no. 1 (October 2011): 92–3.

Challender, Daniel, Baillie, J., Ades, G., Kaspal, P., Chan, B., Khatiwada, A., Xu, L., Chin, S., KC, R., Nash, H. and Hsieh, H. *Manis pentadactyla*. The IUCN Red List of Threatened Species 2014: e.T12764A45222544. http://dx.doi.org/10.2305/IUCN.UK.2014-2.RLTS.T12764A45222544.en.

Challender, Daniel, Stuart Harrop, and Douglas MacMillan. "Toward Informed and Multi-Faceted Wildlife Trade Interventions." *Global Ecology and Conservation* 3, (January 2015): 129–48.

Chao, Sophie. "Forest Peoples: Numbers Across the World," Forest Peoples Programme, 2012 (www.forestpeoples.org/sites/fpp/files/publication/2012/05/forest-peoples-numbers-across-worldfinal_0.pdf).

Chatty, Dawn and Marcus Colchester, eds. *Conservation and Mobile Indigenous Peoples: Displacement, Forced Settlement and Sustainable Development*. New York: Berghahn Books, 2002.

Chatwin, Caroline. "Mixed Messages from Europe on Drug Policy Reform: The Cases of Sweden and the Netherlands," Brookings Institution, 2015 (www.brookings.edu/~/media/Research/Files/Papers/2015/04/global-drug-policy/Chatwin SwedenNetherlands-final.pdf?la=en).

Child, Brian, Brian Jones, David Mazambani, Andrew Mlalazi, and Hasan Moinuddin. "Final Evaluation Report: Zimbabwe Natural Resources Management Program—USAID/Zimbabwe Strategic Objective No.1." Washington, D.C.: USAID, 2003. http://pdf.usaid.gov/pdf_docs/pdact554.pdf.

Chin, Ko-Lin and Sheldon Zhang. *The Chinese Heroin Trade: Cross-Border Drug Trafficking in Southeast Asia and Beyond*. New York: New York University Press, 2015.

"China Ivory Smuggler Fined $1 A Piece." *News24*. March 26, 2013.

Christy, Bryan. "Asia's Wildlife Trade." *National Geographic* 217, no. 1 (January 2010), *National Geographic* http://ngm.nationalgeographic.com/2010/01/asian-wildlife/christy-text.

———. *The Lizard King: The True Crimes and Passions of the World's Greatest Reptile Smugglers*. Guildford: Lyons Press, 2010.

BIBLIOGRAPHY

Clarke, Shelley, Murdoch K. McAllister, E. J. Milner-Gulland, G. P. Kirkwood, Catherine G. J. Michielsens, David J. Agnew, Ellen K. Pikitch, Hideki Nakano and Mahmood S. Shivji. "Global Estimates of Shark Catches Using Trade Records from Commercial Markets." *Ecology Letters* 9, no. 10 (October 2006): 1115–1126.

Clawson, Patrick and Rensselaer W. Lee III. *The Andean Cocaine Industry*. New York: St. Martin's, 1996.

Clements, Tom, Hugo Rainey, Dara An, Vann Rours, Setha Tan, Sokha Thong, William J. Sutherland, and E.J. Milner-Gulland. "An Evaluation of the Effectiveness of Direct Payment Biodiversity Conservation: The Bird Nest Protection Program in the Northern Plains of Cambodia." *Conservation Biology* 157 (January 2013): 50–59.

Clunan, Anne. "US and International Responses to Terrorist Financing," in Jeanne Giraldo and Harold Trinkunas (eds.), *Terrorism Financing and State Responses: A Comparative Perspective* (Stanford: Stanford University Press, 2007): 260–81.

Collins, John. "The Economics of a New Global Strategy." *In Ending the Drug Wars*, compiled by LSE Ideas. London: London School of Economics, 2014, 8–15. http://www.lse.ac.uk/IDEAS/publications/reports/pdf/LSE-IDEAS-DRUGS-REPORT-FINAL-WEB.pdf

Colorado Department of Revenue. "Annual Report 2015." *Department of Revenue*. January 1, 2016. https://www.colorado.gov/pacific/sites/default/files/2015%20Annual%20Report_1.pdf.

Conor, Teresa, "Place, Belonging and Population Displacement: New Ecological Reserves in Mozambique and South Africa," *Development of Southern Africa*, 22(3), 2005: 365–82.

Conrad, Kristen. "Trade Bans: A Perfect Storm for Poaching?" *Tropical Conservation Science* 5, no. 3 (2012): 245–254.

Conti, Domenico. "Want Some Mafia with Your Pizza? How the Mob Is Taking Over Rome's Restaurants." *International Business Times*. February 14, 2014. http://www.ibtimes.com/want-some-mafia-your-pizza-how-mob-taking-over-romes-restaurants-1555647

Convention on International Trade in Endangered Species of Wild Fauna and Flora (CITES), "Monitoring the Illegal Killing of Elephants: Update on Elephant Poaching Trends in Africa to 31 December 2014" (available at: https://cites.org/sites/default/files/i/news/2015/WWD-PR-Annex_MIKE_trend_update_2014_new.pdf).

Cook, Dee, Martin Roberts, and Jason Lowther. "The International Wildlife Trade and Organized Crime: A Review of the Evidence and the Role of the UK." Wolverhampton: University of Wolverhampton and WWF, 2002. http:// www. ibrarian.net/navon/paper/The_InternationalWildlife_Trade_and_Organised_Cr.pdf?paperid=1569954.

Cooke, Kieran. "California is Left High and Dry by Cannabis Growers."

Climate News Network. February 4, 2015. http://climatenewsnetwork.net/california-is-left-high-and-dry-by-cannabis-growers/.

Cooney, Rosie and Paul Jepson. "The International Wild Bird Trade: What's Wrong with Blanket Bans?" *Oryx* 40, no. 1 (January 2006): 18–23.

Cotula, Lorenzo. "Analysis: Land Grab or Development Opportunity." *BBC News* February 22, 2012. http://www.bbc.com/news/world-africa-17099348.

Cribb, Robert. "Conservation in Colonial Indonesia," *Interventions*, 9(1), 2007: 49–61.

Crook, Vicki. "Trade in Anguilla Species, with a Focus on Recent Trade in European Eel A. Anguilla," TRAFFIC, 2010 (available at: https://portals.iucn.org/library/sites/library/files/documents/Traf-114.pdf).

Crosta, Andrea, Kimberly Sutherland, and Mike Beckner. "Blending Ivory: China's Old Loopholes, New Hopes." Los Angeles: Elephant Action League. December 2015 http://elephantleague.org/wp-content-uploads/2015/12/EAL-BLENDING-IVORY-Report-Dec2015.pdf

Cruden, John. "Statement before the Committee on Foreign Affairs, Subcommittee on Terrorism, Nonproliferation, and Trade, U.S. House of Representatives: Poaching and Terrorism: A National Security Challenge" *Hearing*. Department of Justice. April 22, 2015. www.justice.gov/opa/speech/poaching-and-terrorismnational-security-challenge-statement-assistant-attorney-general.

"Cutting Down on Cutting Down." *The Economist*, June 7, 2014.

Da Mao, and Mei Xueqin. "Protecting the Tibetan Antelope: A Historical Narrative and Missing Stories." In *A History of Environmentalism: Local Struggles, Global Histories* edited by Marco Armiero and Lise Sedrez. London: Bloomsbury, 2014, 82–104.

"Dalai Lama Campaigns for Wildlife." *BBC News* April 6, 2005 http://news.bbc.co.uk/2/hi/science/nature/4415929.stm.

Davies, Ella. "'Shocking' Scale of Pangolin Smuggling Revealed." *BBC News* March 13, 2014. http://www.bbc.co.uk/nature/26549963.

Davies, Nick and Oliver Holmes. "China Accused of Defying Its Own Ban on Breeding Tigers to Profit from Body Parts," *Guardian*, September 27, 2016.

Davies Nick and Oliver Holmes. "The Crime Family at the Centre of Asia's Animal Trafficking Network," *Guardian*, September 26, 2016 (www.theguardian.com/environment/2016/sep/26/bach-brothers-elephant-ivory-asias-animal-trafficking-network).

Davis, Ben. *Black Market: Inside the Endangered Species Trade in Asia*. San Rafael: Earth Aware Editions, 2005.

"Deadly Borders: 30 Namibians Killed through Bostwana's Shoot-to-Kill Policy," *The Namibian*, March 9, 2016.

de Castro, Ferreira and Charles Duff. *Jungle: A Tale of the Amazon Rubber-Tappers*. New York: Viking Press, 1935.

BIBLIOGRAPHY

De Suriyani, Luh. "Green Turtle Smuggling Continues." *Jakarta Post*. January 21, 2013. http://www.thebalidaily.com/2013-01-21/green-turtle-smuggling-continues.html.

Defenders of Wildlife, "Combating Wildlife Trafficking from Latin America: The Illegal Trade from Mexico, the Caribbean, Central America and South America and What Can We Do to Address It," 2015 (www.defenders.org/sites/default/files/publications/combating-wildlife-trafficking-from-latin-america-to-the-united-states-andwhat-we-can-do-to-address-it.pdf).

Dellinger, Lauren. "From Dollars to Pesos: A Comparison of the US and Colombian Anti-Money Laundering Initiatives from an International Perspective." *California Western International Law Journal* 38, no. 2 (2008): 419–54.

"Delhi Hub for Trade in Leopard Body Parts: Report." *Business Standard*. September 28, 2012. http:///www.business-standard.com/article/pti-stories/delhi-hub-for-trade-in-leopard-body-parts-report-112092800403_1.html.

Denevan, William "The Pristine Myth: The Landscape of the Americans in 1492," *Annals of the Association of American Geographers*, 82(3), 1992: 369–85

Denyer, Simon. "China Pledges to End Ivory Trading—But Says the U.S. Should, Too." *Washington Post*. June 5, 2015. https://www.washingtonpost.com/world/china-pledges-to-end-ivory-trading—but-says-the-us-should-too/2015/06/05/3b9a1946-0aed-11e5-951e-8e15090d64ae_story.html.

———. "China Tried to Drive a Furry Mammal to Extinction. Maybe That Wasn't Such a Good Idea." *Washington Post*. July 22, 2016. https://www.washingtonpost.com/world/asia_pacific/china-tried-to-drive-a-furry-mammal-to-extinction-maybe-that-wasnt-such-a-good-idea/2016/07/21/97759d34-3ed2-11e6-9e16-4cf01a41decb_story.html.

———. "China's Vow to Shut Down Its Ivory Trade by the End of 2017 Is a 'Game Changer' for Elephants" *Washington Post*, December 30, 2016. https://www.washigntonpost.com/world/asia_pacific/china-cows-to-shut-down-ivory-trade-by-end-of-2017-offering-hope-for-elephants/2016/12/30/9b26a330-ceae-11e6-85cd-e66532e35a44_story.html?utm_term=.070f7b04ff10

Department for the Environment, Food and Rural Affairs, "Declaration: London Conference on the Illegal Wildlife Trade," February 12–13, 2014 (available at: www.gov.uk/government/uploads/system/uploads/attachment_data/file/281289/london-wildlife-conference-declaration-140213.pdf)

Devine, Jennifer. "Counterinsurgency Ecotourism in Guatemala's Maya Biosphere Reserve." *Environment and Planning D: Society and Space* 32, no. 6 (December 2014): 981–1001.

BIBLIOGRAPHY

Díaz, Rodrigo. "Crimen Organizado Opera Tráfico Ilegal de Buche de Totoaba." *Mexicali Digital*. August 4, 2014. http://mexicalidigital.mx/2014/opera-crimen-organizado-trafico-ilegal-de-buche-de-totoaba-19992.html.

Dickman, Amy, Ewan MacDonald, and David MacDonald. "A Review of Financial Instruments to Pay for Predator Conservation and Encourage Human-Carnivore Coexistence." *PNAS* 108, (2011): 13937–44.

Dickson, Barney. "What Is the Goal of Regulating Wildlife Trade? Is Regulation a Good Way to Achieve This Goal." In *Trade in Wildlife: Regulation for Conservation* edited by Sara Oldfield. London: Earthscan, 2003: 23–32.

"Directorate-General for Internal Policies, Policy Department A: Economic and Scientific Policy, European Parliament."Wildlife Crime. 2016. http://www.europarl.europa.eu/RegData/etudes/STUD/2016/570008/IPOL_STU(2016)570008_EN.pdf.

DLA Piper, "Empty Threat: Does the Law Combat Illegal Wildlife Trade? An Eleven-Country Review of Legislative and Judicial Approaches," 2014 (www.dlapiperprobono.com/export/sites/pro-bono/downloads/pdfs/Illegal-Wildlife-Trade-Report-2014.pdf): 131 and 194.

"Doing It My Way." *Economist*. March 2, 2013. http://www.economist.com/news/special-report/21572379-ideological-competition-between-two-diametrically-opposed-economic-models-doing-it-my-way.

Dongol, Kiran. "Nepal's Initiatives to Reduce Illegal Logging." *Ministry of Forests and Soil Conservation*. February 10, 2012, http://www.illegallogging.info/sites/files/chlogging/uploads/Dongol100212.pdf.

Donovan, Deanna G. "Cultural Underpinnings of the Wildlife Trade in Southeast Asia." In *Wildlife in Asia: Cultural Perspectives*, edited by John Knight. London: Routledge, 2004, 88–111.

Dreher, Robert. "On Poaching and Terrorism: Testimony before the U.S. House of Representatives, Committee on Foreign Affairs, Subcommittee on Terrorism, Nonproliferation, and Trade." *Department of the Interior*. April 22, 2015. http://docs.house.gov/meetins/FA/FA18/20150422/103355/HHRG-114-FA18-Wstate-DreherR-20150422.pdf

Dressler, Wolfram et al., "From Hope to Crisis and Back Again?" *Environmental Conservation*, 37(1), 2010: 5–15.

Dressler, Wolfram and Bram Büscher, "Market Triumphalism and the CBNRM 'Crises' at the South African Section of the Great Limpopo Transfrontier Park," *Geoforum*, 39, 2008: 452–65.

Drury, Rebecca. "Hungry for Success: Urban Consumer Demand for Wild Animal Products in Vietnam." *Conservation and Society* 9, no. 3 (September 2011): 247–257.

———. "Identifying and Understanding of Wild Consumer Animal Products in Hanoi, Vietnam: Implications for Conservation Management." PhD thesis, University College London, 2009. http://discovery.ucl.ac.uk/16275/1/16275.pdf.

Duckworth et al. "Why South-East Asia Should Be the World's Priority for Averting Imminent Species Extinctions, and a Call to Join a Developing Cross-Institutional Programme to Tackle This Urgent Issue." *Sapiens* 5, (2012): 1–19.

Duffy, Rosaleen. *A Trip Too Far: Ecotourism, Politics, and Exploitation* (London: Earthscan, 2002).

Duffy, Rosaleen. "Global Environmental Governance and North-South Dynamics: The Case of CITES." *Environment and Planning C: Government and Policy* 31, no. 2 (2013): 222–39.

———. *Nature Crime: How We're Getting Conservation Wrong*. New Haven: Yale University Press, 2010.

———. "Waging A War to Save Biodiversity: The Rise of Militarized Conservation." *International Affairs* 90, no. 4 (July 2014): 819–834.

———. "War, By Conservation." *Geoforum* 69 (October 2015): 238–248.

Durbin, Joanna, Brian Jones, and Marshall Murphree. "Nambian Community-Based Natural Resource Development Programme: Project Evaulation 4–19 May 1997." Windhoek: Integrated Rural Development and Nature Conservation, 1997.

Du Toit, Johan T. and David H.M. Cumming. "Functional Significance of Ungulate Diversity in African Savannas and the Ecological Implications of the Spread of Pastoralism." *Biodiversity and Conservation*. 8 no. 12 (December 1999): 1643–61.

Dutton, Adam, Cameron Hepburn, and David MacDonald. "A Stated Preference Investigation into the Chinese Demand for Farmed vs. Wild Bear Bile." *PLoS One* 6, no. 7 (2011): 1–10.

Dwyer, Michael, Micah Ingalls, and Ian Baird. "The Security Exception: Development and Militarization in Laos's Protected Areas." *Geoforum* 69 (February 2016): 207–217.

EIA, *Under Fire: Elephants in the Front Line*. January 31, 1992. https://eia-international.org/report/under-fire-elephants-in-the-front-line.

Eighth Meeting of the Comité Internacional para la Recuperación de la Vaquita (CIRVA-8), Southwest Fisheries Science Center, November 29–30th, 2016, La Jolla, CA, http://www.iucn-csg.org/wp-content/uploads/2010/03/CIRVA-8-Report-Final.pdf

Eilperin, Juliet. "Distaste Widening for Shark Fin Soup." *Washington Post*. June 5, 2011. https://www.washingtonpost.com/national/distaste-widening-for-sharks-fin-soup/2011/05/18/AG3txgJH_story.html.

"Elephants, People Fight for Life in Nwoya," Observer (Kampala, Uganda), February 14, 2013.

Eliash, Johan. *Climate Change: Financing Global Forests*. London: UK Government, 2008.

BIBLIOGRAPHY

Ellis, Richard. *Tiger Bone and Rhino Horn: The Destruction of Wildlife for Traditional Chinese Medicine*. Washington, D.C.: Island Press, 2013.

Ellis, Stephen. "Of Elephants and Men: Politics and Nature Conservation in South Africa." *Journal of Southern African Studies* 20, no. 1 (March 1994): 53–69.

Emlsie, Richard and Martin Brooks. "African Rhino: Status Survey and Conservation Action Plan." *IUCN/SSC African Rhino Specialist Group*, 1999. https://portals.iucn.org/library/efiles/edocs/1999-049.pdf.

Engel, Stefanie, Stefano Pagiola, Sven Wunder. "Designing Payments for Environmental Services in Theory and Practice: An Overview of the Issues." *Ecological Economics* 65, no. 4 (May 2008): 663–74.

Environmental Affairs Department of the Republic of South Africa. Poaching Statistics. https://www.environment.gov.za/projectprogrammes/rhino-dialogues/poaching_statistics.

Environmental Investigation Agency. "Stop Stimulating Demand! Let Wildlife-Trade Bans Work." London and Washington, D.C.: Environmental Investigation Agency, 2013.

Essen, Erica von and Michael Allen. "Reconsidering Illegal Hunting as Crime of Dissent: Implication for Justice and Deliberative Uptake." *Criminal Law and Philosophy*. (2015). doi:10.1007/s11572-014-9364-8.

Fa, John and Carlos Peres. "Hunting in Tropical Forests," in *Conservation of Exploited Species* edited by John Reynolds, Georgina Mace, Kent Redford, and John Robinson. Cambridge: Cambridge University Press, 2001, 203–241.

Fabricius, Christo and Chris de Wet. "The Influence of Forced Removals and Land Restitutions on Conservation in South Africa." In *Conservation and Mobile Indigenous Peoples: Displacement, Forced Settlement and Sustainable Development* edited by Dawn Chatty and Marcus Colchester. New York: Berghahn Books, 2002, 142–57.

Fackler, Martin. "Uncertainty Buffets Japan's Whaling Fleet." *New York Times*. May 15, 2010. http://www.nytimes.com/2010/05/16/world/asia/16whaling.html.

Fairhead, James, Melissa Leach, and Ian Scoones. "Green Grabbing: A New Appropriation of Nature." *Journal of Peasant Studies* 39, no. 2 (July 2012): 237–261.

Farah, Douglas. *Blood from Stones*. New York: Broadway Books, 2004.

Farooqi, Nauman. "Curbing the Use of Hawala for Money Laundering and Terrorist Financing: Global Regulatory Response and Future Challenges." *International Journal of Business Governance and Ethics* 5, nos. 1–2 (2010): 64–75.

Felbab-Brown, Vanda. "Asia's Role in the Illicit Trade of Wildlife." *Boston Globe*. March 20, 2006.

BIBLIOGRAPHY

———. *Aspiration and Ambivalence: Strategies and Realities of Counterinsurgency and State-Building in Afghanistan*. Washington, D.C.: Brookings Institution, 2013.

———. "Calderón's Caldron: Lessons from Mexico's Battle Against Organized Crime and Drug Trafficking in Tijuana, Ciudad Juárez, and Michoacán." Brookings Institution, (September 2011). http://www.brookings.edu/~/media/research/files/papers/2011/9/calderon-felbab-brown/09_calderon_felbab_brown.pdf.

———. "Counterinsurgency, Counternarcotics, and Illicit Economies in Afghanistan: Lessons for State-Building." In *Convergence: Illicit Networks and National Security in the Age of Globalization* edited by Jacqueline Brewer and Michael Miklaucic. Washington, D.C.: National Defense University Press, 2013, 189–212.

———. "Despite Its Siren Song, High-Value Targeting Doesn't Fit All: Matching Interdiction Patterns to Specific Narcoterrorism and Organized-Crime Contexts." *Brookings Institution*. October 1, 2013. http://www.brookings.edu/~/media/research/files/papers/2013/10/01-matching-interdiction-patterns-to-narcoterrorism-and-organized-crime-contexts-felbabbrown/felbabbrown—matching-interdiction-patterns-to-specific-threat-environments.pdf.

———. "Deterring Non-State Actors." In *U.S. Nuclear and Extended Deterrence: Consideration and Challenges, Brookings Arms Control Series, Paper No. 9*, edited by Steven Pifer, Richard Bush, Vanda Felbab-Brown, Martin Indyk, Michael O'Hanlon, and Kenneth Pollack, Washington, D.C.: Brookings Institution, 2010. http://www.brookings.edu/~/media/research/files/papers/2010/6/nuclear-deterrence/06_nuclear_deterrence.pdf.

———. "Enabling War and Peace: Drugs, Logs, and Wildlife in Thailand and Burma." Washington, D.C.: Brookings Institution, 2015. http://www.brookings.edu/~/media/research/files/papers/2015/12/thailand-burma-drugs-wildlife-felbabbrown/enabling_war_and_peace_final.pdf.

———. "Improving Supply Side Policies: Smarter Eradication, Interdiction, and Alternative Livelihoods and the Possibility of Licensing." London: London School of Economics, 2014. http://www.brookings.edu/~/media/research/files/reports/2014/05/07%20improving%20supply%20side%20policies%20felbabbrown/improvingsupplysidepoliciesfelbabbrown.pdf.

———. "Indonesia Field Report III: The Orangutan's Road: Illegal Logging and Mining in Indonesia." *Brookings Institution*, February 7, 2013. http://www.brookings.edu/research/reports/2013/02/07-indonesia-illegal-logging-mining-felbabbrown.

———. "Indonesia Field Report IV: The Last Twitch? Wildlife Trafficking, Illegal Fishing, and Lessons from Anti-Piracy Efforts." *The Brookings*

Institution. March 26, 2013. http://www.brookings.edu/research/reports/2013/03/25-indonesia-wildlife-trafficking-felbabbrown.

———. "Little to Gloat About." *Cipher Brief*. April 3, 2016. http://www.thecipherbrief.com/article/africa/little-gloat-about-1089.

———. *Narco Noir: Mexico's Cartels, Cops, and Corruption*. Washington, D.C.: Brookings Institution Press, 2016. (Forthcoming.)

———. "No Easy Exit: Drugs and Counternarcotics Policies in Afghanistan." Washington D.C.: Brookings Institution, 2015. http://www.brookings.edu/~/media/Research/Files/Papers/2015/04/global-drug-policy/FelbabBrown—Afghanistan-final.pdf?la=en.

———. "Not as Easy as Falling Off a Log: The Illegal Logging Trade in the Asia-Pacific Region and Possible Mitigation Strategies." Washington D.C.: Brookings Institution, 2011. http://www.brookings.edu/~/media/research/files/papers/2011/3/illegal-logging-felbabbrown/03_illegal_logging_felbabbrown.pdf.

———. "Opium Licensing in Afghanistan: Its Desirability and Feasibility." Washington, D.C.: Brookings Institution 2007. http://www.3brookings.edu/fp/research/felbab-brown200708.pdf.

———. "Organized Criminals Won't Fade Away." *World Today Magazine*. August 2012. http://www.brookings.edu/research/articles/2012/08/drugs-crime-felbabbrown.

———. *Shooting Up: Counterinsurgency and the War on Drugs*. Washington, D.C.: Brookings Institution, 2010.

———. "The Impact of Organized Crime on Governance: The Case Study of Nepal." In *Impact of Organized Crime on Governance* edited by Camino Kavanagh. New York: New York University, 2013. http://cic.nyu.edu/sites/default/files/kavanagh_crime_developing_countries_nepal_study.pdf.

———. "The Political Economy of Illegal Domains in India and China." *International Lawyer* 43, no. 4 (Winter 2009): 1411–1428.

———. "The Not-So-Jolly Roger: Dealing with Piracy off the Coast of Somalia and in the Gulf of Guinea." Washington D.C.: Brookings Institution, 2013. http://www.brookings.edu/~/media/Research/Files/Reports/2014/foresight-africa-2014/02-foresight-piracy-somalia-felbab-brown.pdf?la=en.

Felbab-Brown, Vanda and Anna Newby. "How to Break Free of the Drugs-Conflict Nexus in Colombia." Brookings Institution. December 18, 2015. http://www.brookings.edu/blogs/order-from-chaos/posts/2015/12/16-colombia-drugs-conflict-nexus-felbabbrown-newby.

Felbab-Brown, Vanda and Harold Trinkunas. "UNGASS 2016 in Comparative Perspective: Improving the Prospects for Success." Washington, D.C.: Brookings Institution, 2015. http://www.brookings.edu/~/media/

BIBLIOGRAPHY

Research/Files/Papers/2015/04/global-drug-policy/FelbabBrown-TrinkunasUNGASS-2016-final-2.pdf?la=en.

Felbab-Brown, Vanda et al., "Assessment of the Implementation of the United States Government's Support for Plan Colombia's Illicit Crops Reduction Components," USAID, 2009 (available at: http://pdf.usaid.gov/pdf_docs/PDACN233.pdf).

Fennell, David and David Weaver. "The Ecotourism Concept and Tourism-Conservation Symbiosis," *Journal of Sustainable Tourism*, 13(4), 2005: 373–90

Ferraro, Paul. "Global Habitat Protection: Limitations of Development Interventions and A Role for Conservation Performance Payments," Conservation Biology, 15(4), 2001: 990–1000;

Ferraro, Paul and David Simpson, "The Cost-Effectiveness of Conservation Payments," Discussion Paper 00–31, Resources for the Future, Washington, 2000 (available at: www.rff.org/files/sharepoint/WorkImages/Download/RFF-DP-00–31.pdf).

"Find Land for Displaced, State Urges," *Times of Zambia*, May 6, 2015.

Fitzherbert, Anthony. "The Impact of Afghan Transition Authority's Poppy Eradication Programme on Rural Farmers." *Mercy Corps Mission Report*, October 2003.

Fleishman, Jeffrey. "Daggers Slice Yemen Tradition: Plastic Supplanting Handles Made from Rhino Horns." *The Chicago Tribune*, December 18, 2009.

Forsyth, Tim and Andrew Walker. *Forest Guardians, Forest Destroyers: The Politics of Environmental Knowledge in Northern Thailand* (Seattle: University of Washington Press, 2008).

Franzen, Jonathan. "Last Song for Migrating Birds." *National Geographic* 224, no. 1 (July 2013). http://ngm.nationalgeographic.com/2013/07/songbird-migration/franzen-text.

Frost, Peter and Ivan Bond, "The CAMPFIRE Programme in Zimbabwe: Payments for Wildlife Services," *Ecological Economics*, 65(4), 2008: 776–87.

Fuller, Thomas "US Offers Reward in Wildlife-Trade Fight," *New York Times*, November 13, 2013.

Garric, Audrey. "Pangolins under Threat as Black Market Grows." *Guardian*. March 12, 2013. http://www.theguardian.com/environment/2013/mar/12/endangered-pangolins-illegal-wildlife-trade.

Garshelis, D.L. and R. Steinmetz "*Ursus thibetanus*." *The IUCN Red List of Threatened Species 2008*. http://dx.doi.org/10.2305/IUCN.UK.2008.

Gaworecki, Mike. "Time Running Out to Save World's Most Endangered Porpoise, Environmentalists Warn," *Mongabay*, September 21, 2016 (available at: www.news.mongabay.com/2016/09/time-running-out-tosave-worlds-most-endangered-porpoise-environmentalists-warn/).

BIBLIOGRAPHY

Geisler, Charles. "A New Kind of Trouble: Evictions in Eden." *International Social Science Journal* 55, no. 175 (March 2003): 69–78.

Giraldo, Camilo Meija and James Bargent. "Are Mexican Narcos Moving into Lucrative Fish Bladder Market?" *InSight Crime*, August 6, 2014 (available at: www.insightcrime.org/news-briefs/mexiconarcos-fish-bladder-market).

Giraldo, Jeanne and Harold Trinkunas. "The Political Economy of Terrorism Financing." In *Terrorism Financing and State Responses: A Comparative Perspective* edited by Jeanne Giraldo and Harold Trinkunas. Stanford: Stanford University Press, 2007, 7–20.

"Global Wild Tiger Population Increases, But Still a Long Way to Go," WWF (available at: www.worldwildlife.org/press-releases/globalwild-tiger-population-increases-but-still-a-long-way-to-go-3).

"Global Tiger Recovery Program 2010–2012." Washington, D.C.: World Bank, 2012. http://globaltigerinitiative.org/news-blog/by-tag/global-tiger-recovery-program/.

"Going, Going, Gone … The Illegal Trade in Wildlife in East and Southeast Asia." Washington, D.C.: World Bank, 2005. http://www-wds.worldbank.org/external/default/WDSContentServer/WDSP/IB/2005/09/08/000160016_20050908161459/Rendered/PDF/334670PAPER0Going1goi ng1gone.pdf.

Goldman, Lisa and Stanton Glantz. "Evaluation of Antismoking Advertising Campaigns." *Journal of American Medical Science Association* 279, no. 10 (1998): 772–7.

Goode, Erica. "A Struggle to Save the Scaly Pangolin." *New York Times*. March 30, 2015. http://www.nytimes.com/2015/03/31/science/a-struggle-to-save-the-scaly-pangolin.html.

"Government of Mozambique Announces Major Decline in National Elephant Population." *Wildlife Conservation Society Newsroom*. May 26, 2015. http://newsroom.wcs.org/News-Releases/articleType/ArticleView/articleId/6760/Govt-of-Mozambique-announces-major-decline-in-national-elephant-population.aspx.

Goredema, Charles. "Measuring Money Laundering in Southern Africa." *Institute for Security Studies*. December 1, 2015. https://www.issafrica.org/topics/organised-crime/01-dec-2005-measuring-money-laundering-in-southern-africa-charles-goredema.

Gratwicke, Brian, Elizabeth Bennett, Steven Broad, Sarah Christie, Adam Dutton, Grace Gabriel, Craig Kirkpatrick, and Kristin Nowell. "The World Can't Have Wild Tigers and Eat Them, Too." *Conservation Biology* 22, no. 1 (February 2008): 222–3.

Gray, Dennis. "Asia's Wildlife Hunted for China's Appetite." *Washington Post*. April 6, 2004.

———. "Wildlife at Risk in Southeast Asia: Species Being Used for Food and Medicine." *Washington Post*. April 4, 2004. http://www.nbcnews.com/

BIBLIOGRAPHY

id/4585068/ns/us_news-environment/t/asias-wildlife-hunted-chinas-appetite/#.V_6XEiSD6M8

Grayson, George. *The Cartels*. New York: Praeger, 2014.

Great Elephant Census, "Better Data for a Crisis: Second Tanzania Count Part of Ongoing Population Monitoring" *Great Elephant Census*. September 9, 2015. (available at: www.greatelephantcensus.com/blog/2015/9/8/better-data-for-a-crisis-second-tanzania-count-part-of-ongoing-population-monitoring)

Green, Edmund and Francis Shirley. *The Global Trade in Corals*. Cambridge: World Conservation Monitoring Centre, 1999.

Grindle, Merilee. "Good Enough Governance Revisited." *Development Policy Review* 25, no. 5 (September 2007): 553–74.

Groves-Sunderland, Jacqueline L., Daniel A. Slayback, Michael P.B. Balinga, and Terry Sunderland. "Impacts of Co-Management on Western Chimpanzee (Pan Troglodytes Verus) Habitat and Conservation in Nialama Classified Forest, Republic of Guinea: A Satellite Perspective." *Biodiversity and Conservation* 20, no. 12 (July 2011): 2745–57.

Gurman, Sadie. "Drug Traffickers Seek Haven Amid Legal Marijuana." *Associated Press*. January 28, 2016. http://www.salon.com/2016/01/28/drug_traffickers_seek_safe_haven_amid_legal_marijuana/.

Guynup, Sharon. "U.S. A Major Destination for Trafficked Latin American Wildlife." *Mongabay*. November 25, 2015. http://news.mongabay.com/2015/11/u-s-a-major-destination-for-trafficked-latin-american-wildlife/.

Gwin, Peter. "Rhino Wars." *National Geographic* 221, no. 3 (March 2012). http://ngm.nationalgeographic.com/2012/03/rhino-wars/gwin-text.

Hallam, Christopher, Dave Bewley-Taylor, and Martin Jelsma. "Scheduling in the International Drug Control System." *Series of Legislative Reform of Drug Policies*, No. 25: Amsterdam Transnational Institute, June 2014. https://www.tni.org/files/download/dlr25_0.pdf.

Halliday, Terrence, Michael Levi, and Peter Reuter. "Global Surveillance of Dirty: Assessing Assessments of Regime to Control Money Laundering and Combat the Financing of Terrorism." Champaign: University of Illinois College of Law, 2014. http://faculty.publicpolicy.umd.edu/sites/default/files/reuter/files/report_global_surveillance_of_dirty_money_release_date_30_january_2014.pdf.

Halter, Reese. "Insatiable Demand for African Rhino Horn Spells Extinction." *Huffington Post*. October 6, 2013. http://www.huffingtonpost.com/dr-reese-halter/insatiable-demand-for-afr_b_4055075.html.

Hastie, Jo, Julian Newmann, and Mary Rice. "Back in Business: Elephant Poaching and the Ivory Black Markets of Asia." *Environmental Investigation Agency*. (2002): 1–10.

Hauck, Maria and Neville A. Sweijd. "A Case Study of Abalone Poaching in

BIBLIOGRAPHY

South Africa and Its Impact on Fisheries Management." *ICES Journal of Marine Species: Journal du Conseil* 56, no. 6 (December 1999): 1024–32.

Hauser, Christine. "Numbers of Tigers in the Wild is Rising, Wildlife Groups Say," *New York Times*, April 11, 2016.

Hearne, John and Margaret McKenzie, "Compelling Reasons for Game Ranching in Maputaland." In *Wildlife Conservation by Sustainable Use* edited by Herbert Prins, Jan Geu Grootenhuis, and Thomas T. Dolan. Boston: Kluwer Academic Publishers, 200, 417–438. Boston: Kluwer Academic Publishers, 2000, 417–38.

Helliker, Kirk. "Reproducing White Commercial Agriculture in South Africa." Grahamstown: Rhodes University, 2013. https://www.google.com/url?sa=t&rct=j&q=&esrc=s&source=web&cd=1&cad=rja&uact=8&ved=0ahUKEwjb18_pw7TOAhXC1B4KHeWWAgoQFggcMAA&url=https%3A%2F%2Fwww.ru.ac.za%2Fmedia%2Frhodesuniversity%2Fcontent%2Fsociology%2Fdocuments%2FWhite%2520commercial%2520farms%2520K%2520Helliker%2520seminar%2520paper.doc&usg=AFQjCNELlbG4k3nw_xZxjT-3_bKqCBp6OA.

Higgins, Steven, Charlie Shackleton, and Robbie Robinson. "Changes in Woody Community Structure and Composition under Contrasting Landuse Systems in a Semi-Arid Savanna, South Africa." *Journal of Biogeography* 26, no. 3 (1999): 619–27.

Holmes, Oliver and Nick Davies, "Revealed: The Criminals Making Millions from Illegal Wildlife Trafficking," *Guardian*, September 26, 2016 (https://www.theguardian.com/environment/2016/sep/26/revealed-the-criminals-making-millions-from-illegal-wildlifetrafficking).

Hongfa, Xu and Craig and Kirkpatrick, eds. "The State of Wildlife in China: Information on the Trade in Wild Animals and Plants in China 2007." Beijing: TRAFFIC, 2008, 9. http://www.trafficj.org/publication/09-State_of_Wildlife_China.pdf

Homewood, Katherine and W. Alan Rodgers. *Maasailand Ecology: Pastoralist Development and Wildlife Conservation in Ngorongoro, Tanzania*. Cambridge: Cambridge University Press, 1991.

Honey, Martha. *Ecotourism and Sustainable Development: Who Owns Paradise?* (Washington: Island Press, 1999).

Hornsby, Charles. *Kenya: A History since Independence*. New York: I.B. Tauris, 2013.

Hoven, Wouter van. "Private Game Parks in Southern Africa." In *Institutional Arrangements for Conservation, Development, and Tourism in Eastern and Southern Africa* edited by René van der Duim, Machiel Lamers, and Jakomijn van Wijk. Dordrecht: Springer, 2015: 101–118.

Hübschle, Annette Michaela. "A Game of Horns: Transnational Flows of Rhino Horn." PhD Diss., International Max Planck Research School, 2016, 304.

BIBLIOGRAPHY

Humber, Frances, Brendan Godley, and Annette Broderick. "So Excellent a Fishe: A Global Overview of Legal Marine Turtle Fisheries." *Diversity and Distributions* 20, no. 5 (May 2014): 579–590.

Humphreys, Jasper and M.L.R. Smith. "War and Wildlife: The Clausewitz Connection." *International Affairs* 87, no. 1 (January 2011): 121–142.

Hutton and Barnabas Dikson, *Endangered Species Threatened Convention: The Past, Present and Future of* CITES. London: Earthscan, 2000.

Hutton, Jon and Grahame Webb. "Crocodiles: Legal Trade Snaps Back." In *The Trade in Wildlife* edited by Sara Oldfield. London: Earthscan, 2003, 108–120.

Huxley, Chris. "CITES: The Vision." In *Endangered Species Threatened Convention: The Past, Present, and Future of CITES, the Convention on International Trade in Endangered Species of Wild Fauna and Flora* edited by Jon Hutton and Barnabas Dickson. London: Earthscan, 2000, 3–12.

"If They Were Crooks, Wouldn't They Be Richer?" *Economist*. April 22, 2010.

———. "South Africa Slams the Door on Elephant Tourism." February 25, 2008. http://www.ifaw.org/international/node/25661.

———. "Bidding against Survival: The Elephant Poaching Crisis and the Role of Auctions in the US Ivory Market," August 2014 (available at: www.ifaw.org/sites/default/files/IFAW-Ivory-Auctionsbidding-against-survival-aug-2014_0.pdf).

"India: Tribes Face Harassment and Eviction for, 'Tiger Conservation'." *Survival International*. May 30, 2014. http://www.survivalinternational.org/news/10239.

"Indonesia: Natural Resources and Law Enforcement." *International Crisis Group*. December 20, 2001. http://www.crisisgroup.org/~/media/Files/asia/south-east-asia/indonesia/Indonesia%20Natural%20Resources%20and%20Law%20Enforcement.pdf.

Ingram, Christopher. "Americans Spent More on Legal Weed than on Cheetos and Funyuns Combined Last Year." *Washington Post*. February 2, 2016.

International Fund for Animal Welfare. "Elephants sent into Safari Slavery from Zimbabwe's World Famous Hwange National Park." November 8, 2006. http://www.ifaw.org/united-states/node/11256.

International Consortium on Combatting Wildlife Crime, "Wildlife and Forest Crime Analytic Toolkit" (Vienna: United Nations Office on Drugs and Crime, 2012),

"International Narcotics Strategy Control Report." *U.S. Department of State*. March 2007. http://www.state.gov/j/inl/rls/nrcrpt/2007/vol1/html/80858.htm.

International Union for the Conservation of Nature (IUCN), "Poaching Behind Worst African Elephant Losses in 25 Years—IUCN Report," 2016 (available at: www.iucn.org/news/poaching-behind-worst-africanelephant-losses-25-years-%E2%80%93-iucn-report).

BIBLIOGRAPHY

"Ivory Demand in Hong Kong." WildAid, 2015. http://www.wildaid.org/sites/default/files/resources/WildAid-Hong%20Kong%20Ivory%20Survey,%202015.pdf.

Jachmann, Hugo. "Elephant Poaching and Resource Allocation for Law Enforcement." In *The Trade in Wildlife* edited by Sara Oldfield. London: Earthscan, 2003, 100–107.

"Jailhouse Nation; Justice in America." *Economist*, June 20, 2015.

Jambiya, George, Simon Milledge, and Nangena Mtango. "Night Time Spinach: Conservation and Livelihood Implications of Wild Meat Use in Refugee Situations in North-Western Tanzania." Dar Es Salaam: TRAFFIC, 2007. https://portals.iucn.org/library/sites/library/files/documents/Traf-101.pdf.

James, Alexander N., Kevin J. Gaston, and Andrew Balmford. "Balancing the Earth's Accounts." *Nature* 336, no. 6199 (September 1999): 533–5.

Jamjoom, Mohammed and Gena Somra. "Qat Crops Threaten to Drain Yemen Dry." *CNN*. December 13, 2010. http://www.cnn.com/2010/WORLD/meast/12/02/yemen.water.crisis/.

Jelsma, Martin. "UNGASS 2016: Prospects for Treaty Reform and UN System-Wide Coherence on Drug Policy." Washington, D.C.: Brookings Institution, 2015. http://www.brookings.edu/~/media/Research/Files/Papers/2015/04/global-drug-policy/Jelsma—United-Nations-final.pdf?la=en.

Jenkins, Robert W.G. "The Significant Trade Process: Making Appendix II Work." In *Endangered Species Threatened Convention* edited by Jon Hutton and Barnabas Dickson. London: Earthscan, 2000, 47–56.

Jenkins, Robert W.G., Dietrich Jelden, Grahame J.W. Webb, and S. Charlie Manolis. "Review of Crocodile Ranching Programs." Sanderson: IUCN-SSC Crocodile Specialist Group, 2006. https://cites.org/common/com/ac/22/EFS-AC22-Inf02.pdf.

Jepson, Paul, Richard Ladle, and Sujatnika, "Assessing Market-Based Conservation Governance Approaches: A Socio-Economic Profile of Indonesian Markets for Wild Birds," Oryx, 45(4), October 2011: 482–91.

Johnson, Steve. "State of CBNRM Report 2009." Gabarone: Botswana National CBNRM Forum, 2009. http://emergingsecurityissues.org/sites/default/files/CBNRM%202009%20REPORT.pdf.

Jolly, David. "Whaling Talks in Morocco Fail to Produce Reductions." *New York Times*. June 23, 2010. http://www.nytimes.com/2010/06/24/world/24whale.html?_r=0.

Jolly, David and John Broder. "U.N. Rejects Export Ban on Atlantic Bluefin Tuna." *New York Times*. March 18, 2010. http://www.nytimes.com/2010/03/19/science/earth/19species.html.

Jones, Brian. "Policy Lessons from the Evolution of a Community-Based

Approach to Wildlife Management, Kunene Region, Namibia," *Journal of International Development*, 11(2), 1999: 295–304.

Jones, Brian and Marshall Murphree. "The Evolution of Policy on Community Conservation in Namibia and Zimbabwe." In *African Wildlife and African Livelihoods: The Promise and Performance of Community Conservation* edited by David Hulme and Marshall Murphee. Oxford: James Currey, 2001, 38–58.

Jones, Jonathan. "A Picture of Loneliness." *Guardian*. May 12, 2015.

Journalists for Justice. "Black and White: Kenya's Criminal Racket in Somalia." Nairobi: Journalists for Justice, 2015. http://www.jfjustice.net/userfiles/file/Research/Black%20and%20White%20Kenya's%20Criminal%20Racket%20in%20Somalia.pdf.

Kaimowitz, David, and Douglas Sheil. "Conserving What and For Whom? Why Conservation Should Help Meet Basic Human Needs in the Tropics." *Biotropica* 39, no. 6 (November 2007): S12–17.

Kakala, Taylor Toeka. "Soldiers Trade in Illegal Ivory." *Inter Press Service* July 25, 2013.

Kalron, Nir and Andrea Crosta. "Africa's Wild Gold of Jihad: Al Shabaab and Conflict Ivory." *The Elephant Action League*. https://www.elephantleague.org/project/africas-white-gold-of-jihad-al-shabaab-and-conflict-ivory/.

Karacs, Sarah. "Why Appetite for Shark Fin Soup Continues to Grow Despite Efforts to Stem the Slaughter," *South China Morning Post*, September 20, 2016 (available at: www.scmp.com/news/hong-kong/health-environment/article/2020985/why-politics-stubborntastes-and-resilient).

———. "Shark Fin Soup Still Served at 98 per cent of Hong Kong Restaurants as Restaurants Choose Money over Environmentally Friendly Practices," *South China Morning Post*, January 25, 2016 (available at: www.scmp.com/news/hong-kong/health-environment/article/1905257/shark-fin-soup-still-served-98-cent-hong-kong).

Karanja, Francis, Yemeserach Tessema, and Edmund Barrow. "Equity in the Loita/Purko Naimina Enkiyio Forest in Kenya: Securing Maasai Rights to and Responsibilities for the Forest." *Forest and Social Perspectives in Conservation* no. 11 (2002). https://portals.iucn.org/library/sites/library/files/documents/2000-019-11.pdf.

Karanth, Krithi. "Making Resettlement Work: The Case of India's Bhadra Wildlife Sanctuary." *Biological Conservation* 139, no. 3–4 (October 2007): 315–24.

Karesh, William. "Wildlife Trade and Global Disease Emergence." *Emerging Infectious Diseases* 11, no. 7 (2005): 1000–02.

Karesh, William, Robert A. Cook, Martin Gilbert, and James Newcomb. "Implications of Wildlife Trade on the Movement of Avian Influenza and Other Infectious Diseases." *Journal of Wildlife Diseases* 43, no. 3 (2007): 55–9.

Karsenty, Alain. "Financing Options to Support REDD+ Activities," report

BIBLIOGRAPHY

for the European Commission, DG Climate Action, 2012 (available at: http://ur-forets-societes.cirad.fr/content/download/4123/32260/version/3/file/REDD_study_CIRAD_final.pdf).

"Kauihura Pays Shs 20 Million to Bukwo Victims of Police Brutality," Observer (Kampala, Uganda), January 18, 2016.

Kelly, Alice and Megan Ybarra. "Introduction to the Themed Issue: 'Green Security in Protected Areas.'" *Geoforum* 69 (February 2016): 171–175.

Kennedy, David. *Don't Shoot: One Man, A Street Fellowship, and the End of Violence in Inner-City America*. New York: Bloomsbury, 2011.

Khanal, Dij Raj and Ghana Shyam Pandey. "Illegal Logging and the Issue of Transparency in Nepal." *FECOFUND*, 2011. http://www.illegal-logging.info/uploads/Khanal100212.pdf.

Kievit, Henriette. "Conservation of the Nile Crocodile: Has CITES Helped or Hindered?" in *Endangered Species: Threatened Convention: The Past, Present, and Future of CITES* edited by Jon Hutton and Barnabas Dickson. London: Earthscan, 2000.

Kilmer, Beau. "Marijuana Legalization, Government Revenues, and Public Budgets: Ten Factors to Consider." *RAND*. January 19, 2016. https://www.rand.org/content/dam/rand/pubs/testimonies/CT400/CT449/RAND_CT449.pdf.

Kilmer, Beau, Jonathan Caulkins, Brittany Bond, and Peter Reuter. "Reducing Drug Trafficking Revenues and Violence in Mexico: Would Legalizing Marijuana in California Help?" *RAND*, 2010. http://www.rand.org/content/dam/rand/pubs/occasional_papers/2010/RAND_OP325.pdf.

Kilmer, Beau, Nancy Nicosia, Paul Heaton, and Greg Midgette. "Efficacy of Frequent Monitoring with Swift, Certain, and Modest Sanctions for Violations: Insights from South Dakota's 24/7 Sobriety Project." *American Journal of Public Health* 103 (2013): e37–e43.

Kilmer, Beau and Rosalie Pacula. "Preventing Drug Use." In *Targeting Investments in Children: Fighting Poverty when Resources Are Limited* edited by Phillip B. Levine and David J. Zimmerman, 181–220. Chicago: Chicago University Press, 2010.

Kinoti, Johh. "The Lewa Wildlife Conservancy: IUCN Review of Communities and Their Natural Resources—Sharing Experiences and Learning Lessons in East Africa." IUCN, 2007.

Kirkpatrick, Craig and Lucy Emerton. "Killing Tigers to Save Them: Fallacies of the Farming Argument." *Conservation Biology* 24, no. 3 (June 2010): 665–9.

Kleiman, Mark. "Legal Commercial Cannabis Sales in Colorado and Washington: What Can We Learn?" Brookings Institution, 2015 (available at: http://www.brookings.edu/~/media/Research/Files/Papers/2015/04/global-drug-policy/Kleiman—Wash-and-Co-final.pdf?la=en).

BIBLIOGRAPHY

———. *Against Excess: Drug Policy for Results*. New York: Basic Books, 1992.

———. "Coerced Abstinence: A Neo-Paternalistic Drug Policy Initiative." In *The New Paternalism* edited by Lawrence Mead. Washington, D.C.: Brookings Institution Press, 1998.

———. *When Brute Force Fails: How to Have Less Crime and Less Punishment*. Princeton: Princeton University Press, 2009.

Knights, Peter, Alex Hofford, A. Andersson, and D. Cheng. "The Illusion of Control: Hong Kong's 'Legal' Ivory Trade." *WildAid*, 2015. http://www.wildaid.org/sites/default/files/resources/The%20Illusion%20of%20Control-Full%20Report.pdf.

Koppert, Georgius J.A., Edmond Dounias, Alain. Froment, and Patrick Pasquet. "Consommation alimentaire dans trois populations forestières de la région côtière du Cameroun: Yassa, Mvae et Bakola." In *L'alimentation en forêt tropicale: interactions bioculturelles et perspectives de développement*, edited by Claude Marcel Hladik, Annette Hladik, Hélène Pagezy, Olga F. Linares, Georgius J.A. Koppert and Alain Froment. Paris: UNESCO, 1996, 477–96.

Koziell, Isabella "Diversity, Not Adversity: Sustaining Livelihoods with Biodiversity," Biodiversity and Livelihoods Issues Paper, International Institute for Environment and Development, 2001 (available at: http://pubs.iied.org/pdfs/7822IIED.pdf);

Koziell, Izabella. "The Links between Biodiversity and Poverty." United Nations Development Programme, 2000. http://ec.europa.eu/europeaid/sites/devco/files/publication-biodiversity-in-developmentbrief-1—2001_en.pdf.

Kumar, Ajay Mahapatra, Heidi J. Albers, and Elizabeth J.Z. Robinson. "The Impact of NTFP Sales on Rural Households' Cash Income in India's Dry Deciduous Forest." *Environment Management* 35, no. 3 (2005): 258–65.

Kumar, Jitendra Das and Om Prakash. "Measuring Market Channel Efficiency and Strategy to Improve Income to Local Communities Dependent on Tropical Forests." *Journal of Sustainable Forestry* 15, no. 4 (2002): 27–52.

Laburn, Helen and Duncan Mitchell. "Extracts of Rhinoceros Horn Are Not Antipyretic in Rabbits," *Journal of Basic and Clinical Physiology and Pharmacology*, 8(1/2), 1997: 1–11.

Lavandera, Ed. "Winner of Black Rhino Hunting Auction: My $350,000 Will Help Save the Species." *CNN*, January 17, 2014. http://www.cnn.com/2014/01/16/us/black-rhino-hunting-permit/

Lawson, Sam and Larry McFaul. "Illegal Logging and Related Trade: Indicators of Global Response." Washington, D.C.: Chatham House, 2010, 77. https://www.chathamhouse.org/sites/files/chathamhouse/public/Research/Energy%2C%20Environment%20and%20Development/0710pr_illegallogging.pdf.

BIBLIOGRAPHY

Leader-Williams, Nigel. "Regulation and Protection: Successes and Failures in Rhinoceros Conservation." In *The Trade in Wildlife* edited by Sara Oldfield, London: Earthscan, 2003.

Leakey, Richard. *Wildlife Wars: My Battle to Save Kenya's Elephants*. London: Pan, 2001.

Leclerc, Camille, Céline Bellard, Gloria M. Luque, and Franck Courchamp. "Overcoming Extinction: Understanding Processes of Recovery of the Tibetan Antelope." *Ecosphere* 6, no. 9 (September 2015): 1–14.

Levi, Michael. "The Impact of Anti-Money Laundering Measures within Developing Countries against Proceeds of Corruption." Cardiff: Cardiff School of Social Sciences, 2009. http://www.publicpolicy.umd/.../Levi_final_wold_bank_paper_final_310809.pdf.

Levin, Dan. "China Bans Import of Ivory Carvings for One Year." *New York Times*. February 27, 2015. http://www.nytimes.com/2015/02/27/world/asia/china-bans-import-of-ivory-carvings-for-one-year.html?_r=0.

Levin, Dan. "China Weighs Ban on Manta Ray Gills, Sold in Traditional Market as Modern Panacea." *New York Times*. January 7, 2016.

Levey, Stuart. "How We're Tying Up Terrorists' Cash." *Christian Science Monitor*. December 24, 2008. http://www.csmonitor.com/Commentary/Opinion/2008/1224/p09s01-coop.html.

Li, Peter J. and Gareth Davey. "Culture, Reform Politics, and Future Directions: A Review of China's Animal Protection Challenge." *Society and Animals* 21, no. 1 (2013): 34–53.

Li Zhang and Feng Yi. "Wildlife Consumption and Conservation Awareness in China: A Long Way to Go." *Biodiversity and Conservation* 23, no. 9 (August 2014): 2371–81.

Liddick, Donald. *Crimes Against Nature: Illegal Industries and the Global Environment*. Oxford: Praeger, 2011.

Lin, Jolene. "Tackling Southeast Asia Wildlife Trade." *Singapore Yearbook of International Law* 9 (2005): 191–208.

Lindberg, Kreg. "Economic Aspects of Ecotourism." In *Ecotourism: A Guide for Planners and Managers, Volume II*, edited by Kreg Lindberg, Megan Epler Wood, and David Engeldrum. North Bennington: The Ecotourism Society, 1998: 87–117.

Ling, Justin. "Inside the US-Canadian Smuggling Ring for Narwhal Tusks." *Vice News*. March 18, 2016. https://news.vice.com/article/inside-the-us-canadian-smuggling-ring-for-narwhal-tusks.

Lo, Cheryl and Gavin Edwards. "The Hard Truth: How Hong Kong's Ivory Trade Is Fueling Africa's Poaching Crisis." Hong Kong: World Wildlife Fund, 2015. http://awsassets.wwfhk.panda.org/downloads/wwf_ivory-trade_eng_eversion.pdf.

BIBLIOGRAPHY

Lohmuller, Michael. "How China Fuels Wildlife Trafficking in Latin America." *InSight Crime*, June 10, 2015. http://www.insightcrime.org/news-analysis/how-china-fuels-wildlife-trafficking-latin-america.

———. "Industrial Scale 'Shark Finning' in Ecuador." *InSight Crime*, May 28, 2015.http://www.insightcrime.org/news-briefs/industrial-scale-harvesting-shark-fins-ecuador.

Lombard, Cyril and Pierre du Plessis. "The Impact of the Proposal to List Devil's Claw on Appendix II of CITES." In *The Trade in Wildlife* edited by Sara Oldfield, London: Earthscan, 2003, 132–46.

Lombard, Louisa. "Threat Economies and Armed Conservation in Northeastern Central African Republic." *Geoforum* 69 (November 2015): 218–26.

Lowry, Willy. "Ring of Elephant Poachers Broken Up by Tanzanian Authorities." *New York Times* February 8, 2016. http://www.nytimes.com/2016/02/09/world/africa/ring-of-elephant-poachers-broken-up-by-tanzanian-authorities.html.

Loveridge, John. "A Review of Crocodile Management in Zimbabwe." Harare: University of Zimbabwe, 1996.

Lunstrum, Elizabeth. "Green Militarization: Anti-Poaching Efforts and the Spatial Contours of Kruger National Park." *Annals of the Association of American Geographers* 104, no. 4 (June 2014): 816–32.

Lynam, Antony. "Rain-Forest Guardians." *Wildlife Conservation*, January 2005: 8–9.

MacCoun, Robert and Peter Reuter, *Drug War Heresies: Learning from Other Vices, Times, and Places* (Cambridge: Cambridge University Press, 2001).

Macfie, Elizabeth and Elizabeth Williamson, "Best Practice Guidelines for Great Ape Tourism," Occasional Paper of the IUCN Species Survival Commission No. 28, 2010 (available at: https://portals.iucn.org/library/efiles/edocs/ssc-op-038.pdf).

MacKenzie, Catrina, Colin Chapman, and Raja Sengupta. "Spatial Patterns of Illegal Resource Extraction in Kibale National Park, Uganda." *Environmental Conservation* 39, no. 1 (March 2011): 38–50.

MacKenzie, John. *Empire of Nature: Hunting Conservation and British Imperialism.* Manchester University Press, 1988.

MacMillan, Douglas and Jeonghee Han. "Cetacean By-Catch in the Korean Peninsula—By Chance or By Design?" *Human Ecology* 39, October 2011: 519–33.

MacMillan, Douglas and Sharon Philip. "Can Economic Incentives Resolve Conservation Conflict: The Case of Wild Deer Management and Habitat Conservation in the Scottish Highlands." *Human Ecology* 38, (2010): 485–494.

Maguire, Tom, and Cathy Haenlein. "An Illusion of Complicity: Terrorism and

BIBLIOGRAPHY

the Illegal Ivory Trade in East Africa." London: Royal United Services Institute for Defence and Security Studies, 2015. https://rusi.org/sites/default/files/201509_an_illusion_of_complicity_0.pdf.

Mahurkar, Uday. "King in Shackles," *India Today*, April 23, 2007.

Malkin, Elisabeth. "Before Vaquitas Vanish, a Desperate Bid to Save Them," *New York Times*, February 27, 2017.

Mansfield, David. "An Analysis of Licit Opium Poppy Cultivation: India and Turkey." 2001. http://www.davidmansfield.org/data/Policy_Advice/UK/India-Turkeycultivation.doc.

———. "The Economic Superiority of Illicit Drug Production: Myth and Reality—Opium Poppy Cultivation in Afghanistan." Paper prepared for the International Conference on Alternative Development in Drug Control and Cooperation, Feldafing, Munich, January 7–12, 2002.

Marsh, Bill. "Fretting about the Last of the World's Biggest Cats." *New York Times*, March 6, 2010. http://www.nytimes.com/2010/03/07/weekinreview/07marsh.html.

Marshall, Michael. "Elephant Ivory Could Be Bankrolling Terrorist Groups." *New Scientist*. October 2, 2013.

Martin, Esmond Bradley, Lucy Vigne, and Crawford Allan "On a Knife's Edge: The Rhinoceros Horn Trade in Yemen." Cambridge: TRAFFIC, 1997. http://www.google.com/url?sa=t&rct=j&q=&esrc=s&source=web&cd=1&cad=rja&uact=8&ved=0ahUKEwjTutjxtLLOAhVp64MKHS7-CtoQFggcMAA&url=http%3A%2F%2Fstatic1.1.sqspcdn.com%2Fstatic%2Ff%2F157301%2F9113577%2F1288012352957%2Ftraffic_species_mammals58.pdf%3Ftoken%3Dg0HQoR7lebB4WHQLrvNM4bnPyfU%253D&usg=AFQjCNHP6P_JT9qi_2Bx-OeU_BZYWcSZ2w&sig2=T0yks-xPP-IFquD8rU015g.

Martin, Rowan. "When CITES Works and When it Does Not." In *Endangered Species, Threatened Convention*, edited by Jon Hutton and Barnabas Dickson. London: Earthscan, 2000, 29–37.

Massé, Francis and Elizabeth Lunstrum. "Accumulation by Securitization: Commercial Poaching, Neoliberal Conservation, and the Creation of New Wildlife Frontiers." *Geoforum* 69, (February 2016): 227–237.

Mauer, Bill. "From Anti-Money Laundering to … What? Formal Sovereignty and Feudalism in Offshore Financial Centers." In *Ungoverned Spaces: Alternatives to State Authority in an Era of Softened Sovereignty* edited by Anne L. Clunan and Harold A. Trinkunas. Stanford: Stanford University Press, 2010.

Mayers, James and Sonja Vermeulen. "Power from Trees: How Good Forest Governance Can Help Reduce Poverty." London: International Institute for Environment and Development, 2002. http://pubs.iied.org/pdfs/11027IIED.pdf. Clunnan and Harold Trinkunas, 215–232. Stanford: Stanford University Press, 2010.

BIBLIOGRAPHY

Mbaiwa, Joseph. "The Success and Sustainability of Community-Based Natural Resource Management in the Okavango Delta, Bostwana," *South African Geographical Journal*, 86, no. 1, (2004): 44–53.

Mbaiwa, Joseph, Amanda Stronza, and Urs Kreuter. "From Collaboration to Conservation: Insights from the Okavango Delta, Bostwana." *Society and Natural Resources* 24, no. 4 (January 2011): 400–11.

McCarthy, Cormac. *The Road* (New York: Knopf, 2006): 130.

McClintock, Cynthia. *Peru's Shining Path*. Washington, D.C.: United States Institute of Peace Press, 1998.

———. "Shark Kills Number 100 Million Annually, Research Says." *BBC News*, March 1, 2013. http://www.bbc.com/news/science-environment-21629173.

McGrath, Matt. "'Indiana Jones' Shark Gains Protection at CITES Meeting," *BBC News*, October 3, 2016 (available at: www.bbc.com/news/science-environment-37547103).

McLean, Joanne and Steffen Straede, "Conservation, Relocation, and the Paradigms of Park and People Management: A Case Study of Padampur Village and the Royal Chitwan National Park, Nepal," *Society and Natural Resources*, 16(6), 2003: 509–26.

McNeeley, Jeffrey, Promila Kapoor-Vijay, Lu Zhi, Linda Olsvig-Whittaker, Kashif Sheikh, and Andrew Smith. "Conservation Biology in Asia: The Major Policy Challenges." *Conservation Biology* 23, no. 4 (July 2009): 805–810.

Meibom von, Stephanie et al, "Saiga Antelope Trade: Global Trends with a Focus on Southeast Asia," TRAFFIC-Europe, 2013 (available at: https://portals.iucn.org/library/efiles/edocs/traf-115.pdf).

Migiro, Katy and Stella Dawson. "Kenya: Chic Nairobi Throbs to the Beat of Dirty Money." *Thompson Reuters Foundation News*. December 10, 2013. http://news.trust.org//item/20131209150854-1kirf/.

Milledge, Simon, Ised Gelvas, Antje Ahrends. "Forestry, Governance and National Development: Lessons Learned from a Logging Boom in Southern Tanzania." Cambridge: TRAFFIC, 2007. http://www.trafficj.org/publication/07_Forestry_Governance_and_national.pdf

Millennium Ecosystem Assessment. Ecosystem and Well Being: Synthesis for Decision-Makers. Washington, D.C.: Island Press, 2006.

Milliken, Tom. "Illegal Trade in Ivory and Rhino Horn: An Assessment Report to Improve Law Enforcement under the Wildlife TRAPS Project." Cambridge: TRAFFIC, 2014. https://www.usaid.gov/sites/default/files/documents/1865/W-TRAPS-Elephant-Rhino-report.pdf.

Milliken, Tom, Robert Burn, Fiona Underwood, and Louisa Sangalakula. "Monitoring of Illegal Trade in Ivory and Other Elephant Specimens." Bangkok: TRAFFIC, 2015. https://www.cites.org/sites/default/files/eng/cop/16/doc/E-CoP16-53-02-02.pdf.

BIBLIOGRAPHY

Milliken, Tom and Jo Shaw. "The South Africa—Viet Nam Rhino Horn Trade: A Deadly Combination of Institutional Lapses, Corrupt Wildlife Industry Professionals and Asian Crime Syndicates." Johannesburg: TRAFFIC, 2012. http://www.npr.org/documents/2013/may/traffic_species_mammals.pdf.

Milner-Gulland, E. J. and Nigel Leader-Williams. "A Model of Incentives for the Illegal Exploitation of Black Rhinos and Elephants: Poaching Pays in Luangwa Valley, Zambia." *Journal of Applied Ecology* 29, no. 2 (1992): 388–401.

Miraglia, Paula. "Drugs and Drug Trafficking in Brazil: Trends and Policies," The Brookings Institution, 2015 www.brookings.edu/~/media/Research/Files/Papers/2015/04/global-drug-policy/Miraglia—Brazil-final.pdf?la=en.

Missios, Paul C. "Helped or Hindered?" In *Endangered Species: Threatened Convention: The Past, Present, and Future of CITES* edited by Jon Hutton and Barnabas Dickson. London: Earthscan, 2000.

Mockrin, Miranda, Elizabeth Bennett, and Danielle La Bruna. "Wildlife Farming: A Viable Alternative to Hunting in Tropical Forests?" Washington, D.C.: Wildlife Conservation Society, 2005. http://www.mtnforum.org/sites/default/files/publication/files/2005_mockrin.pdf.

Monitoring the Illegal Killing of Elephants: Update on Elephant Poaching Trends in Africa to 31 December 2014. Conventional on International Trade in Endangered Species of Wild Fauna and Flora (CITES). 2014. https://cites.org/sites/default/files/i/news/2015/WWD-PR-Annex_MIKE_trend_update_2014_new.pdf.

Morales, Zia. "Fighting Wildlife Crime to End Extreme Poverty and Boost Shared Prosperity." *UN Chronicle* 51, no. 2 (September 2014): 23–24.

Moreau, Ron and Sami Yousafzai. "Flowers of Destruction." *Newsweek* 142, no. 2 (July 14, 2003): 33.

Moseley, William "African Evidence on Relation of Poverty, Time Preference, and the Environment," *Ecological Economies*, 38(3), 2001: 317–26.

Moyle, Brendan. "Regulation, Conservation, and Incentives." In *The Trade in Wildlife: Regulation for Conservation* edited by Sara Oldfield. London: Earthscan, 2003: 41–51.

———. "The Black Market in China for Tiger Products." *Global Crime* 10, nos. 1/2 (February 2009): 124–143.

Murombedzi, James. "Devolution and Stewardship in Zimbabwe's CAMPFIRE Programme." *Journal of International Development* 11, no. 2 (April 1999): 287–93.

Murphree, Marshall. "Strategic Pillars of Communal Natural Resource Management: Benefit, Empowerment, and Conservation." *Biodiversity and Conservation* 18, no. 10 (September 2009): 2551–2562.

BIBLIOGRAPHY

Mushava, Everson "Lions, Elephants Used to Kick out Manzou Villagers," *Zimbabwe Standard*, January 18, 2015.

Musumali, Musole, Thor Larsen, and Bjorn Kaltenborn. "An Impasse in Community Based Natural Resource Management in Implementation: The Case of Zambia and Botswana." *Oryx* 41, no. 3 (July 2007): 306–13.

Mwangi, Esther. "The Puzzle of Group Ranch Subdivision in Kenya's Maasailand," *Development and Change*, 38(5), 2007: 889–910.

Nagendra, Harini, Mukunda Karmacharya, and Birendra Karna. "Evaluating Forest Management in Nepal: Views across Space and Time." *Ecology and Society* 10, (2004).

Nash, Stephen. "Sold for a Song: The Trade in Southeast Asian non-CITES Birds." Cambridge: TRAFFIC, 1994. www.traffic.org/species-reports/traffic_species_birds5.pdf.

Nasi, Robert, Andrew Taber, and Nathalie van Vliet. "Empty Forests, Empty Stomachs? Bushmeat and Livelihoods in the Congo and Amazon Basins." *International Forestry Review* 13, no. 3 (January 2011): 355–68.

Nasi, Robert and Tony Cunningham. "Sustainable Management of Non-Timber Forest Resources: A Review with Recommendations for the SBSTTA." Montreal: Secretariat to the Convention on Biological Diversity. 2001.

National Geographic and GlobalScan. "Reducing Demand for Ivory: An International Study," August 2015, http://press.nationalgeographic.com/files/2015/09/NGS2015_Final-August-11-RGB.pdf.

Naylor, Robin Thomas. "The Underworld of Ivory." *Crime, Law, and Social Change* 42, nos. 4/5 (January 2005): 261–95.

Nelson, Robert. "Environmental Colonialism: "Saving" Africa from Africans." *Independent Review* 8, no. 1 (2003): 65–87.

Neme, Laurel. "For Rangers on the Front Lines of Anti-Poaching Wars, Daily Trauma." *National Geographic*, June 27, 2014. http://news.nationalgeographic.com/news/2014/06/140627-congo-virunga-wildlife-rangers-elephants-rhinos-poaching/

"Nepal Winning Battle against One-horned Rhino Poachers," *New York Daily News*, May 11, 2012.

Neumann, Roderick. "Africa's 'Last Wilderness': Reordering Space for Political and Economic Control in Tanzania." *Africa* 71, no. 4 (November 2001): 641–65.

Neumann, Roderick. "Moral and Discursive Geographies in the War for Biodiversity in Africa," *Political Geography*, 23(7), 2004: 813–37.

Newmark, William and John Hough, "Conserving Wildlife in Africa: Integrated Conservation and Development Projects and Beyond," *BioScience*, 50(7), 2000: 585–92.

Ng'wanakilala, Fumbuka. "Tanzania Court Jails Two Chinese Men for Ivory

Smuggling: Media." *Reuters*. March 19, 2016. http://www.reuters.com/article/us-tanzania-poaching-idUSKCN0WL0NK.

Nijman, Vincent. "An Overview of International Wildlife Trade from Southeast Asia." *Biodiversity Conservation* 19 no. 4 (April 2010): 1101–1114.

Nijman, Vincent and Chris Shepherd. "Adding Up the Numbers: An Investigation into Commercial Breeding of Tokay Geckos in Indonesia." *TRAFFIC*. October 2015. www.traffic.org/species-reports/traffic_species_reptiles47.pdf.

Nixon, Ron. "Obama Administration Plans to Aggressively Target Wildlife Trafficking." *New York Times*. February 11, 2015. http://www.nytimes.com/2015/02/12/us/politics/obama-administration-to-target-illegal-wildlife-trafficking.html?_r=0.

Nooren, Hanneke and Gordon Claridge. *Wildlife Trade in Laos: The End of the Game*. Amsterdam: The Netherlands Committee for IUCN, 2001.

Norgrove, Linda and David Hulme. "Confronting Conservation at Mount Elgon, Uganda." *Development and Change* 37, no. 5 (September 2006): 1093–1116.

Norton-Griffiths, Michael. "How Many Wildebeest Do You Need?" *World Economics* 8, no. 2 (June 2007): 41–64.

Norton-Griffiths, Michael and Clive Southey. "The Opportunity Costs of Biodiversity Conservation in Kenya." *Ecological Economics* 12, no. 2 (February 1995): 125–139. http://mng5.com/papers/BiodivCosts.pdf.

Nossiter, Adam, "U.S. Sting That Snared Guinea-Bissau Ex-Admiral Shines Light on Drug Trade." *New York Times*. April 16, 2013. http://www.nytimes.com/2013/04/16/world/africa/us-sting-that-snared-guinea-bissau-ex-admiral-shines-light-on-drug-trade.html.

Nowell, Kristin. "Asian Big Cat Conservation and Trade Control in Selected Range States: Evaluating Implementation and Effectiveness of CITES Recommendations." *TRAFFIC*, 2007. www.felidae.org/KNOWELLPUBL/abc_report.pdf.

Nowell, Kristin. "Species Trade and Conservation—Rhinoceroses: Assessment of Rhino Horn as a Traditional Medicine," TRAFFIC, 2012 (available at: https://cites.org/sites/default/files/eng/com/sc/62/E62-47-02-A.pdf)

Nowell, Kristin and Xu Ling. "Taming the Tiger Trade: China's Markets for Wild and Captive Tiger Products since the 1993 Domestic Trade Ban." *TRAFFIC* 2007. http://www.traffic.org/species-reports/traffic_species_mammals16.pdf

Nuwer, Rachel. "It's Official: Vietnam's Javan Rhino Is Extinct. Which Species Is Next?" *Take Part*. January 2, 2013. http://www.takepart.com/article/2012/12/12/its-official-vietnams-javan-rhino-extinct-and-other-species-will-likely-follow

———. "In Vietnam, Rampant Wildlife Smuggling Prompts Little Concern." *New York Times*. March 30, 2015. http://www.nytimes.com/2015/03/31/science/in-vietnam-rampant-wildlife-smuggling-prompts-little-concern.html.

Nzou, Goodwell. "In Zimbabwe, We Don't Cry for Lions," *New York Times*, August 4, 2015.

Oates, John. "The Dangers of Conservation by Rural Development: A Case Study from the Forests of Nigeria." *Oryx* 29, no. 2 (April 1995): 115–122.

Oborne, Peter and Lucy Edwards. "A Victory for Pushers." *Spectator*, May 31, 2003.

O'Connell, Sandija. "Crocodile Farms: Is It Cruel to Keep These Wild Creatures Captive?" *Independent*. October 4, 2006.

O'Criodain, Colman. "CITES and Community-Based Conservation: Where Do We Go from Here?" In *CITES and CBNRM Proceedings of an International Symposium on the Relevance of CBNRM to the Conservation and Sustainable Use of CITES-listed Species in Exporting Countries, Vienna, May 18–20, 2011*, edited by Max Abensperg-Traun, Dilys Roe, and Colman O'Criodain. London: International Institute for Environment and Development, 2011. http://pubs.iied.org/pdfs/14616IIED.pdf.

Oldekop, Johan, John Homes, Edwin Harris, Karl Evans. "A Global Assessment of the Social and Conservation Outcomes of Protected Areas." *Conservation Biology* 30, no. 1 (February 2016): 133–41.

Olick, Felix. "Govt Hunts for Cash Launderers." *Star*, December 23, 2015. www.allafrica.com/stories/201512230155.html.

Olson, Mancur. *The Logic of Collective Action*. Cambridge: University Press, 1965.

Orenstein, Ronald. *Ivory, Horn, and Blood: Behind the Elephant and Rhinoceros Poaching Crisis*. New York: Firefly Books, 2013.

Ostrom, Elinor. "Beyond Markets and States: Polycentric Governance of Complex Economic Systems." *American Economic Review* 100, no. 3 (June 2010): 641–72.

Ostrom, Elinor. *Governing the Commons: The Evolution of Institutions for Collective Action*. Cambridge University Press, 1990.

"Out of the Shadows: The Origin of SARS." *Economist*. November 1, 2013. http://www.economist.com/blogs/babbage/2013/11/origin-sars.

Packer, Craig. *Lions in the Balance: Man-Eaters, Manes, and Men with Guns*. Chicago: University of Chicago Press, 2015.

Pain, Adam. "Opium Trading Systems in Helmand and Ghor." Kabul: Afghanistan Research and Evaluation Unit, 2006. http://www.yumpu.com/en/document/view/42972723/opium-trading-systems-in-helmand-and-ghor-the-afghanistan-.

BIBLIOGRAPHY

Paine, Robert T. "A Conversation on Refining the Concept of Keystone Species." *Conservation Biology* 9, no. 4 (August 1995): 962–4.

Panday, Raju, Dinesh Jha, Nirajan Thapa, Basant Pokharel and Nanda Aryal. "Forensic Wildlife Parts and Product Identification and Individualization Using DNA Barcoding." *The Open Forensic Science Journal* 7, (2014): 6–13.

Paoli, Letizia, Virginia Greenfield, Molly Charles, and Peter Reuter. "The Global Diversion of Pharmaceutical Drugs." *Addiction* 104, no. 3 (February 2009): 347–54.

"Pardhi Tribe Termed the Biggest Threat to Wildlife." *Express India*. January 7, 2008.

Pearce, Fred, "Big Game Losers," *New Scientist*, 187(2512), 2005: 21.

———. "Laird of Africa," *New Scientist*, 186(2495), 2005: 48–50;

———. "Rhino Rescue Plan Asian Antelopes." *New Scientist*, February 12, 2003. https://www.newscientist.com/article/dn3376-rhino-rescue-plan-decimates-asian-antelopes/

Peluso, Nancy Lee and Peter Vandergeest. "Political Ecologies of War and Forests: Counterinsurgency and the Making of National Natures." *Annals of the Association of American Geographers* 101, no. 3 (March 2011): 587–608.

Peres, Carlos and Barbara Zimmerman, "Perils in the Parks or Parks in Peril? Reconciling Conservation in Amazonian Reserves with and without Use," *Conservation Biology*, 15(3), 2001: 793–97.

"Peru Police Seize Thousands of Dried Seahorses." *BBC News*. August 24, 2012. http://www.bbc.com/news/world-latin-america-19364702.

Piao, Vanessa and Cherie Chan. "Beijing Destroys Confiscated Ivory and Vows to End Trade." *New York Times*. May 29, 2015. http://sinosphere.blogs.nytimes.com/2015/05/29/beijing-destroys-confiscated-ivory-in-effort-to-curb-illegal-trade/.

Pieth, Mark. "Criminalizing the Financing of Terrorism." *Journal of International Criminal Justice* 4, no. 5 (2006): 1074–86.

Pimbert, Michel. "SwedBio and Sida strategic workshop: The role of biodiversity for ecosystem services and its importance for poor people and local livelihoods in developing countries—priorities for the future." In *The Role of Biodiversity for Ecosystem Services and its Importance for People and Local Livelihoods in Developing Countries: Priorities for the Future* edited by Swedish International Biodiversity Program and Swedish Biodiversity Centre. Uppsala: Swedish International Biodiversity Programme, 2003. http://citeseerx.ist.psu.edu/viewdoc/download?rep=rep1&type=pdf&doi=10.1.1.136.5346

Pimentel, David, Lori Lach, Rodolfo Zuniga, and Doug Morrison. "Environmental and Economic Costs of Nonindigenous Species in the United States." *BioScience* 50, no. 1(January 2000): 53–65.

"Poachers Using Science Papers to Target Newly Discovered Species," *The

BIBLIOGRAPHY

Guardian. January 1, 2016. http://www.theguardian.com/environment/2016/jan/01/poachers-using-science-papers-to-target-newly-discovered-species.

"Police Suspect Killing of 10,000 Tibetan Antelopes." *Xinhua-English.* February 21, 2013. http://www.news.xinhuanet.com/english/world/2013-02/21/c_132183480.htm

Posey, Darrell and Graham Dutfield. "Beyond Intellectual Property Rights: Toward Traditional Resource Rights for Indigenous Peoples and Local Communities." Ottwawa: International Development Research Centre, 1996.

Powis, Robert. *The Money Launderers.* Chicago: Probus, 1992.

Pramod, Ganapathiraju, Katrina Nakamura, Tony Pitcher, and Leslie Delagran. "Estimates of Illegal and Unreported Fish in Seafood Imports to the USA." *Marine Policy* 48, no. C (September 2014): 102–113.

Prescott-Allen, Robert, and Christine Prescott-Allen. *What's Wildlife Worth? Economic Contribution of Wild Plants and Animals to Developing Countries.* London: Earthscan, 1982.

Qiang, Dai, Wang Yuezhao and Liang Gang. "Conservation Status of Chinese Giant Salamander (Andrias davidianus)." Translated by Wang Yi. Beijing: Chendu Institute of Biology, 2009. http://www.cepf.net/Documents/final_CIBCAS_giantsalamander_china.pdf.

Qing, Koh Gui. "Exclusive: Risk Ranking: China Revamps Anti-Money Laundering Rules—Sources." *Reuters.* April 17, 2013. http://www.reuters.com/article/us-china-laundering-risk-ratings-idUSBRE93G15A 20130417.

Qing, Yang, Chen Jin, Bai Zhi-Lin, Deng Xiao-Bao, and Liu Zhi-Qiu. "Trade of Wild Animals and Plants in China-Laos Border Areas Status and Suggestion for Effective Management." *Biodiversity Science* 8, no. 3 (2000): 284–256.

Rabinovich, Jorge. "Parrots, Precaution and Project Ele: Management in the Face of Multiple Uncertainties." In *Biodiversity and the Precautionary Principle: Risk and Uncertainty in Conservation and Sustainable Use* edited by Rosie Cooney and Barney Dickson, London: Earthscan, 2005: 173–188.

Rademeyer, Julian. *Killing for Profit.* Cape Town: Zebra Press, 2012.

Ramutsindela, Maano. *Parks and People in Postcolonial Societies: Experiences in Southern Africa.* New York: Kluwer, 2006.

Ramzy, Austin. "Officials Accused of Dining on Endangered Salamander." *New York Times.* January 26, 2015. http://sinosphere.blogs.nytimes.com/2015/01/26/officials-accused-of-dining-on-endangered-salamander/?_r=0.

Ranger, Terence. *Voices from the Rocks: Nature, Culture and History in the Matopos Hills of Zimbabwe.* Oxford: James Currey, 1999.

Ratchford, Marina, Peter LaFontaine, Anya Rushing, and Beth Allgood.

BIBLIOGRAPHY

"Bidding against Survival: The Elephant Poaching Crisis and the Role of Auctions in the U.S. Ivory Market." *International Fund of Animal Welfare*. August 2014. http://www.ifaw.org/sites/default/files/IFAW-Ivory-Auctions-bidding-against-survival-aug-2014_0.pdf

Rao, Madhu, Than Myint, Than Zaw, and Saw Htun. "Hunting Patterns in Tropical Forests Adjoining the Hkakaborazi National Park, North Myanmar." *Oryx* 39, no. 3 (2005): 292–300.

Redford, Kent and Steven Sanderson, "Extracting Humans from Nature," *Conservation Biology*, 14(5), 2000: 1362–64.

Reeve, Roz and Stephen Ellis. "An Insider's Account of the South African Security Force's Role in the Ivory Trade." *Journal of Contemporary African Studies* 13, no. 2 (June 1995): 222–43.

Redford, Kent et al., "Linking Conservation and Poverty Alleviation: Discussion Paper and Good and Best Practice in the Case of Great Ape Conservation," Poverty and Conservation Learning Group Discussion Paper No. 11, International Institute for Environment and Development, 2013 (available at: http://pubs.iied.org/pdfs/G03714.pdf).

Reid, Hannah and Stephen Turner, "The Richtersveld and Makuleke Contractual Parks in South Africa: Win–Win for Communities and Conservation?" in Christo Fabricius et al. (eds.), *Rights, Resources, and Rural Development: Community-Based Natural Resource Management in Southern Africa* (London: Earthscan, 2004): 223–34.

Reilly, Brian. "Game Ranching in South Africa: An Alternative Model Mixing Economics and Conservation." *The Wildlife Professional*, 8, no. 4 (2014): 36–41. http://sustainability.colostate.edu/sites/sustainability.colostate.edu/files/Final%20Private%20Lands%20Package%20TWS%20Summer%202014.pdf.

Reina, Mauricio. "Drug Trafficking and the National Economy." In *Violence in Colombia 1990–2000: Waging War and Negotiating Peace* edited by Charles Berquist, Ricardo Peñaranda, and Gonzalo Sánchez G. Wilmington: A Scholarly Resources Inc. Imprint, 2001.

Renard, Ronald. *Opium Reduction in Thailand, 1970–2000: A Thirty-Year Journey*. Bangkok: UNDCP Silkworm Books, 2001.

Reuter, Peter. *Disorganized Crime: Economics of the Visible Hand*. Cambridge: MIT Press, 1983.

———. "The Continued Vitality of Mythical Numbers." *National Affairs* 75, (1984): 135–47.

———. "The Mismeasurement of Illegal Drug Markets: The Implications of Its Irrelevance." In *Exploring the Underground Economy* edited by Susan Pozo, Kalamazoo: W.E. Upjohn Institute, 1996, 63–80.

Reuter, Peter and Edwin M. Truman. *Chasing Dirty Money*. Washington, D.C.: Institute for International Economics, 2004.

"Rhino Death Toll at 80," *The Namibian*, January 7, 2016.

BIBLIOGRAPHY

Rihoy, Elizabeth, ed. "The Commons Without the Tragedy? Strategies for Community Based Natural Resource Management in Southern Africa: Proceedings of the Regional Natural Resources Management Programme Annual Conference, Kasane, Botswana, April 3–6 1995." Lilongwe: SADC Wildlife Technical Coordination Unit, 1995.

Rivalan, Philippe, Virginie Delmas, Elena Angulo, Leigh Bull, Richard Hall, Franck Courchamp, Alison Rosser, and Nigel Leader-Williams. "Can Bans Stimulate Wildlife Trade?" *Nature* 477, no. 7144 (2007): 529–30.

Robbins, Paul. "The Rotten Institution: Corruption in the Natural Resource Management." *Political Geography* 19, no. 4 (May 2000): 423–43.

Robbins, Paul, Kendra McSweeney, Anil Kumar Chhangani, and Jennifer Rice. "Conservation as it is: Illicit Resource Use in a Wildlife Reserve in India." *Human Ecology* 37, (2009): 356–82.

Robins, Steven and Kees van der Waal. "'Model Tribes' and Iconic Conservationists? The Makuleke Restitution Case in Kruger National Park." *Development and Change* 39, no. 1 (April 2008): 53–72.

Robinson, John and Elizabeth Bennett, eds. *Hunting for Sustainability in Tropical Forests*. New York: Columbia University Press, 2000.

———. "Will Alleviating Poverty Solve the Bushmeat Crisis?" *Oryx* 36, no. 4 (October 2002): 332.

Robinson, John and Kent Redford, eds., *Neotropical Wildlife Use and Conservation*. Chicago: Chicago University Press, 1991.

Rocky Mountains High Intensity Drug Trafficking Area 3, "The Legalization of Marijuana in Colorado: The Impact," Vol. 2, 2014 www.rmhidta.org/html/august%202014%20legalization%20 of%20mj%20in%20colorado%20the%20impact.pdf).

Roe, Dilys et al. (eds.), "Community Management of Natural Resources in Africa," International Institute for the Environment and Development, '2009 http://pubs.iied.org/pdfs/17503IIED.pdf)

Roe, Dilys, Man Fancourt, Chris Sandbrook, Mxolisi Sibanda, Alessandra Giuliani, and Andrew Gordon-Maclean. "Which Components or Attributes of Biodiversity Influence Which Dimensions of Poverty?" *Environmental Evidence* 3, no. 3 (February 2014): 1–15.

Roe, Dilys, Teresa Mulliken, Simon Milledge, Josephine Mremi, Simon Mosha, and Maryanne Grieg-Gran. "Making a Killing or Making a Living? Wildlife Trade, Trade Controls, and Rural Livelihoods." *Biodiversity and Livelihoods* 6, (March 2002). file:///C:/Users/vfelb/AppData/Local/Temp/traffic_pub_trade4.pdf.

Roelf, Wendell, "South Africa's Scorpion Crime Fighters to Be Disbanded," *Reuters*, February 12, 2008 (www.reuters.com/article/us-safrica-scorpions-idUSL1223 114320080212).

Romero, Simon. "Colombia Lists Civilian Killings in Guerrilla Toll." *New York*

BIBLIOGRAPHY

Times. October 29, 2008. http://www.nytimes.com/2008/10/30/world/americas/30colombia.html?mtrref=www.bing.com&gwh=64C03A7BDBF1983DFB098E9BF503DB01&gwt=pay.

Ross, James Perran, ed. "Crocodiles: Status Survey and Conservation Action Plan." Gland: International Union of Nature, 1998. http://iucncsg.org/ph1/modules/Publications/action_plan1998/plan1998a.htm.

Rubio, Mauricio. "Violence, Organized Crime, and the Criminal Justice System in Colombia." *Journal of Economic Issues* 32, no. 2: 605–610.

Rujivanarom, Pratch. "Villagers Fight a Losing Battle in 2015," The Nation (Thailand), December 28, 2015.

Rushby, George. *No More the Tusker*. London: W.H. Allen, 1965.

Saint Louis, Catherine. "New Challenge for Police: Finding Pot in Lollipops and Marshmallows." *New York Times*. May 16, 2015. http://www.nytimes.com/2015/05/17/us/new-challenge-for-police-finding-pot-in-lollipops-and-marshmallows.html?mwrsm=Email&_r=0.

Sakamoto, Masayuki. "Black and Grey: Illegal Ivory in Japanese Markets." Japan Wildlife Conservation Society, 2004. http://www.ifaw.org/sites/default/files/Black%20and%20Grey%20Illegal%20ivory%20in%20Japan%20Markets%20-%202004.pdf.

Salisbury, Claire. "Communities and Cutting-Edge Tech Keep Cambodia's Gibbons Singing." *Mongabay*. February 22, 2016. https://www.google.com/?gws_rd=ssl#q=Claire+Salisbury%2C+%E2%80%9CCommunities+and+Cutting-Edge+Tech+Keep+Cambodia%E2%80%99s+Gibbons+Singing%2C%E2%80%9D+Mongabay.com%2C+.

Samii, Cyrus et al., "Effects of Payment for Environmental Services (PES) on Deforestation and Poverty in Low and Middle Income Countries: A Systematic Review CEE 3–015b," Collaboration for Environmental Evidence, 2014 (available at: www.environmentalevidence.org/wp-content/uploads/2015/01/Samii_PES_Review-formatted-for-CEE.pdf).

Sample, Ian and John Gittings. "In China, The Civet Is a Delicacy—and May Have Caused SARS." *Guardian*. May 23, 2003. https://www.theguardian.com/world/2003/may/24/china.sars.

"Sansar Chand, Notorious Tiger Poacher, Dead," Times of India, March 19, 2014 (www.timesofindia.indiatimes.com/city/jaipur/Sansar-Chand-notorious-tiger-poacher-dead/articleshow/32261903.cms).

Saunders, Jade and Jens Hein. "EUTR, CITES and Money Laundering: A Case Study on the Challenges of Coordinated Enforcement in Tackling Illegal Logging." London: Chatham House, 2015. http://efface.eu/sites/default/files/EFFACE_EUTR%20CITES%20and%20money%20laundering%20A%20case%20study%20on%20the%20challenges%20to%20coordinated%20enforcement%20in%20tacking%20illegal%20logging.pdf.

Schaller, George. *Tibet Wild: A Naturalist's Journeys on the Roof of the World*. Washington, D.C.: Island Press, 2012.

BIBLIOGRAPHY

Schama, Simon. *Landscape and Memory*. London: Fontana Press, 1996.

Scherer, Steve. "Mafia Thrives on Italy's Legalized Gambling Addiction." *Reuters*. March 11, 2015. http://www.reuters.com/article/usitaly-mafia-slots idUSKBN0M720R20150312.

Schmidt-Soltau, Kai. "Conservation led Resettlement in Central Africa: Environmental and Social Risks." *Development and Change* 34, no. 3. (2003) 536.

Schmidt-Soltau, Kai and Daniel Brockington, "Protected Areas and Resettlement: What Scope for Voluntary Relocation." *World Development* 35, no. 12 (November 2007): 2182–202.

Schneider, Jacqueline. *Sold Into Extinction: The Global Trade in Endangered Species*. Santa Barbara: Praeger, 2012.

Schoppe, Sabine. *Status, Trade Dynamics, and Management of the Southeast Asian Box Turtle in Indonesia*. Kuala Lumpur: TRAFFIC, 2009. http://www.traffic.org/species-reports/traffic_species_reptiles19.pdf.

Scoones, Ian. "Why Cecil the Lion Offers Lessons for Land Reform and the Role of Elites," *The Conversation*, August 5, 2015.

Sellar, John. "Policing the Trafficking of Wildlife: Is There Anything to Be Learned from Law Enforcement Responses to Drug and Firearm Trafficking?" *Global Initiative Against Organized Crime*. February 2014. http://www.globalinitiative.net/download/global-initiative/Global%20Initiative%20%20Wildlife%20Trafficking%20Law%20Enforcement%20-%20Feb%202014.pdf.

"Secretary of Interior: U.S. Supports Sustainable Trophy Hunts." *The Associated Press*, January 19, 2016. htttp://www.yahoo.com/news/secretaryinterior-us-supports-sustainable-trophy-hunts-215748549.html?ref=gs

"Seeing the Wood." *Economist*. September 25, 2010: 13.

Sheikh, Pervaze A. "Illegal Logging: Background and Issues." Washington D.C.: Congressional Research Service, 2008. http://nationalaglawcenter.org/wp-content/uploads/assets/crs/RL33932.pdf.

Shepard, Edward and Paul Blackley. "Medical Marijuana and Crime: Further Evidence from the Western States." *Journal of Drug Issues*, 46, no. 2 (April 2016): 122–34.

Shepherd, Chris. "Export of Live Freshwater Turtles and Tortoises from Northern Sumatra and Riau, Indonesia: A Case Study." In *Asian Turtle Trade: Proceedings of a Workshop on Conservation and Trade of Freshwater Turtles and Tortoises in Asia* edited by Peter Paul Van Dijk, Bryan L. Stuart, and Andres G.J. Rhodin. Chelonian Research Monographs No. 2, 2000.

———. "The Bird Trade in Medan, North Sumatra: An Overview." *Birding ASIA* 5, (2006): 16–24.

Shepherd, Chris, and Nolan Magnus. "Nowhere to Hide: The Trade in Sumatran Tiger." *TRAFFIC*. 2004. www.traffic.org/species-reports/traffic_species_mammals15.pdf.

BIBLIOGRAPHY

Shepherd, Chris and Vincent Nijman. "The Trade in Bear Parts from Myanmar: An Illustration of the Ineffectiveness of Enforcement of International Wildlife Trade Regulations." *Biodiversity Conservation* 17, no. 1 (2008): 35–42.

———. "The Wild Cat Trade in Myanmar." *TRAFFIC*. 2008. www.traffic.org/species-reports/traffic_species_mammals40.pdf.

Shova, Thapa, and Klaus Hubacek. "Drivers of Illegal Resource Extraction: An Analysis of Bardia National Park, Nepal." *Journal of Environmental Management* 92, no. 1 (January 2011): 156–64.

Simon, Dan. "Mexican Cartels Running Pot Farms in U.S. National Forest." *CNN*. August 8, 2008. http://www.cnn.com/2008/CRIME/08/08/pot.eradication/.

Sina, Stephen et al., "Wildlife Crime," Study for the ENVI Committee, Directorate-General for Internal Policies, Policy Department A: Economic and Scientific Policy, European Parliament, 2016 (www.europarl.europa.eu/RegData/etudes/STUD/2016/570008/IPOL_STU(2016)570008_EN.pdf): 9.

Smith, David. "Elephant Deaths Rise in Tanzania After Shoot-to-Kill Poachers Policy is Dropped." *Guardian*. December 31, 2013. https://www.theguardian.com/world/2013/dec/31/elephant-deaths-rise-tanzania-shoot-to-kill-poachers.

Smith, Robert, David Roberts, Rosaleen Duffy, and Freya St John. "New Rhino Conservation Project in South Africa to Understand Landowner Decision-Making." *Oryx* 47, no. 3 (September 2013): 323.

Smythe, Christie. "HSBC Judge Approves $1.9B Drug-Money Laundering Accord." *Bloomberg*. July 3, 2012. http://www.bloomberg.com/news/articles/2013-07-02/hsbc-judge-approves-1-9b-drug-money-laundering-accord.

Sollund, Ragnhild and Jennifer Maher. "The Illegal Wildlife Trade: A Case Study Report on the Illegal Wildlife Trade in the United Kingdom, Norway, Colombia and Brazil." Oslo: University of Oslo, 2015. http://efface.eu/sites/default/files/EFFACE_Illegal%20Wildlife%20Trade_revised.pdf.

"Somalia Famine 'Killed 260,000 People'." *BBC News*, May 2, 2013 (available at: www.bbc.com/news/world-africa-22380352).

Song, Nguyen Van. "Tracking the Trade: Vietnam's Illegal Wildlife Business." Hanoi: Hanoi Agricultural University, 2003. http://www.eepsea.org/[ub/pb/132573.pdf

"South African Group Reports Slight Drop in Rhino Poaching." *Associated Press*. January 2, 2016. http://bigstory.ap.org/article/fadef7f9221c48908636ac0dc731e87f/south-african-group-reports-slight-drop-rhino-poaching.

"South African Judge Lifts Domestic Ban on Rhino Horn Trade." *Guardian*. November 26, 2015. http://www.theguardian.com/environment/2015/nov/26/south-african-judge-lifts-domestic-ban-on-rhino-horn-trade.

BIBLIOGRAPHY

South, Nigel and Tanya Wyatt. "Comparing Illicit Trades in Wildlife and Drugs: An Exploratory Study." *Deviant Behavior* 32, no. 6 (June 2011): 538–561.

Spinage, Clive "Social Change and Conservation Misrepresentation in Africa." *Oryx* 32, no. 4 (October 1998): 265–76.

Starkey, Malcolm Paul. "Commerce and Subsistence: The Hunting, Sale and Consumption of Bushmeat in Gabon." PhD diss. Cambridge University, 2004. https://www.repository.cam.ac.uk/handle/1810/251940.

Stronza, Amanda. "The Economic Promise of Ecotourism for Conservation." *Journal of Ecotourism* 6, no. 3 (March 2009): 210–30.

Srivastava, Mihir. "Tracking the Tiger Killers." *India Today*. May 28, 2010. http://indiatoday.intoday.in/story/tracking-the-tiger-killers/1/99228.html.

Steele, Paul "Ecotourism: An Economic Analysis," *Journal of Sustainable Tourism*, 3(1), 1995: 29–44.

Steenkamp, Conrad and Jana Uhr. "The Makuleke Land Claim: Power Relations and Community-Based Natural Resource Management." London: International Institute for Environment and Development, 2000, 2 http://pubs.iied.org/pdfs/7816IIED.pdf.

Steinberg, Jonny. "The Illicit Abalone Trade in South Africa." Pretoria: Institute for Security Studies, 2005. http://www.dlist.org/sites/default/files/doclib/Abalone%20Trade%20in%20South%20Africa.pdf.

Stiles, Daniel. "Elephant Ivory Trafficking in California, USA." Natural Resources Defense Council, 2015 http://docs.nrdc.org/wildlife/files/wil_15010601a.pdf.

———. "The Ivory Trade and Elephant Conservation." *Environmental Conservation* 31, no. 4 (2004): 309–321.

———. "Elephant and Ivory Trade in Thailand," TRAFFIC, Petaling Jaya, Selangor, Malaysia, 2009 (available at: www.traffic.org/species-reports/traffic_species_mammals50.pdf).

Stiles, Daniel, Ian Redmond, Doug Cress, Christian Nelleman, and Rannveig Knutsdatter Formo. "Stolen Apes: The Illicit Trade in Chimpanzees, Bonobos, Gorillas, and Orangutans." UNEP 2013. https://cld.bz/book-data/KY3u76i/basic-html/page-1.html.

Stoner, Sarah, and Natalia Pervushina. "Reduced to Skin and Bones Revisited: An Updated Analysis of Tiger Seizures from 12 Tiger Range Countries (2000–2012)." *TRAFFIC*, 2013. http://portals.iucn.org/library/sites/library/files/documents/Trf-137.pdf.

Strang, John, Thomas Babor, Jonathan Caulkins, Benedikt Fischer, David Foxcroft, and Keith Humphreys. "Drug Policy and the Public Good: Evidence for Effective Interventions." *Lancet* 379, no. 9810 (January 2012): 71–83.

BIBLIOGRAPHY

Stuart, Bryan, Jady Smith, Kate Davey, Prom Din, and Steven G. Platt. "Homalospine Watersnakes: The Harvest and Trade from Tonle Sap, Cambodia." *TRAFFIC Bulletin* 18, no. 3 (2000): 115–24.

Stueck, Wendy, "Convicted Canadian Narwhal Tusk Smuggler Extradited to US on Money-Laundering Charges," *Globe and Mail*, March 17, 2016.

"Sudan Army Accused of Ivory Trade," *Al Jazeera*. March 14, 2005.

Supplement to the 2009 National Drug Control Strategy, Office of National Drug Control Policy (ONDCP). http://www.whitehousedrugpolicy.gov/publications/policy/ndcs09/ndcs09_data_supl/ds_drg_rltd_tbls.pdf.

Suzman, James. "Etosha Dreams: An Historical Account of the Hai//Om Predicament." *Journal of Modern African Studies* 42, no. 2 (June 2004): 221–38.

Swanson, Timothy. "A Tale of Rent Seeking, Corruption, Stockpiling and (Even) Tragedy: Re-telling the Tale of the Commons." *International Review of Environmental and Resource Economics* 1, no. 1 (2007): 111–150.

Swiderska, Krystyna, with Dilys Roe, Linda Siegele, and Maryanne Grieg-Gran. "The Governance of Nature and the Nature of Governance: Policy that Works Biodiversity and Livelihoods." London: International Institute for Environment and Development, 2008. http://pubs.iied.org/pdfs/14564IIED.pdf.

Sy, Amadou and Amy Copley. "Understanding the Economic Effects of the 2014 Ebola Outbreak in West Africa." *Brookings Institution*. October 1, 2014. http://www.brookings.edu/blogs/africa-in-focus/posts/2014/10/01-ebola-outbreak-west-africa-sy-copley.

"Sympathy for the Misunderstood Shark," *Bloom Hong Kong*, April 20, 2015 (available at: www.bloomassociation.org/en/bloom-hongkong-sympathy-for-the-misunderstood-shark/).

't Sas Rolfes, Michael. "Assessing CITES: Four Case Studies." in *Endangered Species: Threatened Convention: The Past, Present, and Future of CITES* edited by Jon Hutton and Barnabas Dickson. London: Earthscan, 2000: 69–87.

———. "The Rhino Poaching Crisis: A Market Analysis." February 2012. http://www.rhinoresourcecenter.com/pdf_files/133/1331370813.pdf.

Tangley, Laura. "The Sustainable Extraction of Rainforest Products in Guatemala." *US News and World Report* 124, no. 15 (1998): 40–11 and 44.

Taylor, Edward, George Dyer, Micki Stewart, Antonio Yunez-Naude, and Sergio Ardila. "The Economics of Ecotourism: A Galápagos Islands Economy-Wide Perspective." *Economic Development and Cultural Change* 51, no. 4 (July 2003): 978–97.

Taylor, Russell and Marshall Murphree, "Case Studies on Successful Southern African NRM Initiatives and Their Impact on Poverty and Governance: Zimbabwe-Masoka and Gairezi." Washington D.C.: International Resources Group, 2007. https://rmportal.net/library/content/frame/case-studies-on-successful-southern-african-nrm-initiatives-and-their-impacts-on-poverty-and-governance-zimbabwe-masoka-and-gairezi/at_download/file. The link above automatically downloads the document. If you'd prefer to use the landing page with the download link, it's https://rmportal.net/library/content/frame/case-studies-on-successful-southern-african-nrm-initiatives-and-their-impacts-on-poverty-and-governance-zimbabwe-masoka-and-gairezi/view

Tejaswi, Pillenahalli Basavarajappa. "Non-Timber Forest Products (NTFPs) for Food and Livelihood Security: An Economic Study of Tribal Economy in Western Ghats of Karnataka, India." M.Sc. thesis. University of Ghent, 2008, 2. http://ageconsearch.umn.edu/bitstream/54184/2/Thesis_Tejaswi__18th_Aug_2008_final[1].pdf, 2.

Terborgh, John. *Requiem for Nature* (Washington: Island Press, 1999)

Thapa, Surya. "Neither Forests Nor Trees." *Nepali Times*, January 7, 2011.

"The Elephants Fight Back." *The Economist*, November 21, 2015.

The International Consortium on Combatting Wildlife Crime (ICCWC). *Wildlife and Forest Crime Analytic Toolkit*, Revised Edition. UNODC. Vienna 2012. https://www.unodc.org/documents/Wildlife/Toolkit_e.pdf.

"The Trade in Marine Turtle Products in Vietnam." *TRAFFIC*. 2004. www.traffic.org/species-reports/traffic_species_reptiles23.pdf.

Thin Green Line Foundation, "2009–2016: Ranger Roll of Honor, In Memoriam" (www.europarc.org/wp-content/uploads/2016/07/2009-2016-Honour-Roll-1.pdf).

Thompson, Michael, Suzanne Serneels, Dickson Ole Kaelo, and Pippa Trench. "Maasai Mara—Land Privatization and Wildlife Decline: Can Conservation Pay Its Way?" In *Staying Maasai? Livelihoods, Conservation, and Development in East African Rangelands* edited by Kathrine Homewood, Patti Krisjanson, and Pippa Trench. New York: Springer Press, 2009: 77–110

Thorbjarnarson, John and Alvaro Velasco. "Venezuela's Caiman Harvest Program: A Historical Perspective and Analysis of Its Conservation Benefits." Wildlife Washington D.C.: Wildlife Conservation Society, 1998.

Thorson, Erica and Chris Wold. "Back to Basics: An Analysis of the Object and Purposes of CITES and a Blueprint for Implementation." Portland: International Environmental Law Project, 2010. http://www.lclark.edu/live/files/4620.

Thoumi, Francisco. *Illegal Drugs, Economy, and Society in the Andes*. Baltimore: John Hopkins University Press, 2004.

Thoumi, Francisco and Marcela Anzola. "Extra-Legal Economy, Dirty Money, Illegal Capital Inflows and Outflows and Money Laundering in Colombia." 2009.

Tillen, James and Laura Billings. "Anti-Money Laundering 2015: In 23 Jurisdictions Worldwide; Getting the Deal Through." 2015. https://www.amt-law.com/res/news_2015_pdf/150625_4396.pdf.

BIBLIOGRAPHY

Titeca, Kristof. "Central Africa: Ivory Beyond the LRA—Why a Broader Focus Is Needed in Studying Poaching." *AllAfrica*. September 17, 2013. http://allafrica.com/stories/201309170982.html.

Tongson, Edgardo and Marisel Dino. "Indigenous People and Protected Areas: The Case of the Sibuyan Mangyan Tagabukid, Philippines." In *Getting Biodiversity Projects to Work: Towards More Effective Conservation and Development* edited by Thomas McShane and Michael Wells. New York: Columbia University Press, 2004: 181–207.

TRAFFIC, "The State of Wildlife Trade in China: Information on the Trade in Wild Animals and Plants in China 2008," 2010 (www.awsassets.wwfcn.panda.org/downloads/wwf_state_of_wildlife_trade_report_2010____1.pdf): 7.

TRAFFIC. February 2003. www.traffic.org/species-reports/traffic_species_mammals20.pdf

TRAFFIC "What's Driving the Wildlife Trade: A Review of Expert Opinion on Economic and Social Drivers of the Wildlife Trade and Trade Control Efforts in Cambodia, Indonesia, Lao PDR, and Vietnam." Washington, D.C.: World Bank 2008. www.traffic.org/generalreports/traffic_pub_gen24.pdf.

Transparency International. "Global Corruption Index 2015." http://www.transparency.org/cpi2015#results-table.

Trench, Pippa, John Rowley, Marthe Diarra, Fernand Sano, and Boubacar Keita. "Beyond Any Drought: Root Causes of Chronic Vulnerability in the Sahel." Sahel Working Group, 2007. http://www.livestock-emergency.net/userfiles/file/assessment-review/Trench-et-al-2007.pdf.

"Tribespeople Illegally Evicted from 'Jungle Book' Tiger Reserve." *Survival International*. January 14, 2015. http://www.survivalinternational.org/news/10631.

Tumnukasetchai, Piyanut. "AMLO Follows Money Trail to Bring Down Siamese Rosewood Smuggling Cartel." *Nation* (Thailand). July 21, 2014.

Uhm, Daan van. "Illegal Wildlife Trade to the EU and Harms to the World." edited by Toine Spapens, Rob White, and Win Huisman. *Environmental Crime and the World*. London: Routledge, 2016: 43–66.

United Nations Office on Drugs and Crime. "Opium Poppy Cultivation in Southeast Asia." 2008, 1. https://www.unodc.org/documents/crop-monitoring/East_Asia_Opium_report_2008.pdf.

Ulmer, Jeffrey, and Darrell Steffensmeier. "The Age and Crime Relationship: Social Variation, Social Explanations." In *The Nurture Versus Biosocial Debate in Criminology: On the Origins of Criminal Behavior and Criminality* edited by Kevin Beaver, J.C. Barnes, and Brian Boutwell. London: Sage, 2014, 377–397.

Underwood, Fiona, Robert Burn, and Tom Milliken. "Dissecting the Illegal

BIBLIOGRAPHY

Ivory Trade: An Analysis of Ivory Seizure Data." *PLOS One*, October 18, 2013.

United Nations Environmental Programme, "Elephants in the Dust: The African Elephant Crisis," March 2013 (available at: www.unep.org/pdf/RRAivory_draft7.pdf).

United Nations Environmental Program. "Year Book 2014 Emerging Issues Update: Illegal Trade in Wildlife." *United Nations Environmental Program (UNEP)*. 2014. http://www.unep.org/yearbook/2014/PDF/chapt4.pdf.

———. "Republique Democratique du Congo: Evaluation Environnementale Post-Conflit." November 2012. http://postconflict.unep.ch/publications/UNEP_DRC_PCEA_full_FR.pdf.

UNEP-INTERPOL. *The Rise of Environmental Crime*. 2016. http://pfbc-cbfp.org/news_en/items/unep-interpol-enen.html.

United Nations Food and Agriculture Organization (FAO), "The State of World Fisheries and Aquaculture." 2016 (available at: www.fao.org/3/a-i5555e.pdf).

United Nations International Drug Control Program. "The Social and Economic Impact of Drug Abuse and Control." Vienna: UNDCP, 1994, 29.

United Nations Office on Drugs and Crime. "Afghanistan Opium Survey 2013", December 2013, 12. https://www.unodc.org/documents/crop-monitoring/Afghanistan/Afghan_Opium_survey_2013_web_small.pdf.

———. "Opium Amounts to Half of Afghanistan's GDP in 2007, Reports UNODC." November 16, 2007. http://www.unodc.org/india/afghanistan_gdp_report.html.

———. "Transnational Organized Crime in the Fishing Industry." 2011. http://www.unodc.org/documents/human-trafficking/Issue_Paper_-_TOC_in_the_Fishing_Industry.pdf.

———. "World Wildlife Crime Report: Trafficking in Protected Species." Vienna: UNODC, 2016. https://www.unodc.org/documents/data-and-analysis/wildlife/World_Wildlife_Crime_Report_2016_final.pdf.

United States Department of Health and Human Services, "The Health Consequences of Smoking—50 Years of Progress: Report of the Surgeon General," 2014 (available at: www.surgeongeneral.gov/library/reports/50-years-of-progress/sgr50-chap-2.pdf): 15–41.

United Nations Environmental Programme (UNEP), "Elephants in the Dust: The African Elephant Crisis," March 2013 (available at: www.unep.org/pdf/RRAivory_draft7.pdf).

Vallianos, Christina, "Pangolins: On the Brink," (San Francisco: WildAid, 2016).

van der Walle, Nicholas, *African Economies and the Politics of Permanent Political Crisis, 1979–1999* (Cambridge: Cambridge University Press, 2001).

Van Dijk, Peter Paul, Bryan L. Stuart, and Andres G.J. Rhodin. eds. *Asian

BIBLIOGRAPHY

Turtle Trade: Proceedings of a Workshop on Conservation and Trade of Freshwater Turtles and Tortoises in Asia. Chelonian Research Monographs No. 2, 2000.

Varisco, Daniel Martin. "From Rhino Horns to Dagger Handles: A Deadly Business." *Animal Kingdom* 92, no. 3 (June 1989): 44–49.

Veit, Peter and Catherine Benson, "When Parks and People Collide," Carnegie Council for Ethics and International Relations, 2004 (availableat: www.carnegiecouncil.org/publications/archive/dialogue/2_11/section_2/4449.html/:pf_printable): 13–14.

Veríssimo, Diogo, Daniel Challender, and Vincent Nijman. "Wildlife Trade in Asia: Start with the Consumer." *Asian Journal of Conservation Biology* 1, no. 2 (2012): 49–50.

Velásquez Gomar, José Octavio and Lindsay Stringer. "Moving toward Sustainability? An Analysis of CITES' Conservation Policies." *Journal of Environmental Policy* 21, no. 4(August 2011): 240–258.

Vidal, John "WWF Accused of Facilitating Human Rights Abuses of Tribal People in Cameroon," *Guardian*, March 3, 2016.

Vigne, Lucy, Esmond Martin, and Benson Okita-Ouma. "Increased Demand for Rhino Horn in Yemen Threatens Eastern Africa's Rhinos." *Pachyderm* 43, (December 2007): 73–86.

Vigne, Lucy and Esmond Martin. "Increasing Rhino Awareness in Yemen and a Decline in the Rhino Horn Trade." *Pachyderm* 53, (2013): 51–58.

———. "Yemen Strives to End Rhino Horn Trade." *Swara* 16, no. 6 (1993): 20–24.

———. "Yemen's Attitudes toward Rhino Horn and Jambiyas." *Pachyderm* 44, (2008): 45–53.

Vira, Varun, Thomas Ewing, and Jackson Miller. "Out of Africa: Mapping the Global Trade in Illicit Elephant Ivory." *C4ADS*. 2014. http://www.wwf.se/source.php/1578610/out%20of%20africa.pdf.

Vira, Varun and Thomas Ewing. "Ivory's Curse: The Militarization and Professionalization of Poaching in Africa." *C4ADS*. April 2014. http://www.rhinoresourcecenter.com/pdf_files/139/1398477046.pdf?view.

von Meibom, Stephanie, Alexey Vaisman, Neo Liang Song Horng, Julia Ng, and Xu Hongfa. "Saiga Antelope Trade: Global Trends with a Focus on Southeast Asia." *TRAFFIC*, 2013. https://portals.iucn.org/library/efiles/edocs/traf-115.pdf.

Wadley, Jago, Pallavi Shah, and Sam Lawson. "Behind the Veneer: How Indonesia's Last Rainforests Are Being Felled for Flooring." Washington D.C.: Environmental Investigation Agency, 2006. http://www.eia-international.org/cgi/reports/reports.cgi?t=template&a=117.

Wakefeld, Melanie, Barbara Loken, and Robert Hornik. "Use of Mass Media Campaigns to Change Health Behavior." *Lancet* 376, no. 9748 (October 2010): 1261–71.

BIBLIOGRAPHY

Walker, John Frederick. *Ivory Ghosts: The White Gold of History and the Fate of Elephants*. New York: Grove Press, 2010.

Walsh, John and Geoff Ramsey. "Uruguay's Drug Policy: Major Innovations, Major Challenges," The Brookings Institution, April 2015 (available at: http://www.brookings.edu/media/Research/Files/Papers/2015/04/global-drug_policy/Walsh—Uruguay-final.pdf?la=en)

Warchol, Greg. "The Transnational Illegal Wildlife Trade." *Criminal Justice Studies* 17, no. 1 (March 2004): 57–73.

Warchol, Greg, Linda L. Zupan, and Willie Clark. "Transnational Criminality: An Analysis of the Illegal Wildlife Market in Southern Africa." *International Criminal Justice Review* 13, no. 1 (2003): 1–27.

Ward, Christopher and William Byrd. "Afghanistan's Opium Drug Economy." *World Bank Report No. SASPR-5*. Washington, D.C.: World Bank, 2004. http://siteresources.worldbank.org/INTAFGHANISTAN/Publications-Resources/20325060/AFOpium-Drug-Economy-WP.pdf.pdf.

Washington Office on Latin America. "La Comisión Internacional Contra La Impunidad en Guatemala: Un Estudio de Investigación de WOLA Sobre La Experiencia de la CICIG." March 2015. http://www.wola.org/sites/default/files/CICIG%203.25.pdf.

Wassener, Bettina. "Environmental Cost of Shark Finning is Getting Attention in Hong Kong." *New York Times*, June 20, 2010. http://www.nytimes.com/2010/06/21/business/global/21iht-green.html.

Wasser, Rachel M. and Priscilla Bei Jiao. "Understanding Motivations: The First Step toward Influencing China's Unsustainable Wildlife Consumption." *TRAFFIC*. 2010. www.traffic.org/general-reports/traffic_pub_gen33.pdf.

Wasser, Samuel, Lisa Brown, Celia Mailand, Samrat Mondol, William Clark, and Cathy Laurie. "Genetic Assignment of Large Seizures of Elephant Ivory Reveals Africa's Major Poaching Hotspots." *Science* 349, no. 6243 (June 2015): 1–7.

Wasser, Samuel, Joyce Poole, Phyllis Lee, Keith Lindsay, Andrew Dobson, John Hart, Iain Douglas-Hamilton, George Wittemyer, Petter Granli, Bethan Morgan, Jody Gunn, Susan Alberts, Rene Beyers, Patrick Chiyo, Harvey Croze, Richard Estes, Kathleen Gobush, Ponjoli Joram, Alfred Kikoti, Jonathan Kingdon, Lucy King, David Macdonald, Cynthia Moss, Benezeth Mutayoba, Steve Njumbi, Patrick Omondi and Katarzyna Nowak. "Elephants, Ivory, and Trade." *Science* 327, no. 5971 (March 12, 2010): 1331–2.

Weaver, David. "Asian Ecotourism: Patterns and Themes," *Tourism Geographies*, 4(2), 2002: 153–72.

Weaver, Kent. "Getting People to Behave: Research Lessons for Policy Makers." *Public Administration Review* 75 no. 6 (December 2015): 806–816.

BIBLIOGRAPHY

"Weight Watchers." *Economist*. April 13, 2013.

Wells, Michael, Katrina Brandon, and Lee Hannah. "People and Parks: Linking Protected Area Management with Local Communities." Washington, D.C.: World Bank, 1992.

Wells, Miriam. "Don't Forget That Your Weekend Fun is Killing People, Cocaine Users Told." *Vice News*, December 2, 2015. https://news.vice.com/article/dont-forget-that-your-weekend-fun-is-killing-people-cocaine-users-told.

Wells, Susan M. and Jonathan G. Barzdo. "The International Trade in Marine Species: Is CITES a Useful Control Mechanism?" *Coastal Management*, 19, 1, January 1991.

Western, David and Michael Wright (eds), *Natural Connections: Perspectives in Community-Based Conservation* (Washington: Island Press, 1994).

West, Steven and Kerry O'Neal. "Project D.A.R.E. Outcome Effectiveness Revisited." *American Journal of Public Health* 94 (2004): 1027–9.

Western, David "Amboseli National Park: Enlisting Land Owners to Conserve Migrating Wildlife," *Ambio*, 11(5), 1982: 302–8.

Wheeler, Jane, and Domingo Hoces R. "Community Participation, Sustainable Use, and Vicuña Conservation in Peru." *Mountain Research and Development* 17, no. 3 (July 1997): 283–7.

WildAid, "Pangolins: On the Brink," 2016 (available at: www.wildaid.org/sites/default/files/resources/WildAid-Pangolins%20on%20the%20Brink-2016.pdf).

Wildlife Protection Society of India. *Rhino Poaching 2009–2014*. http://www.wpsi-india.org/crime_maps/rhino_poaching.php.

WildAid, "Pangolins: On the Brink," 2016 (available at: www.wildaid.org/sites/default/files/resources/WildAid-Pangolins%20on%20the%20Brink-2016.pdf).

Wildlife Protection Society of India. Elephant Poaching 2009–2014. http://www.wpsi-india.org/crime_maps/elephant_poaching.php.

Wildlife Protection Society of India. "WPSI's Tiger Poaching Statistics." http://www.wpsi-india.org/statistics/index.php.

"WII to Start Tiger Census This Month." *Times of India*, October 5, 2009. http://timesofindia.indiatimes.com/city/lucknow/WII-to-start-tiger-census-this-month/articleshow/5087896.cms.

White, Natascha. "The 'White Gold of Jihad': Violence, Legitimisation and Contestation in Anti-Poaching Strategies." *Journal of Political Ecology* 21 (2014): 452–474. http://jpe.library.arizona.edu/volume_21/White.pdf.

White House, "National Strategy for Combating Wildlife Trafficking" (available at: www.whitehouse.gov/sites/default/files/docs/nationalstrategy-wildlifetrafficking.pdf).

"Whoops Apocalypse," *Economist*. February 27, 2016. http://www.economist.com/news/finance-and-economics/21693603-american-regulators-wield-big-stick-not-always-fairly-whoops-apocalypse.

Wilkie, David S., Gilda A. Morelli, Josefien Demmer, Malcolm Starkey, Paul Telfer, and Matthew Steil. "Parks and People: Assessing Human Welfare Effects of Establishing Protected Areas for Biodiversity Conservation." *Conservation Biology* 20, no. 1 (February 2006): 247–9.

Wilkie, David, and Julia Carpenter. "Bushmeat Hunting in the Congo Basin: An Assessment of Impacts and Options for Mitigation." *Biodiversity and Conservation* 8, no. 7 (July 1999): 927–955.

Wilkie, David, John Sidle, Georges Boundzanga, Phillippe Auzel, and Stephen Blake. "Defaunation, not Deforestation: Commercial Logging and Market Hunting in Northern Congo." In *The Cutting Edge: Conserving Wildlife in Logged Tropical Forests* edited by Robert Fimbel, Alejandro Grahal, and John Robinson. New York: Columbia University Press. 2001, 375–399.

Willis Katherine et al., "How 'Virgin' Is Virgin Rainforest?" *Science*, 304(5669), 2004: 402–3.

Wilson, Edward. *The Future of Life* (New York: Vintage, 2002).

Windle, James. *Suppressing the Poppy: A Comparative Historical Analysis of Successful Drug Control*. London: I.B. Tauris, 2016.

Winkler, Daniel. "The Mushrooming Fungi Market in Tibet Exemplified by *Cordyceps Sinensis* and *Tricholoma Matsutake*." *Journal of the International Association of Tibetan Studies* 4 (December 2008): 1–47.

Winter, Allison. "Report Finds 42,000 Turtles Harvested Each Year by Legal Fisheries." *Environmental News Network*. February 21, 2014. http://www.enn.com/top_stories/article/47074.

Witte, Kim et al., "A Theoretically-Based Evaluation of HIV/AIDS Prevention Campaigns along the Trans-Africa Highway in Kenya," *Journal of Health Communication*, 3(4), 1998: 345–63.

Wittemyer, George, David Daballen, and Iain Douglas-Hamilton. "Poaching Prices: Rising Ivory Prices Threaten Elephants." *Nature* 476 (August 2011): 282–3.

Wittemyer, George, Joseph M. Northrup, Julian Blanc, Iain Douglas-Hamilton, Patrick Omondi, and Kenneth P. Burnham. "Illegal Killing for Ivory Drives Global Decline in African Elephants." *Proceedings of National Academy of Sciences USA* 111, no. 36 (September 2014): 13117–13121.

World Bank, "Turning the Tide—Saving Fish and Fishers: Building Sustainable and Equitable Fisheries and Governance," 2005 (http://siteresources.worldbank.org/ESSDNETWORK/Publications/20631963/seaweb_FINAL_pt.1.pdf).

———, "Going, Going, Gone … The Illegal Trade in Wildlife in East and Southeast Asia," Discussion Paper, 2005 (available at: wwwwds.worldbank.org/external/default/WDSContentServer/WDSP/IB/2005/09/08/00

BIBLIOGRAPHY

0160016_20050908161459/Rendered/PDF/334670 PAPER0Going1going1gone.pdf): 6.

United Nations Office on Drugs and Crime. "World Drug Report 2005". https://www.unodc.org/pdf/WDR_2005/volume_1_web.pdf.

———; *United Nations Office on Drugs and Crime (UNODC)* http://www.unodc.org/documents/wdr2015/World_Drug_Report_2015.pdf

World Rainforest Movement. "Laos: FSC Certified Timber Is Illegal." http://www.illegal-logging.info/item_single.php?it_id=1683&it=news.

World Resources Institute. "Global Biodiversity Strategy." 1992.

World Travel Tourism Council, "Travel and Tourism, Economic Impact: Tanzania." (www.wttc.org/-/media/files/reports/economic%20impact%20research/countries%202015/tanzania2015.pdf): 3.

World Wildlife Fund. "Souvenir Alert Highlights Deadly Trade in Endangered Species" (available at: www.wwf.org.uk/news/scotland/n_0000000409.asp).

World Wildlife Fund. "Fighting Illicit Wildlife Trafficking: A Consultation with Governments." 2012. http://wwf.panda.org/about_our_earth/species/problems/illegal_trade/wildlife_trade_campaign/wildlife_trafficking_report/

Wright, Timothy F., Catherine A. Toft, Ernesto Enkerlin-Hoeflich, Jaime Gonzalez-Elizondo, Mariana Albornoz, Adriana Rodríguez-Ferraro, Franklin Rojas-Suárez, Virginia Sanz, Ana Trujillo, Steven R. Beissinger, Vicente Berovides A., Xiomara Gálvez A., Ann Brice, Kim Joyner, Jessica Eberhard, James Gilardi, S. E. Koenig, Scott Stoleson, Paulo Martuscelli, J. Michael Meyers, Katherine Renton, Angélica M. Rodríguez, Ana Sosa-Asanza, Francisco J. Vilella and James W. Wiley. "Nest Poaching in Neotropical Parrots." *Conservation Biology* 15, no. 3 (2001): 710–20.

Wu, S.B., N. Liu, Y. Zhang, and G.Z. Ma. "Assessment of Threatened Status of Chinese Pangolin *(Manis Pentadactyla)*." *Chinese Journal of Applied Environmental Biology* 10, no. 4 (2004): 456–461.

Wyatt, Tanya. "Exploring the Organization of Russia Far East's Illegal Wildlife Trade: Two Case Studies of the Illegal Fur and Illegal Falcon Trades." *Global Crime* 10, nos. 1–2 (March 2009): 144–54.

Wyler, Liana Sun, and Pervaze A. Sheikh. "International Illegal Trade in Wildlife: Threats and U.S. Policy." Washington, D.C.: Congressional Research Service, 2008. https://www.fas.org/sgp/crs/misc/RL34395.pdf.

Yang, Xiyun. "Tiger Deaths Raise Alarms About Chinese Zoos." *New York Times*. March 18, 2010.

Ybarra, Megan. "'Blind Passes' and the Production of Green Security through Violence on the Guatemalan Border." *Geoforum* 69 (June 2015): 194–206.

Yong, Nicholas. "Drop in Illegal Wildlife Trade Here." *Straits Times*. May 2, 2009.

BIBLIOGRAPHY

Youngers, Colleta and Eileen Rosin, eds. *Drugs and Democracy in Latin America*. Boulder: Lynne Rienner, 2005.

Yves, Mike. "Vietnam Craves Rhino Horn; Costs More than Cocaine." *Associated Press* April 4, 2012.

Zabel, Astrid and Karin Holm-Muller. "Conservation Performance Payments for Carnivore Conservation in Sweden." *Conservation Biology* 22, no. 2 (April 2008): 247–51.

Zak, Annie. "Synthetic Rhino Horn Maker: It's Not Cloning, Exactly." *Puget Sound Business Journal*. February 6, 2015.

Zarate, Juan Carlos. *Treasury's War: The Unleashing of a New Era of Financial Warfare*. New York: PublicAffairs, 2013.

Zhang, Li. "China Must Act Decisively to Eradicate the Ivory Trade," *Nature*, 527(135), 2015.

Zhang, Li et al., "Wildlife Trade, Consumption, and Conservation Awareness in Southwest China," *Biodiversity and Conservation*, 17(6), 2008: 1513.

Zhang Li et al., "Culture, Reform Politics, and Future Directions," Society and Animals, 21(1), 2013: 34–54.

Zhang, Li and Feng Yi. "Wildlife Consumption and Conservation Awareness in China: A Long Way to Go," *Biodiversity and Conservation*, 23(9), 2014: 2371–81.

Zimmerman, Mara. "The Black Market for Wildlife: Combatting Transnational Organized Crime in the Illegal Wildlife Trade." *Vanderbilt Journal of Transnational Law* 36, (November 2003): 1656–89.

INDEX

24/7 Sobriety Project: 231–2

abalone: use in organized crime smuggling, 73–4
Abu Sayyaf: drug trade activity of, 81
Adam, William: view of ecotourism, 170
Afghanistan: 24, 76, 192, 210, 258; drug-control schemes in, 201–2; drug market of, 114; government of, 77; Nangarhar Province, 82; opium poppy farming in, 70, 82, 106, 114, 153
African Parks: 191
African Union (AU): 18
African Wildlife Foundation: 223
Allen, Paul: 59
alligators: trade of, 94, 136
Andean vicuña: recovery of, 137–8, 192, 255
Andorra: 215
animal rights groups: 6, 49, 136–7
antelopes: managed hunting of, 138; saiga, 238
anti-money-laundering (AML): 207–8, 242, 251–2; informal cash transfer mechanisms, 210–11; legislation, 214; measures, 205–6, 208–9, 211–17, 255, 266; prosecutions, 206, 211–12, 215; reporting, 209; shortcomings of, 215–18
Arabic (language): 236
Argentina: 57; Chaco region of, 140
Association of Southeast Asian Nations Wildlife Enforcement Network (ASEAN-WEN): 111, 117
Australia: medicinal opiates supplied by, 152
Austria: government of, 191
avian influenza: spread of, 6, 147

baboons: gelada, 183
Bach Mai 'Boonchai': associates of, 213; criminal activity of, 73, 98, 116
Bach Van Limh: associates of, 213; criminal activity of, 73, 98, 116
badgers: collapse in population of, 67
Baiga (tribe): forced eviction of, 188

INDEX

Bangladesh: 77
banteng: 273
Barzdo, Jonathan: 145
Bathangyl (ethnic group): cultural practices regarding chimpanzees, 68
bears: 62; brown, 67
biodiversity: 22, 26, 30, 43, 66, 68–9, 79, 92, 169, 173; habitat destruction as threat to, 69; preservation of, 1, 3–5, 9, 102, 159, 163, 265, 269; violent conflict in hotspots, 77
bird flu: outbreak of (2004), 227
Birdlife International: role in formation of Conservation Initiative on Human Rights, 47
Bloom Association: surveys conducted by, 221
bobcats: collapse in population of, 67
Bodo Security Forces (BdSF): targeting of poachers, 83
Botswana: 276; ivory stock of, 140; managed elephant population of, 140; military of, 117; Okavango Delta, 172, 174; tourism industry of, 70; use of shoot-to-kill policies in, 127
Brazil: 24, 48, 57; Bolsa Familia, 200; illegal parrot trade in, 8, 108; Pantanal, 127
Buddhism: 98, 229
Burkina Faso: Nazinga Game Ranch, 187
bushmeat: 77, 179; annual hunting rates of, 60; demand for, 42–3, 144; link of Ebola outbreaks to consumption of, 72; nutritional value of, 51; relationship with food security, 52

caiman: trade of, 136
Calderón, Felipe: 205
Cambodia: 76, 198; conditional cash transfer schemes in, 202; NTFPs as percentage of household income, 90–1; Tonle Sap, 62
Cameroon: 140, 187
Canada: 50, 255; Royal Canadian Mounted Police, 120, 213
Cecil the Lion: killing of (2015), 275
Center for International Forestry Research (CIFOR): 174
Central African Republic: 77, 79, 101–2; conservation efforts in, 83; Dzanga-Ndoki National Park, 77; ivory hunting in, 93
Chad: 77
Chaimat, Jay Daoreung: associates of, 213
Chand, Sansar: death of (2014), 94
Chepang (indigenous group): alleged poaching activity of, 96
chimpanzees: cultural practices focusing on, 68
China, People's Republic of: 7, 13, 24–5, 34, 38, 53, 63, 74, 145, 147, 180, 212, 215–16, 226–7, 230, 236, 269–72, 274; Batang, 220; consumption of wildlife products in, 56, 72, 88–9, 252, 254; corruption in, 120, 221; demand for rhino horn in, 17, 109, 254; destruction of stocks of ivory, 20; domestic ivory market of, 14–15, 100, 145–6, 186, 270–3; ecological compensation schemes in, 202; efforts to combat global poaching and wildlife trafficking, 14; Fuzhou, 115; government of, 92, 271; Guangdong Province, 99, 125;

INDEX

Guangzhou, 221; Hong Kong, 16, 25, 54, 60, 63, 94, 100, 146, 220–3, 238, 269–72, 274; Kanding, 220; Kunming, 99; Linxia, 100; Litang, 220; plateau pika extirpation campaigns in, 66; prosecution of illegal hunting in, 125; Sichuan Province, 220; tiger farms in, 15, 146; tiger populations of, 61; tiger product trade in, 15; wildlife market in, 7, 112; Yunnan Province, 91–2

Chumlong Lemtognthai: role in Vixay Keosavang's smuggling operation, 94–5

civets: consumption of, 72

climate change: 1

Coalition of Congolese Patriotic Resistance (PARECO): poaching activity of, 77

cocaine: 113; efforts to combat use of, 224; use of coca leaves in, 69–70, 132, 192, 199–200, 224

cocaleros: 104

Colombia: 24, 50, 97, 104, 107, 192, 207, 215, 258; banking system of, 211; deforestation in, 224; farming of illegal crops in, 70; impact of drug eradication efforts in, 69–70; interdiction in, 107; Medellín, 106; monetary compensation for destruction of legal crop, 199–200; Tayrona National Park, 80

community-based natural resource management (CBNRM): 19, 21, 24, 46–7, 49, 102, 166, 197–8, 203–4, 219, 242, 263–4; aims of, 162–3, 192; ecotourism as focus of, 195; formulations of, 48, 163–4, 167, 193–4, 236–7, 250, 255, 259, 269, 276–7; maximalist variants of, 22; support for, 23, 250–1

Conservation Initiative on Human Rights: formation of, 47

Conservation International: role in formation of Conservation Initiative on Human Rights, 47

Convention on the International Trade in Endangered Species of Wild Fauns and Flora (CITES): 15, 32–5, 41, 60–1, 100, 125, 131; Appendix 1, 17, 33, 143; Appendix II, 33, 105; Appendix III, 33; ban on rhino horn trade (1977), 109; Conference of the Parties (1989), 14; Conference of the Parties (2008), 17; Conference of the Parties (2016), 14, 17, 50, 98, 227, 256, 270; Secretariat, 33–4; signatories to, 33; Standing Committee, 33–4

corruption: 10, 17, 26, 49, 82, 100–2 109, 111, 114, 118, 141, 143, 196, 256–7; amongst ranger forces, 82, 109, 118–21, 248–9, 262; efforts to reduce, 27, 121–2, 221; institutional, 107, 116; judicial, 74; networks, 74, 93, 113; patronage, 119–20; political, 73–4, 88; systemic, 120

crocodile: CITES ban on trade in all species sourced from wild (1975), 136; legal trade of, 19, 136–7; skins, 136

Cuba: 152

Cyprus: 215

dacoits: insurgency activity of, 83

Dalai Lama: 98

Dayak (ethnic group): 273–5

INDEX

deforestation: 44, 70, 200; as cause of desertification, 183

Democratic Forces for the Liberation of Rwanda (FDLR): poaching activity of, 77

Democratic Republic of Congo: 52; Bathangyi population of, 68; military of, 79; rhino horn smuggling in, 96; Rwenzori Mountains National Parks, 68; Virunga National Park, 116

Denmark: 120

Djibouti: 24, 96

drug economy: estimations of size, 57

drug policy: 3, 22–4, 34–5, 43, 48, 104, 230–1, 242, 257–8; activism, 205; alternative livelihood schemes, 175–8; decriminalization, 40–1; demand reduction, 224, 252; eradication, 69–70, 80, 92, 201–2; hard drugs, 34; harm reduction, 72, 134; legalization, 21–2, 34, 43, 50, 105, 131–5, 152–5, 246; non-violent arrests, 153–4; potential for gray market, 155; prevention programs, 229–30, 253; prohibition, 104–5; seizure, 41; treatment for addicts, 133–4; use reduction programs, 230–1

drug production: 31

drug trafficking/trade: 3, 23, 31, 65, 73–4, 86, 97, 209–10; arrests of traffickers, 48; as cause of environmental harm, 69–70; legal, 131; links to wildlife trade, 73; use by militant groups, 76, 81

DuYueshen: 71, 124

Dutch dieseas: concept of, 71

eagles: collapse in population of, 67

Ebola: 1; link to bushmeat consumption of, 72; spread of, 229

ecotourism: 9, 70, 139, 145, 187, 190, 276; as focus of CBNRM approaches, 195; ecolodges, 168–9; employment, 168–9, 171–2; growth of, 170; impact on poaching rates, 170–3; revenues, 168, 190, 193; trophy hunting as part of, 172; use in local community involvement, 162, 166–9

Egmont Group: 207

Ejército de Liberación Nacional (National Liberation Army) (ELN): initial efforts to ban illegal drug production, 81

elephants: 8–10, 47, 49, 63; as source of crop damage, 53; culling of, 140; decline of population, 4, 12, 14; illegal hunting of, 7, 11–12; management of populations, 139–40; poaching of, 12, 77, 95, 105, 111, 115, 123, 264

environmental crime: rise in, 57–8

Environmental Investigation Agency: 94

environmental policies: association with colonial oppression, 44

environmentalism: 45–6

Eritrea: 96, 191

Escobar, Pablo: 70, 94

Ethiopia: 24, 45, 96, 179, 198; Addis Ababa, 141; Awash, 179–80; Bale Mountains National Park, 183; Constitution of (1991), 180; government of, 190; Lake Abiata-Shala National Park, 141–2, 179–80, 183; Nechisar National Park,

INDEX

191; land ownership issues in, 179–81; Nechisar, 179; Simien Mountains National Park, 179, 183, 189
European Commission: 18
European Parliament: 74; study of wildlife crime enforcement in EU, 117
European Union (EU): 55, 105, 117, 151; ban on import of wild birds, 147; legal importing of wildlife in, 62; penalties for wildlife crime in, 125–6

falcons: hunting of, 77; prairie, 67; saker, 77
Feng Yi: survey of Chinese customer view of wildlife products (2012), 223
ferrets: black-footed, 67
Fiji: sea turtle poaching activity in, 61
Financial Action Task Force (FATF): 207, 216
food security: 52; impact of illegal fishing on, 68–9
Forest Stewardship Council (FSC): certification system of, 150–2
Forsyth, Tim: *Forest Guardians, Forest Destroyers*, 184
fox: Tibetan, 67
Fox, Vicente: support for drug legalization policies, 153
France: medicinal opiates supplied by, 152
Frankfurt Zoological Society: personnel of, 45
Fuerzas Armadas Revolucionarias de Colombia (FARC): 128; drug trade activity of, 76, 81; illegal logging/mining activity of, 76; infrastructure provided by, 84; initial efforts to ban illegal drug production, 81

Gabon: destruction of stocks of ivory, 20
Galapagos Islands: 63
Gambella (ethnic group): forced displacement of, 180
Germany: 14; importing of TCM in, 54
Global Financial Integrity: assessment of illicit money entering Kenya (2013), 211–12
global warming: 11
globalization: 7
Gond (tribe): forced eviction of, 188
Great Limpopo Transfrontier Park: establishment of (2001), 189
green militarization: concept of, 79–80
Grzimek, Bernhard: President of Frankfurt Zoological Society, 45
Guatemala: bio-prospecting in, 69
Guinea: 72
Gulf of Thailand: seahorse fishing in, 60
Gutiérrez Rebollo, José de Jésus: 124

habitat conservation: 108
habitat destruction: 11, 178–9; threat to biodiversity, 69
Haqqani Network: illegal logging/mining activity of, 76
hawala: 210; networks, 210
Heath, Brian: Manager of Laikipia Conservancy, 138; Manager of Mara Conservancy, 121–3, 141
hepatitis: spread of, 72
heroin: 72, 113, 152; use as painkiller, 42–3

INDEX

Himalayas: 185
HIV/AIDS: spread of, 72, 154
Hmong (ethnic group): inaccurate depictions of, 184
Home Depot: supply of FSC certified products, 150
Honduras: land theft from locals in, 79–80
HSBC: AML programs of, 215; money laundering activities of, 215
Hübschle, Annette Michaela: study of rhino horn trafficking, 84–5, 159
human security: 1, 4, 249; threats related to poaching, 68
human trafficking: 28
hunting: 89–90, 131, 149–50, 246, 257; fees, 141; illegal, 108; impact of illegal wildlife trade on, 91; low-tech, 92–3; managed, 138–9; professional, 93; subsistence, 21, 67, 149; sustainable, 24, 44

ibex: Ethiopian, 191
illegal grazing: law enforcement efforts against, 178–9
illegal/illicit economics: 42, 47–8, 51, 65, 71, 73, 75, 87, 113–14; consumption and demand, 38–9, 56; detectability of flows, 37–8; dispersion of supply, 36; displacement of, 113–14; seconomic effects of, 70–1; large-scale, 75; measurement of, 56–7, 59; technological/institutional requirements of, 36–7; presence of militant groups in, 76–7; production/transhipment, 36–7; profit mark-up, 98–9; sponsorship of, 84–6

illegal fishing: 55; impact on food security, 68–9
illegal logging: 2, 24, 142, 164–5, 181, 258, 274–5; small-scale, 265; use by militant groups, 76
illegal mining: use by militant groups, 76
India: 24, 55, 180, 186; Assam, 83; Bhadra Wildlife Sanctuary, 188; British Raj Criminal Tribes Act (1871), 90; Gir Forest National Park, 45; Gujarat, 90; Kanha Tiger Reserve, 188; Maharashtra, 90; Manas National Park, 83; opium poppy cultivation in, 152; Similpal Tiger Reserve, 188; use of shoot-to-kill policies in, 127; Uttar Pradesh, 83
Indonesia: 5–7, 24, 45, 62, 145, 157, 210, 222, 272; Ambon, 96, 112; Bali, 9, 111–12; Borneo, 273; Christian population of, 144; Flores, 6; Forestry Ministry, 111; importing of TCM in, 55; Guiting-Guiting Natural Park, 171–2; Jakarta, 6; Jatinegara, 6; Java, 61; Kalimantan, 96, 112, 151, 182, 273–5; Kutai National Park, 273–4; Labuan Bajo, 6; logging industry of, 143; Moluccas, 165; Papua, 165; Samarinda, 151, 182; sea turtle poaching activity in, 61; Seram, 170–1; Sibuyan Island, 171; Sulawesi, 96, 144; Sumatra, 61; Tangkoko Reserve, 144; tiger killings in, 61
Infield, Mark: view of ecotourism, 170
International Commission Against Impunity in Guatemala (La

INDEX

Comisión Internacional Contra La Impunidad en Guatemala) (CICIG): 124
International Fund for Animal Welfare (IFAW): 50, 220
International Police Organization (INTERPOL): 57; efforts to combat wildlife trafficking, 117
International Union for the Conservation of Nature: Report of (2016), 12
Irish Republican Army (IRA): 207
Islam: 210, 237
Islamic State in Afghanistan: ban on poppy cultivation on Nangarhar Province, 82
Islamism: militant, 77
Italy: 48; legalization of gambling in, 154
ivory: 21, 25, 49, 77, 100, 145, 148–9, 157, 223, 226, 232, 269, 276–7; banning of global trade of (1989), 14–16; destruction of global stocks, 17, 20; domestic markets, 14–15; hunting, 93; pricing of, 13; products, 52; seizure of, 41, 119–20; smuggling of, 52, 125; stocks from natural deaths, 139–40; whale, 120

jaguars: poaching of, 127
jambiya: 234–5, 237–8; agate, 238; use of rhino horn in, 234–5
Janjaweed: poaching of elephants by, 77
Japan: 25, 143, 148, 227; demand for rhino horn in, 109, 254; domestic ivory market of, 14–15; illegal fishing in, 55; importing of TCM in, 54; ivory trade in, 100

Kambaata (ethnic group): 183
Karen (ethnic group): inaccurate depictions of, 184
Kazakhstan: 238
Karki, Babu Krishna: 117
Kenya: 12, 24, 80–1, 99, 108–9, 118, 120, 126, 129, 164, 198, 210–12; Act on Wildlife Conservation (2013), 199; Amboseli National Park, 46, 188; borders of, 122; CBNRM efforts in, 194; corruption in, 118–120; destruction of stocks of ivory, 20; economy of, 212; Forestry Service, 123; Kinguri Council, 121; Laikipia Conservancy, 138, 255; Laikipia District, 194; Lewa Conservancy, 194; Maasai population of, 122, 141, 164, 196; Maasai Mara National Reserve, 54, 121, 141, 196; Mara Conservancy, 10, 121–5, 128–9, 138; Ministry of Energy, 119; Mombasa, 120, 129; Narok County Council 121; Narok District, 121, 196; Ol Pejeta Conservancy, 12; rhino horn smuggling in, 96; Tsvao National Park, 54, 118
Kenya Wildlife Society: 141; compensation scheme established by, 199
Kenyan Defense Forces: presence in southern Somalia, 82
keystone species: concept of, 66
Khmer Rouge: illegal logging/mining activity of, 76
Kipling, Rudyard: *Jungle Book, The*, 188
Komodo dragon: 6; trade of, 94
Kruger, Paul: 44

397

INDEX

Laos: 73, 213, 272; government of, 146–7; illegal exporting of pangolin skins from, 62; military of, 80; Nakai-Nam Theun National Protected Area, 92; NTFPs as percentage of household income, 90–1; tiger farms in, 15, 146; tiger populations of, 61; Vientiane, 100

law enforcement: 2–5, 11, 14, 19–20, 22–5, 27, 37–8, 41, 43, 47, 66, 74, 79, 82, 86–7, 93, 98, 101–4, 106, 108, 129, 142, 157, 170, 178–9, 181–2, 204, 210, 261; agencies, 49, 56, 68, 72, 75, 116, 123–4; cooperation with prosecutors, 124; costs of, 75; *in situ*, 28, 260–1, 268–9; interdiction, 107; problematic consequences of, 112–14; prohibition, 104–5, 112–13; regulatory theory, 104–6; rogue, 124; shoot-to-kill policies, 127–8; special interdiction units (SIUs), 123–4

leopards: hunting of, 77; legal trading of, 62; snow, 77

Li Zhang: 55; survey of Chinese customer view of wildlife products (2012), 223

Liberia: 72

lions: 63

llamas: 137

Lo Hsing Han: 71

local community involvement: 2, 22, 29, 48, 68, 119, 121–2, 141, 178–9, 196–7; alternative livelihood schemes, 161, 166–7, 174–8; buffer zone creation, 184–7; conditional cash transfers, 198–204; conservation project design, 162–3; development of, 121, 138, 162; government support for, 27; habitat/wildlife management, 173–4, 180–2; payments for ecosystem services (PES), 200–1; social injustice, 165–6; use of ecotourism, 162, 166–9, 171–3; voluntary resettlement, 187–9

Loera, Joaquín Archivaldo Guzmán (El Chapo): 70–1

London Declaration on Illegal Wildlife Trade (2014): 18

Lord's Resistance Army (LRA): poaching of elephants by, 77

Lusaka Agreement Task Force: establishment of (1999), 111

lynx: Iberian, 67

Maasai (ethnic group): livestock owned by, 122; mobilization efforts of, 45, 141; territory inhabited by, 122, 141, 164, 196

macaques: 144; legal trading of, 62

Macau: 274

Makuleke (ethnic group): impact of CBNRM approaches on, 194

Malawi: Liwonde National Park, 128

Malaysia: 94, 147, 238; bushmeat hunting rates in, 60; importing of TCM in, 54; prosecution of wildlife smuggling in, 125; Sabah Province, 60; Sarawak Province, 60

manta ray: demand for gills of, 225

Maoism: 78

marijuana (cannabis): 21, 186, 225; illegal cultivation of, 69; legalization of 34, 43, 105, 132–3, 152, 156; medical, 42–3, 69, 133; recreational, 43, 133; smuggling of, 154

INDEX

Masai (ethnic group): 45
Medllín Cartel: drug trafficking networks of, 97
methamphetamines: 77, 113
Mexico: 8, 24, 48–9, 53, 62–3, 76, 97, 215; Chiapas, 154; drug cartel activity in, 48, 153–5, 209; Guerrero, 154; Michoacán, 154; organized crime in, 63, 76; sea turtle poaching activity in, 61; Zetas, 124
Microsoft Corporation: 59
Montesinos, Vladimiro: 124
Morocco: 24
Mozambique: 45, 84, 140, 189; Civil War (1977–92), 77; decline of elephant population in, 12; poaching groups in, 84
Mugabe, Grace: alleged role in forcible expulsion of local communities from private ranch (2015), 187; family of, 187
Mugabe, Robert: family of, 187
Myanmar (Burma): 71, 77, 79, 82–3, 100, 107, 210, 222, 258, 272; military of, 91; Mong Law wildlife market, 99; NTFPs as percentage of household income, 90–1; poppy cultivation in, 48, 92, 165; Shan State, 91; Special Region No. 4, 91–2; wildlife market in, 7, 99

Nagas (ethnic group): hunting traditions of, 90
Namibia: 12, 24, 46–7, 256, 276; CBNRM efforts in, 193–4, 197; ivory stocks of, 17, 140, 256; Kunene region, 194; managed elephant population of, 140
narwhals: smuggling of tusks, 120, 213

National Democratic Alliance Army (NDAA): 91–2
National Rifle Association (NRA): opposition to total federal ban on ivory trade, 16
Nationalist Socialist Council of Nagaland: 82–3; illegal wildlife trading activity of, 77
Nature Conservancy: role in formation of Conservation Initiative on Human Rights, 47
Nepal: 24; Bardia National Park, 78, 181; buffer zone creation in, 185–7; Chitwan National Park, 181; Civil War (1996–2006), 78; ecotourism in, 169; Kathmandu, 96, 169; Lhasa, 98; military of, 117; Terai, 169, 185–6; Sherpa community of, 186; wildlife smuggling networks in, 98; *yarchagumba* trafficking in, 78
Netherlands: harm reduction measures in, 134; legalization of possession and sale of cannabis in, 34; Rotterdam, 120
Ning Hua: 55
non-governmental organisations (NGOs): 11, 22, 56–7, 111, 135, 144, 150, 170, 187–9, 216–17, 226, 239, 264, 267; conservation, 65, 90, 103; donors, 251; efforts to train rangers, 117; environmental, 10, 19–20, 22–4, 41, 46–7, 50, 58, 71, 102, 111, 115, 131, 146, 191, 206, 221–2, 227, 256, 265, 271–2; humanitarian, 216; monitoring, 59; wildlife conservation, 16
non-timber forest products (NTFP): 67, 90

INDEX

Norte del Valle Cartel: money laundering activity of, 215
Norway: 188
nuclear smuggling: 36, 39

Obama, Barack: administration of, 110
Olson, Mancur: 65
Operation Crash: 110–11
Operation Terminate: allegations of extreme violence during, 127
opium: resin, 152; use as painkiller, 42–3, 152
orang-utans: 6, 38
organized crime: 75, 123–4; links to wildlife trade/trafficking, 73–4; popular perceived relationship with poaching, 164; smuggling activity, 73–4
Oromo (ethnic group): 184; forced displacement of, 180; logging activity of, 183

Pacific Institute: personnel of, 112
Pakistan: 211; drug cultivation in, 114
Pallas's cats: collapse in population of, 67
pangolins: 111; illegal trade of, 98–9; legal trading of, 62; skins, 62
parakeets: 89
Pardhis (ethnic group): hunting traditions of, 90
parrots: 89; blue-headed, 108; breeding in captivity, 136; trafficking of, 7, 107–8
Peña Nieto, Enrique: 205
Peru: 107, 192; conservation efforts in, 137–8, 255; farming of illegal crops in, 70; impact of drug eradication efforts in, 69–70
Philippines: 81; destruction of stocks of ivory, 20
plateau pika: as target of extirpation campaigns, 66–7
poaching: 2, 4, 8, 12–13, 18, 26, 31, 47, 55, 61, 63, 65–6, 73, 77, 83–4, 86–7, 95, 100, 103–5, 109, 111, 120, 138–40, 144, 149–50, 165, 179, 194, 224, 264–5, 273; as targets of shoot-to-kill policies, 127–9; efforts to halt/reduce, 18, 29–30, 78, 103, 109–10, 114–15, 122–3, 149–50, 155, 170–1, 198–9, 242–3, 261; impact of ecotourism on, 170–3; involvement of militant groups/militaries in, 77–82; ivory, 12, 14–15; killing of rangers, 116–17; kingpins, 84–5, 97, 114–15, 189; laundering, 19, 100; meat, 54; middle layer, 114–16, 128; networks, 244, 248–9; popular perceived relationship with organized crime, 164; potential effect on ecosystems, 66; prosecution of, 124–6; removal of militant groups from sensitive ecosystems, 78–9; small-scale, 75; subsistence, 67–8; threats to human security, 68; use by war refugees, 77–8
polecats: steppe, 67
poppy: cultivation of, 48, 70, 82, 106, 114, 132, 153, 165, 176–7, 225; eradication of, 92; poppy straw method, 153
Portugal: 48
prairie dogs: as target of extirpation campaigns, 66

INDEX

private ranches: 143; put and take, 143–4; unregistered ownership via, 148

Profauna: 170; encouragement of zero-catch quotas, 111

qat: cultivation of, 69

rabbit: European, 66–7

rangers: 11, 101, 118, 128–9, 179; corruption amongst, 82, 109, 118–21, 248–9, 262; equipment used by, 82, 111; forest, 9; killing of, 116–17; park, 10, 24, 116; patronage networks, 119–20; training of, 117

Rathkeale Rovers: criminal activity of, 158

reducing emissions from deforestation and forest degradation (REDD+): 200–1

reptile skins: 53; trade of, 89

Republic of Congo: Parc National Nouabalé-Ndoki, 229

Reuter, Peter: *Disorganized Crime: Economics of the Visible Hand*, 97

Resistência Nacional Moçambique (RENAMO): poaching activity of, 79; trading of rhino horn and ivory, 77

Resources Himalaya: 186

Revolutionary United Front (RUF): illegal logging/mining activity of, 76

rhinoceros: 10, 63, 93, 254; black, 47, 109, 118, 138, 234, 255; decline of population, 4, 12–13, 109; illegal hunting of, 7, 94–5; Indian, 12; Javan, 12–13; live trade of, 139; northern white, 234; poaching of, 12–13, 83, 109, 115, 123, 264; preserved stock, 138–9, 143; recovery of, 17; rhino horn, 17–18, 25, 54, 77, 84, 96, 99, 101, 109, 113, 146, 157–9, 226, 234–6, 266, 276; Sumatran, 12, 273; white, 138–9

Rocky Mountain High Intensity Drug Trafficking Task Force: findings of, 152

Russian Federation: 22, 53, 75, 226, 238; opposition to drug harm reduction policies in, 72; Siberian tiger population of, 13

Rwanda: Genocide (1994), 77

salamander: giant, 147

Saudi Arabia: 180

Save the Elephants: 223

Selamatkan Yaki Project: 144

Seleka: alleged poaching activity of, 77

Sendero Luminoso (Shining Path): drug trade activity of, 76, 81; infrastructure provided by, 84; initial efforts to ban illegal drug production, 81

severe acute respiratory syndrome (SARS): 1; economic impact of, 72; outbreak of (2003), 6, 72, 227

al-Shabaab: 9; alleged elephant poaching activity of, 80–1; territory controlled by, 81

Shan Sun: 55

sharks: 6, 52, 63, 221; harvesting of fins for shark-fin soup, 63, 220–2

Sierra Leone: 72, 76

Simien Mountains National Park Integrated Development Project: proposals for, 191

Sinaola Cartel: drug trafficking

networks of, 97; money laundering activity of, 215

Singapore: 238, 274; importing of TCM in, 55; legal importing of wildlife in, 62; popularity of shark-fin soup in, 220–1; prosecution of wildlife smuggling in, 125

snow finches: disappearance of, 67

Somalia: 24, 96, 211, 260; Famine (2011), 216; Juba State, 81–2; Kismayo, 81–2, 120; Puntland, 9

Somaliland: 9–10; Hargesia, 10

songbirds: trade of, 53–4

South Africa: 12, 25, 45–6, 103, 143, 157, 190, 195; Apartheid, 45–6, 117, 195; Cape Town, 73–4; Constitution of, 17; Directorate of Special Operations (Scorpions), 124; ecotourism in, 9, 168; elephant population of, 140; game parks in, 138; ivory stock of, 140; Johannesburg, 98; Kruger National Park, 44, 127–8, 138, 140, 189, 194–5; Makuleke population of, 194; managed elephant population of, 140; National Prosecuting Authority, 124; organized crime in, 73–4; poaching groups in, 84; private ranches in, 148; rhino horn smuggling in, 96; rhino poaching in, 115; rhino population of, 109, 138, 148; Western Cape, 85

South African Defence Force: personnel of, 117

South China Sea: seahorse fishing in, 60

South Korea: demand for rhino horn in, 109, 254; importing of TCM in, 54

South Sudan: 52; rhino horn smuggling in, 96

Spatial Monitoring and Reporting Tool (SMART): use of, 111

Spix's macawa: 94

Sri Lanka: destruction of stocks of ivory, 20

Stevenson-Hamilton, Lieutenant-Colonel James: First Warden of Kruger National Park, 44

Sudan: 140, 190

Suharto: 7; regime of, 182

Survival International: 187

Swaziland: 17; stock of rhino horn, 157, 256

Sweden: 198, 202

Switzerland: Geneva, 33

Taiwan: demand for rhino horn in, 109; importing of TCM in, 54; popularity of shark-fin soup in, 220–1

Taliban: drug trade activity of, 76, 81; hunting activity of, 77; illegal logging/mining activity of, 76; initial efforts to ban illegal drug production, 81

Tanzania: 24, 104, 123, 129, 140, 142, 164; borders of, 122; corruption in, 120, 196; decline of elephant population in, 12; ecotourism in, 70; GDP per capita, 70; Maasai mobilisation in, 45; military of, 127; refugee camps in, 78; Serengeti National Park, 111; trophy hunting in, 139; trophy ivory exports from, 16; use of shoot-to-kill policies in, 127

Thailand: 45, 73, 176, 213, 222; alternative livelihood schemes in, 176–7; consumption of wildlife products in, 89; corruption in, 120; Hmong population of, 184; ivory trade in, 100; Karen population of, 184; logging industry of, 143; Manyara National Park, 45; opium poppy cultivation in, 176; prosecution of wildlife smuggling in, 125; resettlement conflicts in, 188; Tarangire National Park, 45; tiger farms in, 15; wildlife market in, 7, 112

Tharu (ethnic group): alleged poaching activity of, 96; resettlement of, 181

Thin Green Line Foundation: estimation of killing of rangers (2009–16), 116

Tibet: plateau pika extirpation campaigns in, 66

Tibetan ground-tits: disappearance of, 67

tigers: 8–10, 63; bones, 54; conservation efforts, 62; decline of population, 4; farming of, 15, 146; killing of, 61; poaching of, 13; Siberian, 13; tiger product trade, 15, 146–7

Tockay gecko: farming of, 145; trading of, 61

tortoises: 111; freshwater, 61; land, 112; Madagascar ploughshare, 94; Sulawesi land, 112; trading of, 61

totoaba fish: conservation efforts, 63; trade in bladders of, 63, 95

Traditional Chinese Medicine (TCM): 7, 11, 78, 112, 232–3, 238–9, 266–8, 274; as cultural practice, 54; demand for, 8, 54, 267; importing of, 54–5; marketing of, 156–7

TRAFFIC: 59; studies conducted by, 145, 174, 238

Trans-Pacific Trade Partnership (TTP): negotiation of, 14

Transparency International: Global Corruption Index, 120

trophy hunting: 16–17, 21, 26, 94–5, 101, 257; as part of ecotourism, 172; corruption in, 17; licensed, 147–8; revenues from, 139, 142, 193, 275–6

tuberculosis: drug-resistant, 72

Turkey: use of poppy straw method in, 153

turtles: 112; freshwater, 61; illegal capture of, 111; marine hawksbill, 61; sea, 61, 111; trading of, 61

Uganda: 52, 198; military of, 79; Mburo National park, 188; Mount Elgon National Park, 188

União Nacional para a Independência Total de Angola (UNITA): elephant poaching activity of, 77

United Arab Emirates (UAE): 180; Dubai, 120

United for Wildlife: 206

United Kingdom (UK): 14, 215; corruption in, 120; government of, 201; Heathrow, 120

United Nations (UN): 207, 221; Convention Against Illicit Traffic in Narcotic Drugs and Psychotropic Substances (1988), 34; Convention Against Transnational Organized Crime, 18; Convention on Psychotropic Substances (1971), 34; Convention on Terrorism, 207; corruption in, 120; Educational,

INDEX

Scientific and Cultural Organization (UNESCO), 62, 189–90; Environment Programme (UNEP), 57, 68; Financial Crimes Enforcement Network (FinCEN), 207, 215; International Economic Emergency Powers Act, 207; Office on Drugs and Crime (UNODC), 57, 59, 70; Security Council (UNSC), 14; Single Convention on Narcotic Drugs (1961), 34; *World Drug Report*, 59; *World Wildlife Crime Report: Trafficking in Protected Species, The* (2016), 59

United States of America (USA): 7, 14, 36, 43–4, 53, 100, 111, 148, 150–1, 153, 155, 188, 220, 230, 255, 276; 9/11 Attacks, 207; anti-smoking campaigns in, 228; ban on importing wild birds, 136; ban on ivory trade, 16; Congress, 206; Congressional Research Service, 58; consumption of wildlife products in, 89; Death Valley, 44; demand for TCM in, 8; destruction of stocks of ivory, 20; displacement of Native Americans in, 44; drug policy in, 40–1; drug trafficking in, 48; efforts to combat global poaching and wildlife trafficking, 14, 18; Global Anti-Poaching Act (2015), 18; government of, 95; habitat conservation of, 108; House of Representatives, 18, 206; importing of TCM in, 54; Intelligence Authorization Act (2016), 18; Lacey Act, 151; Los Angeles, CA, 16; *National Strategy for Combating Wildlife Trafficking* (2014), 18; PATRIOT Act (2001), 208; penalties for wildlife trafficking in, 125–6; Project Hope, 231; San Francisco, CA, 16; Senate, 18; State Department, 18; Treasury Department, 215; US Agency for International Development (USAID), 174; US Fish and Wildlife Agency, 110; US National Strategy for Combating Wildlife Trafficking, 206; US Racketeer Influenced and Corrupt Organizations Act (RICO), 214; Yellowstone National Park, 44

United Wa State Army: drug and wildlife trafficking activity of, 77

Uruguay: legalization of cannabis in (2013), 34

Uzbekistan: 238

Vannaseng: alleged wildlife trafficking operations of, 98, 116

Vietnam: 13, 125, 147, 212, 222, 230, 232, 236; consumption of wildlife products in, 89, 159, 229, 254; corruption in, 120; Cuc Phuong National Park, 169; demand for rhino horn in, 17, 109; Hanoi, 98; illegal wildlife trade in, 58; importing of TCM in, 54; Pu Mat National Park, 173–4; Tam Dao National Park, 98; tiger populations of, 61; tiger farms in, 15, 146; wildlife market in, 7, 112

village development committee (VDC): members of, 185

Vinasakhone: alleged wildlife trafficking operations of, 98, 116

Vixay Keosavang: 97; imprisonment of, 94, 126; smuggling activity of, 94–5; withdrawal from wildlife trafficking (2014) 94, 98, 116

Walker, Andrew: *Forest Guardians, Forest Destroyers*, 184
war on drugs: criticisms of, 21, 133
Weaver, Kent: 233
Wells, Susan: 145
WildAid: 223
Wildlife Conservation Society: 206; role in formation of Conservation Initiative on Human Rights, 47
wildlife markets: 6–7, 99; specialized, 7
wildlife products: 11, 52–3, 58, 82, 89, 95, 147, 155, 230, 232; consumer views of, 223; consumption of, 39–40, 87–9, 137, 220–3; costs, 227; demand reduction efforts, 219–20, 224–5, 229, 252–4; depletable resources, 40; farmed supply of, 142, 144; green certification, 233–4, 246–7; non-depletable resources, 40; legal supply/trade of, 98, 246–8, 254–5, 262–3; one-off sales, 157–9; seizure of, 119–20; subsistence, 90; taxation of, 144–5; value of, 42, 55, 90, 108, 139, 227, 244; wildlife protection messaging, 227–9
wildlife trade/trafficking: 1–2, 8, 13, 23, 25, 29, 31–2, 34–5, 39, 65–6, 71–2, 86, 97, 100, 102, 107–8, 113, 125, 127, 133, 144–5, 217–18, 243, 245, 248, 258–9, 269, 273, 275, 277; certification systems, 150–1;
consumption and demand, 38–9, 56, 135; DNA testing, 148–9; efforts to halt/reduce, 14, 18, 30, 110–11, 167, 242–3; estimation of scale of, 57, 59–60; impact on hunting, 91; laundering via legal systems, 145–7, 157–8, 248; legal, 131–2, 135–6, 140–1, 143–6, 161, 227, 241–2; links to drug trade/organized crime, 73–4; middlemen, 92–3, 128; networks, 27, 96, 165, 210, 244–5, 248–9; penalties for, 125–6; pet trade, 53; regulation of, 101–2; role of diaspora communities in, 98
wolves: Ethiopian, 183, 191
Wong, Anson (Wong Keng Liang): 97; background of, 93–4
World Agroforestry Center (ICRAF): 174
World Bank: 58
Worldwide Fun for Nature (WWF): 58, 171–2; role in formation of Conservation Initiative on Human Rights, 47

Xi Jinping: anti-corruption efforts of, 221, 271

Yang Feng Glan: arrest of, 115
yarchagumba: legalization of collection of, 78; use in TCM, 78
Yemen: 211, 234–5; demand for wildlife products in, 159, 236, 254; mining of agate in, 238; Unification (1990), 235; use of *jambiyas* in, 234–5, 237–8; water scarcity in, 69
Yogayakarta: market of, 6

Zambia: 140, 142, 187–8; agricultural practices in, 53–4; poaching of elephants, 111

zebras: Hartmann's, 193, 255; managed hunting of, 138; meat, 138; skins, 138

Zimbabwe: 12, 46, 142, 275; CAMPFIRE program, 192–3, 197, 276; Gonarezhou National Park, 79; ivory stocks of, 17, 256; military of, 79; trophy ivory exports from, 16; wild resources as percentage of household income in, 67

Zulu (language): 44

Zuma, Jacob: Scorpion investigation into, 124